ELECTRICAL INSULATION
FOR ROTATING MACHINES

Other Books in the IEEE Press Series on Power Engineering

ELECTRICAL INSULATION FOR ROTATING MACHINES

Design, Evaluation, Aging, Testing, and Repair

GREG C. STONE

EDWARD A. BOULTER

IAN CULBERT

HUSSEIN DHIRANI

POWER
ENGINEERING

IEEE Press Series on Power Engineering
Mohamed E. El-Hawary, *Series Editor*

IEEE PRESS

A JOHN WILEY & SONS, INC., PUBLICATION

For general information on our other products and services please contact our Customer Care Department within the U.S. at 877-762-2974, outside the U.S. at 317-572-3993 or fax 317-572-4002.

Wiley also publishes its books in a variety of electronic formats. Some content that appears in print, however, may not be available in electronic format.

Library of Congress Cataloging-in-Publication Data is available.

ISBN 0-471-44506-1

Printed in the United States of America.

10 9 8 7 6 5 4 3 2

CONTENTS

PREFACE

This book arose out of the conviction that both designers and users of large motors and generators would appreciate a single reference work about the electrical insulation systems used in rotating machines. We also wanted to document how and why the insulation systems in current use came to be. Since rotating machine insulation is not the most glamorous field of study in the engineering world, it is sometimes treated as an afterthought. The result has been a gradual loss of knowledge as innovators in the field have retired, with few new people specializing in it. We hope that the archiving of the information in this book will slow this gradual loss of knowledge and be a useful starting point for future innovations.

This book is unique in that two of the authors (Alan Boulter and Ian Culbert) have a machine design background, whereas the other two have experience as primarily users of machines. With luck, both users and manufacturers of machines can find their interests represented here.

Collectively, three of us (Greg Stone, Ian Culbert, and Hussein Dhirani) want to thank Ontario Hydro (now Ontario Power Generation) for enabling us to become specialists in this field. It seems that the current business climate enables few engineers to become as specialized as we were allowed to be. We would also like to thank John Lyles, Joe Kapler, and Mo Kurtz, all former employees of Ontario Hydro, who taught us much of what we know. EPRI, and Jan Stein in particular, are acknowledged for allowing the three of us to have a "dry run" at this book when they sponsored the writing of a handbook in the 1980s.

We thank Resi Lloyd, who worked valiantly to put a consistent style on the various chapters, created many of the figures, and brought the book together.

Hussein Dhirani thanks his family for allowing him to sneak away to the office on some weekends to work on the book on the dubious pretext of better productivity. Hussein is grateful for the generosity of the many who shared their knowledge, from tradesmen working on generators, to designers poring over drawings, to staff in sister utilities discussing common problems, to supplier organizations explaining the intricacies of their insulation systems design. The understanding and support of Derek Sawyer and Bill Wallace at Ontario Power Generation is particularly appreciated.

Ian Culbert thanks Ontario Power Generation for allowing him to participate in a number of internal and EPRI projects from which he gained much of the information he contributed to this book. He also appreciates the opportunities his former employers Reliance Electric and Parsons Peebles gave him to learn how to design, test, and troubleshoot motors.

Finally, Greg Stone wants to thank his original partners at Iris Power Engineering—Blake Lloyd, Steve Campbell, and Resi Lloyd—for allowing his attention to wander from day-to-day business matters to something as esoteric as contributing to a book. Of course, Greg's wife, Judy Allan, is thanked because she never did question the premise that most folks use vacations for writing books and papers.

GREG C. STONE
EDWARD A. BOULTER
IAN CULBERT
HUSSEIN DHIRANI

Toronto, Ontario, Canada
October 2003

ELECTRICAL INSULATION
FOR ROTATING MACHINES

CHAPTER 1

ROTATING MACHINE INSULATION SYSTEMS

In the hundred years since motors and generators were invented, a vast range of electrical machine types have been created. In many cases, different companies called the same type of machine or the same component by completely different names. Therefore, to avoid confusion, before a detailed description of motor and generator insulation systems can be given, it is prudent to identify and describe the types of electrical machines that are discussed in this book. The main components in a machine, as well as the winding subcomponents, are identified and their purposes described.

Although this book concentrates on machines rated at 1 kW or more, much of the information on insulation system design, failure, and testing can be applied to smaller machines, linear motors, servomotors, etc. However, these latter machines types will not be discussed explicitly.

1.1 TYPES OF ROTATING MACHINES

Electrical machines rated at about 1 HP or 1 kW and above are classified into two broad categories: (1) motors, which convert electrical energy into mechanical energy (usually rotating torque) and (2) generators (also called alternators), which convert mechanical energy into electrical energy. In addition, there is another machine called a synchronous condenser that is a specialized generator/motor generating reactive power. Consult any general book on electrical machines for a more extensive description of machines and how they work [1.1, 1.2, 1.3].

Motors or generators can be either AC or DC, that is, they can use/produce alternating current or direct current. In a motor, the DC machine has the advantage that its output rotational speed can be easily changed. Thus, DC motors and generators were widely used in industry in the past. However, with variable speed motors now easily made by combining an

Electrical Insulation for Rotating Machines. By Stone, Boulter, Culbert, and Dhirani
ISBN 0-471-44506-1 © 2004 Institute of Electrical and Electronics Engineers

AC motor with an electronic "inverter-fed drive" (IFD), DC motors in the 100's of kW range and above are becoming less common.

Machines are also classified according to the type of cooling used. They can be directly or indirectly cooled, using air, hydrogen, and/or water as a cooling medium.

This book concentrates on AC induction and synchronous motors, as well as synchronous generators. Other types of machines exist, but these motors and generators constitute the vast majority of electrical machines rated more than 1 kW presently used around the world.

1.1.1 AC Motors

Nearly all AC motors have a single-phase (for motors less than about 1 kW) or three-phase stator winding through which the input current flows. For AC motors, the stator is also called the armature. AC motors are usually classified according to the type of rotor winding. The rotor winding is also known as a field winding in most types of machines. A discussion of each type of AC motor follows.

Squirrel Cage Induction (SCI) Motor (Figure 1.1). The rotor produces a magnetic field by transformer-like AC induction from the stator (armature) winding. This is by far the most common type of AC motor made, with millions manufactured every year. SCI motors can range in size from a fraction of a horsepower motor (< 1 kW) to tens of thousands of horsepower (greater than 30 MW). The predominance of the squirrel cage induction motor is attributed to the simplicity and ruggedness of the rotor. In an SCI motor, the speed of the rotor is usually 1% or so slower than the "synchronous" speed of the rotating magnetic field in the air gap created by the stator winding. Thus, the rotor speed "slips" behind the speed of the air gap magnetic flux [1.1, 1.2]. The SCI motor is used for almost every conceivable application, including fluid pumping, fans, conveyor systems, grinding, mixing, and power tool operation.

Wound Rotor Induction Motor. The rotor is wound with insulated wire and the leads are brought off the rotor via slip rings. In operation, a current is induced into the rotor from the stator, just as for an SCI motor. However, in the wound rotor machine it is possible to limit the current in the rotor winding by means of an external resistance or slip-energy recovery system. This permits some control of the rotor speed. Wound rotor induction motors are relatively rare due to the extra maintenance required for the slip rings. IFD SCI motors are often a more reliable, cheaper alternative.

Synchronous Motor. This motor has a direct current flowing through the rotor (field) winding. The current creates a DC magnetic field, which interacts with the rotating magnetic field from the stator, causing the rotor to spin. The speed of the rotor is exactly related to the frequency of the AC current supplied to the stator winding (50 or 60 Hz). There is no "slip." The speed of the rotor depends on the number of rotor pole pairs (a pole pair contains one north and one south pole) times the AC frequency. There are two main ways of obtaining a DC current in the rotor. The oldest method, still popular, is to feed current onto the rotor by means of two slip rings (one positive, one negative). Alternatively, the "brushless" method uses a DC winding mounted on the stator to induce a current in an auxiliary three-phase winding mounted on the rotor to generate AC current, which is rectified (by "rotating" diodes) to DC. Synchronous motors require a small "pony motor" to run the rotor up to near synchronous speed. Alternatively, an SCI type of winding on the rotor can be used to drive the motor up to speed, before DC current is permitted to flow in the main

Figure 1.1. Photograph of a SCI rotor being lowered into the squirrel cage induction motor stator.

rotor winding. This winding is referred to as an amortisseur or damper winding. Because of the more complicated rotor and additional components, synchronous motors tend to be restricted to very large motors today (greater than 10 MW) or very slow speed motors. The advantage of a synchronous motor is that it usually requires less "inrush" current on start-up in comparison to a SCI motor, and the speed is more constant. Also, the operating energy costs are lower since, by adjusting the rotor DC current, one can improve the power factor of the motor, reducing the need for reactive power and thus the AC supply current. Refer to the section on synchronous generators below for further subdivision of the types of synchronous motor rotors. Two-pole synchronous motors use round rotors, as described in Section 1.1.2.

Permanent Magnet Motors. These motors have rotors made of a special permanently magnetized material. That is, no DC or AC current flows in the rotor, and there is no rotor winding. In the past, such motors were always rated at < 50 HP, since they can be hard to shut down. However, some large permanent magnet motors have been recently used in marine applications, due to their simplicity.

1.1.2 Synchronous Generators

Although induction generators do exist, particularly in wind turbine generators, they are relatively rare compared to synchronous generators. Virtually all generators used by electrical utilities are of the synchronous type. In synchronous generators, DC current flows through the rotor (field) winding, which creates a magnetic field from the rotor. At the same time, the rotor is spun by a steam turbine (using fossil or nuclear fuel), gas turbine, diesel engine or, a hydroelectric turbine. The spinning DC field from the rotor induces current to flow in the stator (armature) winding. As for motors, the following types of synchronous generators are determined by the design of the rotor, which is primarily a function of the speed of the driving turbine.

Round Rotor Generators (Figure 1.2). Also known as cylindrical rotor machines, round rotors are most common in high-speed machines, that is, machines in which the rotor revolves at about 1000 rpm or more. Where the electrical system operates at 60 Hz, the rotor speed is usually either 1800 rpm or 3600 rpm. The relatively smooth surface of the rotor reduces "windage" losses, that is, the energy lost to moving the air (or other gas) around in the air gap between the rotor and the stator—the fan effect. This loss can be substantial at high speeds in the presence of protuberances from the rotor surface. The smooth cylindrical shape also lends itself to a more robust structure under the high centrifugal forces that occur in high-speed machines. Round rotor generators, sometimes called "turbogenerators," are usually driven by steam turbines or gas turbines (jet engines). Turbogenerators using round ro-

Figure 1.2. Phototgraph of a small round rotor. The retaining rings are at each end of the rotor body.

tors have been made in excess of 1500 MW. (1000 MW is a typical load for a city of 500,000 people in an industrialized country). Such a machine may be 10 m in length and about 5 m in diameter, with a rotor on the order of 1.5 m in diameter. Such large generators almost always have a horizontally mounted rotor and are hydrogen-cooled (see Section 1.1.3).

Salient Pole Generators (Figure 1.3). Salient pole rotors usually have individual magnetic field poles that are mounted on a rim, with the rim in turn fastened to the rotor shaft by a "spider"—a set of spokes. Since the magnetic field poles protrude from the rim with spaces between the poles, the salient pole rotor creates considerable air turbulence in the air gap between the rotor and the stator as the rotor rotates, resulting in a relatively high windage loss. However, since the rotational speed is usually significantly less than 1000 rpm, the loss is considered moderate. Salient pole machines typically are used with hydraulic turbines, which have a relatively low rpm (the higher the penstock, i.e., the larger the fall of the water, the faster the speed). To generate 50 or 60 Hz current in the stator, a large number of field poles are needed (recall that the generated AC frequency is the number of pole pairs times the rotor speed in revolutions per second). Fifty pole pairs are not uncommon on a hydrogenerator, compared to one or two pole pairs on a turbogenerator. Such a large number of pole pairs requires a large rotor diameter in order to mount all the poles. Hydrogenerators have been made up to about 800 MW. The rotor in a large hydrogenerator is almost always vertically mounted, and may be more than 10 m in diameter.

Pump/Storage Generator. This is a special type of salient pole machine. It is used to pump water into an upper reservoir during times of low electricity demand. Then, at times of high demand for electricity, the water is allowed to flow from the upper reservoir to the lower reservoir, where the machine operates in reverse as a generator. The reversal of the ma-

Figure 1.3. Photograph of a salient pole rotor for a large, low-speed motor. (Courtesy TECO-Westinghouse.)

chine from the pump to generate mode is commonly accomplished by changing the connections on the machine's stator winding to reverse rotor direction. In a few cases, the pitch of the hydraulic turbine blades is changed. In the pump motor mode, the rotor can come up to speed by using a SCI-type winding on the rotor (referred to as an amortisseur or damper winding), resulting in a large inrush current, or by using a "pony" motor. If the former is used, the machine is often energized by an inverter-fed drive (IFD) that gradually increases the rotor speed by slowly increasing the AC frequency to the stator. Since the speed is typically less than a few hundred rpm, the rotor is of the salient pole type. Pump storage units have been made up to 500 MW.

1.1.3   Classification by Cooling

Another important means of classifying machines is by the type of cooling medium they use: water, air, and/or hydrogen gas. One of the main heat sources in electrical machines is the DC or AC current flowing through the stator and rotor windings. These are usually called I^2R losses, since the heat generated is proportional to the current squared times the resistance of the conductors (almost always copper in stator windings, but sometimes aluminum in SCI rotors). There are other sources of heat: magnetic core losses, windage losses, and eddy current losses. All these losses cause the temperature of the windings to rise. Unless this heat is removed, the winding insulation deteriorates and the machine fails due to a short circuit.

Indirect Air Cooling. Motors and modern generators rated less than about 100 MVA are almost always cooled by air flowing over the rotor and stator. This is called indirect cooling since the winding conductors are not directly in contact with the cooling air due to the presence of electrical insulation on the windings. The air itself may be continuously drawn in from the environment, that is, not recirculated. Such machines are termed open-ventilated, although there may be some effort to prevent particulates (sand, coal dust, pollution, etc.) and/or moisture from entering the machine using filtering and indirect paths for drawing in the air. These open-ventilated machines are referred to as weather-protected or WP.

A second means of obtaining cool air is to totally enclose the machine and recirculate air via a heat exchanger. This is often needed for motors that are exposed to the elements. The recirculated air is most often cooled by an air-to-water heat exchanger in large machines, or cooled by the outside air via radiating metal fins in small motors or a tube-type cooler in large ones. Either a separate blower motor or a fan mounted on the motor shaft circulates the air. IEC and NEMA standards describe the various types of cooling methods in detail [1.4, 1.5].

Although old, small generators may be open-ventilated, the vast majority of hydrogenerators and turbogenerators (rated less than about 50 MVA) have recirculated air flowing through the machine. Virtually all hydrogenerators use recirculated air, with the air often cooled by air-to-water heat exchangers. For turbogenerators rated up to a few hundred megawatts, recirculated air is now the most common form of cooling.

Indirect Hydrogen Cooling. Almost all large turbogenerators use recirculated hydrogen as the cooling gas. This is because the smaller and lighter hydrogen molecule results in a lower windage loss and better heat transfer than air. It is then cost effective to use hydrogen in spite of the extra expense involved, due to the few percent gain in efficiency. The dividing line for when to use hydrogen cooling is constantly changing. In the 1990s, there was a definite trend to reserve hydrogen cooling for machines rated more than 300 MVA, whereas in the past, hydrogen cooling was sometimes used on steam and gas turbine generators as small as 50 MVA [1.6, 1.7].

Directly Cooled Windings. Generators are referred to as being indirectly or convention-ally cooled if the windings are cooled by flowing air or hydrogen over the surface of the windings and through the core, where the heat created within the conductors must first pass through the insulation. Large generator stator and rotor windings are frequently "directly" cooled. In direct-cooled windings, water or hydrogen is passed internally through the con-ductors or through ducts immediately adjacent to the conductors. Direct water-cooled stator windings pass very pure water through hollow copper conductors strands, or through stain-less steel tubes immediately adjacent to the copper conductors. Since the cooling medium is directly in contact with the conductors, this very efficiently removes the heat developed by I^2R losses. With indirectly cooled machines, the heat from the I^2R losses must first be trans-mitted through the electrical insulation covering the conductors, which forms a significant thermal barrier. Although not quite as effective in removing heat, in direct hydrogen-cooled windings the hydrogen is allowed to flow within hollow copper tubes or stainless steel tubes, just as in the water-cooled design. In both cases, special provisions must be taken to ensure that the direct water or hydrogen cooling does not introduce electrical insulation problems. See Sections 1.4 and 8.13.

Direct water cooling of hydrogenerator stator windings is applied to machines larger than about 500 MW. There are no direct hydrogen-cooled hydrogenerators. In the 1950s, turbo-generators as small as 100–150 MVA had direct hydrogen or direct water stator cooling. Modern turbogenerators normally only use direct cooling if they are larger than about 200 MVA.

Direct cooling of rotor windings in turbogenerators is common whenever hydrogen is present, or in air-cooled turbogenerators rated more than about 50 MVA. With the exception of machines made by ASEA, only the very largest turbo and hydrogenerators use direct water cooling of the rotor.

1.2 PURPOSE OF WINDINGS

The stator winding and rotor winding consist of several components, each with their own function. Furthermore, different types of machines have different components. Stator and ro-tor windings are discussed separately below.

1.2.1 Stator Winding

The three main components in a stator are the copper conductors (although aluminum is sometimes used), the stator core, and the insulation. The copper is a conduit for the stator winding current. In a generator, the stator output current is induced to flow in the copper con-ductors as a reaction to the rotating magnetic field from the rotor. In a motor, a current is in-troduced into the stator, creating a rotating magnetic field that forces the rotor to move. The copper conductors must have a cross section large enough to carry all the current required without overheating.

Figure 1.4 is the circuit diagram of a typical three-phase motor or generator stator wind-ing. The diagram shows that each phase has one or more parallel paths for current flow. Mul-tiple parallels are often necessary since a copper cross section large enough to carry the entire phase current may result in an uneconomic stator slot size. Each parallel consists of a number of coils connected in series. For most motors and small generators, each coil consists of a number of turns of copper conductors formed into a loop. The rationale for selecting the number of parallels, the number of coils in series, and the number of turns per coil in any par-

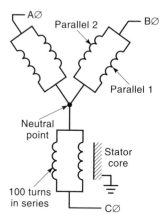

Figure 1.4. Schematic diagram for a three-phase, Y-connected stator or winding, with two parallel circuits per phase.

ticular machine is beyond the scope of this book. The reader is referred to any book on motors and generators, for example references 1.1 to 1.3.

The stator core in a generator concentrates the magnetic field from the rotor on the copper conductors in the coils. The stator core consists of thin sheets of magnetic steel (referred to as laminations). The magnetic steel acts as a low-reluctance (low magnetic impedance) path for the magnetic fields from the rotor to the stator, or vice versa for a motor. The steel core also prevents most of the stator winding magnetic field from escaping the ends of the stator core, which would cause currents to flow in adjacent conductive material. Chapter 6 contains more information on cores.

The final major component of a stator winding is the electrical insulation. Unlike copper conductors and magnetic steel, which are active components in making a motor or generator function, the insulation is passive. That is, it does not help to produce a magnetic field or guide its path. Generator and motor designers would like nothing better than to eliminate the electrical insulation, since the insulation increases machine size and cost, and reduces efficiency, without helping to create any torque or current [1.8]. Insulation is "overhead," with a primary purpose of preventing short circuits between the conductors or to ground. However, without the insulation, copper conductors would come in contact with one another or with the grounded stator core, causing the current to flow in undesired paths and preventing the proper operation of the machine. In addition, indirectly cooled machines require the insulation to be a thermal conductor, so that the copper conductors do not overheat. The insulation system must also hold the copper conductors tightly in place to prevent movement.

As will be discussed at length in Chapters 3 and 4, the stator winding insulation system contains organic materials as a primary constituent. In general, organic materials soften at a much lower temperature and have a much lower mechanical strength than copper or steel. Thus, the life of a stator winding is limited most often by the electrical insulation rather than by the conductors or the steel core. Furthermore, stator winding maintenance and testing almost always refers to testing and maintenance of the electrical insulation. Section 1.3 will describe the different components of the stator winding insulation system and their purposes.

1.2.2 Insulated Rotor Windings

In many ways, the rotor winding has the same components as the stator, but with important changes. In all cases, copper, copper alloy, or aluminum conductors are present to act as a conduit for current flow. However, the steady-state current flowing through the rotor winding is usually DC (in synchronous machines), or very low frequency AC (a few Hz) in induction machines. This lower frequency makes the need for a laminated stator core less critical.

The conductors in rotor windings are often embedded in the laminated steel core or surround laminated magnetic steel. However, round rotors in large turbogenerator and high-speed salient pole machines are usually made from forged magnetic steel, since laminated magnetic steel rotors cannot tolerate the high centrifugal forces.

Synchronous machine rotor windings, as well as wound rotor induction motors, contain electrical insulation to prevent short circuits between adjacent conductors or to the rotor body. As will be discussed in Chapters 3 and 5, the insulating materials used in rotor windings are largely composites of organic and inorganic materials, and thus have poor thermal and mechanical properties compared to copper, aluminum, or steel. The insulation then often determines the expected life of a rotor winding.

1.2.3 Squirrel Cage Induction Motor Rotor Windings

SCI rotor windings are unique in that they usually have no explicit electrical insulation on the rotor conductors. Instead, the copper, copper alloy, or aluminum conductors are directly installed in slots in the laminated steel rotor core. (Smaller SCI rotors may have the aluminum conductors cast in place.) In normal operation, there are only a few volts induced on the rotor conductors, and the conductivity of the conductors is much higher than that of the steel core. Because the current normally only flows in the conductors, electrical insulation is not needed to force the current to flow in the right paths. Reference 1.9 describes the practical aspects of rotor design and operation in considerable detail.

The only time that significant voltage can appear on the rotor conductors is during motor starting. This is also the time that extremely heavy currents will flow in the rotor windings. Under some conditions during starting, the conductors make and break contact with the rotor core, leading to sparking. This is normally easily tolerated. However, some SCI motors operate in a flammable environment, and this rotor sparking may ignite an explosion. Therefore, some motor manufacturers do insulate the conductors from the rotor core to prevent the sparking [1.10]. Since such applications are rare, for the purposes of this book, we assume that the rotor is not insulated.

Although SCI rotor windings are generally not insulated, for completeness, Section 9.4 does discuss such rotors, and Chapters 12 and 13 present some common tests for SCI rotor winding integrity.

1.3 TYPES OF STATOR WINDING CONSTRUCTION

Three basic types of stator winding structures are employed over the range from 1 kW to more than 1000 MW:

1. Random-wound stators
2. Form-wound stators using multiturn coils
3. Form-wound stators using Roebel bars

In general, random-wound stators are typically used for machines less than several hundred kW. Form-wound coil windings are used in most large motors and many generators rated up to 50 to 100 MVA. Roebel bar windings are used for large generators. Although each type of construction is described below, some machine manufacturers have made hybrids that do not fit easily into any of the above categories; these are not discussed in this book.

1.3.1 Random-Wound Stators

Random-wound stators consist of round, insulated copper conductors (magnet wire or winding wire) that are wound continuously (by hand or by a winding machine) through slots in the stator core to form a coil (Figure 1.5). Figure 1.5 shows that most of the turns in the coils can be easily seen. Each turn (loop) of magnet wire could, in principle, be placed randomly against any other turn of magnet wire in the coil, independent of the voltage level of the turn, thus the term "random." Since a turn that is connected to the phase terminal can be adjacent to a turn that is operating at low voltage (i.e., at the neutral point), random-wound stators usually operate at voltages less than 1000 V. This effectively limits random-wound stators to machines less than several hundred kW or HP.

1.3.2 Form-Wound Stators—Coil Type

Form-wound stators are usually intended for machines operating at 1000 V and above. Such windings are made from insulated coils that have been preformed prior to insertion in the slots in the stator core (Figure 1.6). The preformed coil consists of a continuous loop of magnet wire shaped into a coil (sometimes referred to as a diamond shape), with additional insulation ap-

Figure 1.5. Photograph of the end-winding and slots of a random-wound stator. (Courtesy TECO-Westinghouse.)

(a)

(b)

Figure 1.6. (a) Photograph of a form-wound motor stator winding. (Courtesy TECO-Westinghouse.) (b) A single form-wound coil being inserted into two slots.

plied over the coil loops. Usually, each coil can have from two to 12 turns, and several coils are connected in series to create the proper number of poles and turns between the phase terminal and ground (or neutral); see Figure 1.4. Careful design and manufacture are used to ensure that each turn in a coil is adjacent to another turn with the smallest possible voltage difference. By minimizing the voltage between adjacent turns, thinner insulation can be used to separate the turns. For example, in a 4160 volt stator winding (2400 V line-to-ground), the winding may have 10 coils connected in series, with each coil consisting of 10 turns, yielding 100 turns between the phase terminal and neutral. The maximum voltage between adjacent turns is 24 V. In contrast, if the stator were of a random-wound type, there might be up to 2400 V between adjacent turns, since a phase-end turn may be adjacent to a neutral-end turn. This placement would require an unacceptably large magnet wire insulation thickness.

1.3.3 Form-Wound Stators—Roebel Bar Type

In large generators, the more the power output, the larger and mechanically stiffer each coil usually is. In stators larger than about 50 MW, the form-wound coil is large enough that there are difficulties in inserting both legs of the coil in the narrow slots in the stator core without risking mechanical damage to the coil during the insertion process. Thus, most large generators today are not made from multiturn coils, but rather from "half-turn" coils, often referred to as Roebel bars. With a Roebel bar construction, only one half of a "coil" is inserted into the slot at a time, which is considerably easier than inserting two sides of a coil in two slots simultaneously. With the Roebel bar approach, electrical connections to make the "coils" are needed at both ends of the bar (Figure 1.7).

1.4 STATOR WINDING INSULATION SYSTEM FEATURES

The stator winding insulation system contains several different components and features, which together ensure that electrical shorts do not occur, that the heat from the conductor I^2R losses are transmitted to a heat sink, and that the conductors do not vibrate in spite of the magnetic forces. The basic stator insulation system components are the:

- Strand (or subconductor) insulation
- Turn insulation
- Groundwall (or ground or earth) insulation

Figures 1.8 and 1.9 show cross sections of random-wound and form-wound coils in a stator slot, and identify the above components. Note that the form-wound stator has two coils per slot; this is typical. Figure 1.10 is a photograph of the cross section of a multiturn coil. In addition to the main insulation components, the insulation system sometimes has high-voltage stress-relief coatings and end-winding support components.

The following sections describe the purpose of each of these components. The mechanical, thermal, electrical, and environmental stresses that the components are subjected to are also described.

1.4.1 Strand Insulation

In random-wound stators, the strand insulation can function as the turn insulation, although extra sleeving is sometimes applied to boost the turn insulation strength in key areas. Many

Figure 1.7. Photo of a turbogenerator stator winding using Roebel bars.

form-wound machines employ separate strand and turn insulation. The following mainly addresses the strand insulation in form-wound coils and bars. Strand insulation in random-wound machines will be discussed as turn insulation. Section 1.4.8 discusses strand insulation in its role as transposition insulation.

There are both electrical and mechanical reasons for stranding a conductor in a form-wound coil or bar. From a mechanical point of view, a conductor that is big enough to carry

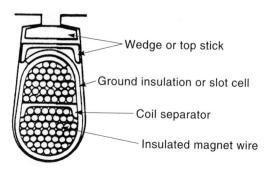

Figure 1.8. Cross section of a random stator winding slot.

the current needed in the coil or bar for a large machine will have a relatively large cross-sectional area. That is, a large conductor cross section is needed to achieve the desired ampacity. Such a large conductor is difficult to bend and form into the required coil/bar shape. A conductor formed from smaller strands (also called subconductors) is easier to bend into the required shape than one large conductor.

From an electrical point of view, there are reasons to make strands and insulate them from one another. It is well known from electromagnetic theory that if a copper conductor has a large enough cross-sectional area, the current will tend to flow on the periphery of the conductor. This is known as the skin effect. The skin effect gives rise to a skin depth through which most of the current flows. The skin depth of copper is 8.5 mm at 60 Hz. If the conductor has a cross section such that the thickness is greater than 8.5 mm, there is a tendency for the current *not* to flow through the center of the conductor, which implies that the current is not making use of all the available crossection. This is reflected as an effective AC resistance that is higher than the DC resistance. The higher AC resistance gives rise to a larger I^2R loss than if the same cross section had been made from strands that are insulated from one another to prevent the skin effect from occurring. That is, by making the required cross section from strands that are insulated from one another, all the copper cross section is used for current flow, the skin effect is negated, and the losses are reduced.

In addition, Eddy current losses occur in solid conductors of too large a cross section. In the slots, the main magnetic field is primarily radial, that is, perpendicular to the axial direction. There is also a small circumferential (slot leakage) flux that can induce eddy currents to flow. In the end-winding, an axial magnetic field is caused by the abrupt end of the rotor and stator core. This axial magnetic field can be substantial in synchronous machines that are under-excited. By Ampere's Law, or the 'right hand rule', this axial magnetic field will tend to cause a current to circulate within the cross section of the conductor (Figure 1.11). The larger the cross sectional area, the greater the magnetic flux that can be encircled by a path on the periphery of the conductor, and the larger the induced current. The result is a greater I^2R loss from this circulating current. By reducing the size of the conductors, there is a reduction in stray magnetic field losses, improving efficiency.

The electrical reasons for stranding require the strands to be insulated from one another. The voltage across the strands is less than a few tens of volts; therefore, the strand insulation can be very thin. The strand insulation is subject to damage during the coil manufacturing process, so it must have good mechanical properties. Since the strand insulation is immedi-

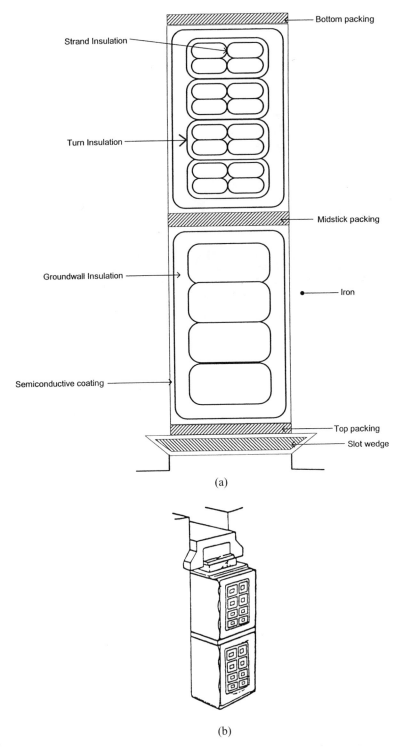

Strand Insulation

Turn Insulation

Groundwall Insulation

Iron

Semiconductive coating

Midstick packing

Top packing

Slot wedge

(a)

(b)

Figure 1.9. Cross sections of slots containing (a) form-wound multiturn coils; (b) directly cooled Roebel bars.

Figure 1.10. Cross-section of a multiturn coil, with three turns and three strands per turn.

ately adjacent to the copper conductors that are carrying the main stator current, which produces the I^2R loss, the strand insulation is exposed to the highest temperatures in the stator. Therefore, the strand insulation must have good thermal properties. Section 3.8 describes in detail the strand insulation materials that are in use. Although manufacturers ensure that strand shorts are not present in a new coil, they may occur during service due to thermal or mechanical aging (see Chapter 8). A few strand shorts in form-wound coils/bars will not cause winding failure, but will increase the stator winding losses and cause local temperature increases due to circulating currents.

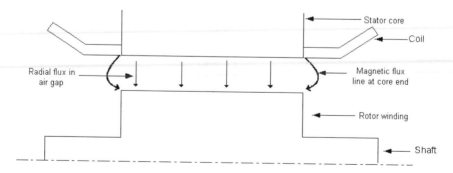

Figure 1.11. Side view of a generator showing the radial magnetic flux in the air gap and the bulging flux at the core end, which results in an axial flux.

1.4.2 Turn Insulation

The purpose of the turn insulation in both random- and form-wound stators is to prevent shorts between the turns in a coil. If a turn short occurs, the shorted turn will appear as the secondary winding of an autotransformer. If, for example, the winding has 100 turns between the phase terminal and neutral (the "primary winding"), and if a dead short appears across one turn (the "secondary"), then 100 times normal current will flow in the shorted turn. This follows from the transformer law:

$$n_p I_p = n_s I_s \tag{1.1}$$

where n refers to the number of turns in the primary or secondary, and I is the current in the primary or secondary. Consequently, a huge circulating current will flow in the faulted turn, rapidly overheating it. Usually, this high current will be followed quickly by a ground fault due to melted copper burning through any groundwall insulation. Clearly, effective turn insulation is needed for long stator winding life.

The power frequency voltage across the turn insulation in a random-wound machine can range up to the rated phase-to-phase voltage of the stator because, by definition, the turns are randomly placed in the slot and thus may be adjacent to a phase-end turn in another phase, although many motor manufacturers may insert extra insulating barriers between coils in the same slot but in different phases and between coils in different phases in the end-windings. Since random winding is rarely used on machines rated more than 600 V (phase-to-phase), the turn insulation can be fairly thin. However, if a motor is subject to high-voltage pulses, especially from modern inverter-fed drives (IFDs), interturn voltage stresses that far exceed the normal maximum of 600 Vac can result. These high-voltage pulses give rise to failure mechanisms, as discussed in Section 8.7.

The power frequency voltage across adjacent turns in a form-wound multiturn coil is well defined. Essentially, one can take the number of turns between the phase terminal and the neutral and divide it into the phase–ground voltage to get the voltage across each turn. For example, if a motor is rated 4160 Vrms (phase–phase), the phase–ground voltage is 2400 V. This will result in about 24 Vrms across each turn, if there are 100 turns between the phase end and neutral. This occurs because coil manufacturers take considerable trouble to ensure that the inductance of each coil is the same, and that the inductance of each turn within a coil is the same. Since the inductive impedance (X_L) in ohms is:

$$X_L = 2\pi f L \tag{1.2}$$

where f is the frequency of the AC voltage and L is the coil or turn inductance, the turns appear as impedances in a voltage divider, where the coil series impedances are equal. In general, the voltage across each turn will be between about 10 Vac (small form-wound motors) to 250 Vac (for large generator multiturn coils).

The turn insulation in form-wound coils can be exposed to very high transient voltages associated with motor starts, IFD operation, or lightning strikes. Such transient voltages may age or puncture the turn insulation. This will be discussed in Section 8.7. As described below, the turn insulation around the periphery of the copper conductors is also exposed to the rated AC phase–ground stress, as well as the turn–turn AC voltage and the phase coil-to-coil voltage.

Before about 1970, the strand and the turn insulation were separate components in multiturn coils. Since that time, many stator manufacturers have combined the strand and turn in-

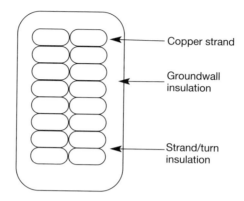

Figure 1.12. Photo of the cross section of a coil where the turn insulation and the strand insulation are the same.

sulation. Figure 1.12 shows the strand insulation is upgraded (usually with more thickness) to serve as both the strand and the turn insulation. This eliminates a manufacturing step (i.e., the turn taping process) and increases the fraction of the slot cross section that can be filled with copper. However, some machine owners have found that in-service failures occur sooner in stators without a separate turn insulation component [1.11].

Both form-wound coils and random-wound stators are also exposed to mechanical and thermal stresses. The highest mechanical stresses tend to occur in the coil forming process, which requires the insulation-covered turns to be bent through large angles, which can stretch and crack the insulation. Steady-state, magnetically induced mechanical vibration forces (at twice the power frequency) act on the turns during normal machine operation. In addition, very large transient magnetic forces act on the turns during motor starting or out-of-phase synchronization in generators. These are discussed in detail in Chapter 8. The result is the turn insulation requires good mechanical strength.

The thermal stresses on the turn insulation are essentially the same as those described above for the strand insulation. The turn insulation is adjacent to the copper conductors, which are hot from the I^2R losses in the winding. The higher the melting or decomposition temperature of the turn insulation, the greater the design current that can flow through the stator.

In a Roebel bar winding, no turn insulation is used and there is only strand insulation. Thus, as will be discussed in Chapter 8, some failure mechanisms that can occur with multi-turn coils will not occur with Roebel bar stators.

1.4.3 Groundwall Insulation

Groundwall insulation is the component that separates the copper conductors from the grounded stator core. Groundwall insulation failure usually triggers a ground fault relay, taking the motor or generator off-line.* Thus the stator groundwall insulation is critical to the

---

*If the fault occurs electrically close to the neutral, many types of relays will not detect the ground fault. This allows a current to flow from the copper to the stator core, which may damage the stator core. Special third-harmonic ground fault relays are available to detect this type of condition.

proper operation of a motor or generator. For a long service life, the groundwall must meet the rigors of the electrical, thermal, and mechanical stresses that it is subject to.

Electrical Design. The ground insulation in a random-wound stator, in its most basic form, is the same as the turn insulation. So the magnet wire insulation serves as both the turn and ground insulation. The turn insulation is designed to withstand the full phase–phase applied voltage, usually a maximum of 600 Vac. However, especially in motors rated at more than 250 volts, random-wound stators usually also have sheets of insulating material lining the slots, to provide additional ground insulation (Figure 1.8). They may also have sheets of insulating material separating coils in different phases. Such insulating materials are discussed in Section 3.7. The thermal capability of the liners and separators are less stringent than required of the turn insulation since the liners are not in immediate contact with the copper conductors. Mechanically, however, the liners must have excellent abrasion resistance to withstand the magnetic forces, which cause the turns to vibrate, and good tear resistance to withstand the manufacturing operation.

The groundwall insulation in form-wound multiturn coils and Roebel bars requires considerably more discussion. Coils or bars connected to the phase end of the winding will have the full rated phase–ground voltage across it. For example, a stator rated at 13.8 kV (phase–phase) will have a maximum of 8 kV (13.8/$\sqrt{3}$) between the copper conductors and the grounded stator core. This high voltage requires a substantial groundwall insulation thickness. The high groundwall voltage only occurs in the coils/bars connected to the phase terminals. The coils/bars connected to the neutral have essentially no voltage across the groundwall during normal operation. Yet, virtually all machines are designed to have the same insulation thickness for both phase-end and neutral-end coils. If the coils all had different groundwall thicknesses, then, to take advantage of the smaller width of a neutral end bar or coil, the stator slot would be narrower. All the slots would be of different sizes and problems would occur when a neutral bar/coil had to be placed on top of a phase-end bar in the same slot. It is simply easier to make the slots all the same size. An advantage to this design approach is that since all coils/bars have the same groundwall thickness, changing connections to reverse the line and neutral ends may extend the life of a winding. That is, the coils formerly at the neutral are now subjected to high voltage, and vice versa. Such a repair may be useful if purely electrical failure mechanisms, such as those described in Sections 8.5 and 8.6, are occurring.

Other aspects of the electrical design are discussed in Section 1.4.4.

Thermal Design. The groundwall insulation in indirectly cooled form-wound machines is the main path for transmitting the heat from the copper conductors (heat source) to the stator core (heat sink). Thus, the groundwall insulation should have as low a thermal resistance as possible, to prevent high temperatures in the copper. To achieve a low thermal resistance requires the groundwall materials to have as high a thermal conductivity as possible, and for the groundwall to be free of voids. Such air voids block the flow of heat, in the same way that two layers of glass separated by a small air space inhibits the flow of heat through a window. Therefore, the insulation must be able to operate at high temperatures (in the copper) and be manufactured in such a way as to minimize the formation of air pockets within the groundwall.

Mechanical Design. There are large magnetic forces acting on the copper conductors. These magnetic forces are primarily the result of the two magnetic fields from the current flowing in the top and bottom coils/bars in each slot. These fields interact, exerting a force that makes the individual copper conductors as well as the entire coil or bar vibrate (primarily) up and down in the slot. The force, F, acting on the top coil at 120 Hz for a 60 Hz current in the radial direction for 1 meter length of coil is given by [1.12]:

$$F = \frac{kI^2}{d} \text{ kN/m} \tag{1.3}$$

where I is the rms current through the Roebel bar, or $I = nI_0$, with I_0 being the rms coil current times the number of turns in the coil; d is the width of the stator slot in meters; and k is 0.96. The force is expressed in kN of force acting per meter length of coil/bar in the slot. If the current in a stator bar is

$$I = A \sin \omega t$$

where ω is $2\pi f$, f is the 50 or 60 Hz power frequency, and t is time. Then (1.3) becomes

$$F = \frac{kA^2(1 - \cos 2\omega t)}{(2d)}$$

Thus, there is a net force to the bottom of the slot. Around this "DC" force is an oscillating force at twice the power frequency, i.e., 100 Hz or 120 Hz. There is also a 100 or 120 Hz force in the circumferential direction caused by the rotor's magnetic field interacting with the current in the stator coil/bar. This circumferential force is only about 10% of the radial force [1.12].

The groundwall insulation must also help to prevent the copper conductors from vibrating in response to the magnetic forces. If the groundwall were full of air pockets, the copper conductors might be free to vibrate. This would cause the conductors to bang against the remaining groundwall insulation, as well as allowing the copper strands and turns to vibrate against one another, leading to insulation abrasion. If an incompressible insulating mass exists between the copper and the coil surface, then the conductors cannot move.

1.4.4   Groundwall Partial Discharge Suppression

In form-wound bars and coils rated greater than about 4 kV, partial discharges can occur within the groundwall insulation or between the surface of the coil or bar and the stator coil. These partial discharges (PDs), which are sometimes colloquially (but incorrectly) called coronas,* are created by the high-voltage stress that occurs in the groundwall. If an air pocket (also called a void or a delamination) exists in the groundwall, the high electric stress will break down the air, causing a spark. This spark will degrade the insulation and, if not corrected, repeated discharges will eventually erode a hole through the groundwall, leading to failure. Therefore, efforts are needed to eliminate voids in the groundwall to prevent stator winding failure. In addition, a partial discharge suppression system is needed to prevent PD in any air gaps between the surface of the coils and bars and the core. The following is a discussion of the physics of the PD process within voids in the groundwall. Section 1.4.5 discusses the requirement for a PD suppression system on coil and bar surfaces.

---

*According to the IEEE Dictionary (IEEE Standard 100-1996), corona is a form of partial discharge. The term corona is reserved for the visible partial discharges that can occur on bare metal conductors operating at high voltage, which ionize the surrounding air. Since PD within the groundwall is not visible, it should not be termed corona.

Electric breakdown of an insulation is analogous to mechanical failure of a material. For example, the tensile strength of a material depends on the nature of the material (specifically, the strength of the material's chemical bonds) and the cross-sectional area of the material. Mechanical failure occurs when the chemical bonds rupture under the mechanical stress. Tensile stress (kPa) is defined in terms of force (weight) supported (in kN or pounds) per unit cross-sectional area (m^2) or, in British units, pounds per square inch (psi). The larger the cross section, the more force a material (for example, a steel wire) can support before it breaks. Different materials have vastly different tensile strengths. The tensile strength of steel exceeds that of copper, which, in turn, is hundreds of times greater than the tensile strength of paper.

Electric breakdown strength is also a property of an insulating material. Electric breakdown is not governed by voltage alone. Rather, it depends on the electric field, just as the tensile stress on a copper wire is not solely determined by the force it is supporting but by the force per cross-sectional area. Electric stress, E, in a parallel plate geometry is given by

$$E = \frac{V}{d} \qquad \text{(kV/mm)} \qquad (1.4)$$

where V is the voltage across the metal plates in kV and d is the distance between the plates in mm. Note that as for the tensile stress, there is an element of dimensionality. If the voltage is gradually increased across the metal plates, there will be a voltage at which electric breakdown occurs, i.e., at which a spark will cross between the plates. Using Equation 1.4, one can then calculate the electric strength of the insulation material. Breakdown involves a process in which the negatively charged electrons orbiting the atoms within the insulation are ripped away from the molecules because they are attracted to the positive metal plate. This is called ionization. The electrons accelerate toward the positive metal plate under the electric field, and often collide with other atoms, ionizing these also. A cloud of positive ions is left behind that travels gradually to the negative metal plate. The electrons and ions short the voltage difference between the two metal plates. The result is the electric breakdown of the insulation. Two examples of the electric breakdown of air are lightning and the static discharge from a person who has acquired a charge by walking across a carpet and then reaches for a grounded doorknob.

Like mechanical (tensile) strength, each material has its own characteristic electric breakdown strength. For air at room temperature and one atmosphere (100 kPa) pressure, and in low humidity, the electric strength is about 3 kV/mm. The electric strength of gas insulation depends on the gas pressure and humidity. For example, the breakdown strength of air at 300 kPa is about 9 kV/mm, that is, for the same distance between the plates, the breakdown voltage is three times higher than at atmospheric pressure (100 kPa). This relationship is known as Paschen's Law [1.14]. The breakdown strength of air and hydrogen is about the same. However, since hydrogen-cooled generators often operate at 300 kPa or more, the breakdown strength of hydrogen under this pressure is 9 kV/mm. As we will see later, this allows hydrogen-cooled generators to operate at higher voltages than air-cooled machines, which operate at atmospheric pressure. The intrinsic breakdown strength of most solid insulating materials such as epoxy and polyester composites is on the order of 300 kV/mm. That is, solid materials used as stator winding insulation are about 100 times stronger than air. More details on electric breakdown and the physics behind it are in Reference 1.14.

The presence of air (or hydrogen) pockets within the groundwall can lead to the electric breakdown of the air pockets, a process called a partial discharge (PD). To understand this process, consider the groundwall cross section in Figure 1.13. For electric breakdown to oc-

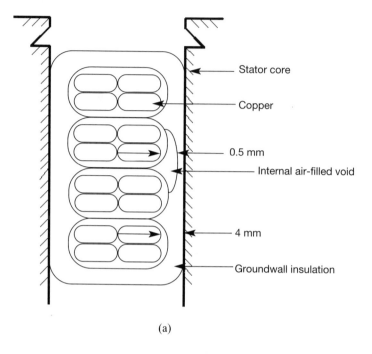

(a)

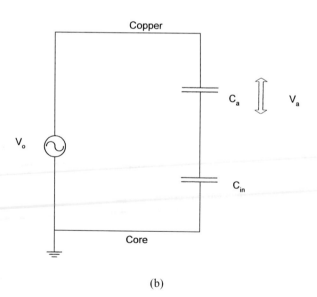

(b)

Figure 1.13. (a) Cross section of a coil with an air pocket next to the turn insulation. (b) An electrical equivalent circuit.

cur in the air pocket, there must be a high electric stress across it. Using a simple capacitive voltage divider circuit (Figure 1.13b), one can calculate the voltage across the air pocket. The capacitance of the air pocket, to a first approximation, can be calculated assuming it is a parallel plate capacitor, i.e.,

$$C_a = \frac{\varepsilon A}{d_a}$$ (1.5)

where ε is the permittivity of the insulating material, A is the cross-sectional area of the void, and d_a is the thickness of the void (0.5 mm in the example in Figure 1.13a). The permittivity is often represented as:

$$\varepsilon = \varepsilon_r \varepsilon_0$$ (1.6)

where ε_r is called the relative dielectric constant and ε_0 is the permittivity of free space, equal to 8.85×10^{-12} F/m. The dielectric constant for air is 1.0. For most stator winding insulation materials, the dielectric constant is about 4. Thus, assuming a unity cross-sectional area, the capacitance of the air pocket in Figure 1.13a can be calculated.

The air pocket is in series with another capacitor, which represents the capacitance (C_{in}) of the solid insulating material. Using Equation 1.5, assuming a dielectric constant of 4, and assuming that the thickness of the insulation capacitor is 4 mm, the insulation capacitances can be calculated to a first approximation.

Using simple circuit theory, the voltage across the air pocket can be calculated:

$$V_a = \frac{C_{in} V_0}{C_a + C_{in}}$$ (1.7)

where V_0 is the applied AC voltage (8 kV rms if the coil is at the phase terminal) and C_{in} is the solid insulating material capacitance. Using the above equations, the dimensions in Figure 1.13a, recognizing that A and ε_0 will cancel out, and assuming the dielectric constants are 1 and 4 for air and the insulation, respectively, one can calculate that the voltage across the air pocket is 33% of the applied voltage. For a V_0 of 8 kV rms (rated phase–ground voltage for a phase-end coil in a 13.8 kV stator), the voltage across the air pocket is about 2.6 kV. From Equation 1.4, this implies the electric stress within the air pocket is 5.2 kV/mm. This far exceeds the 3 kV/mm electric strength of the air, and thus electric breakdown will occur within the air. The resulting spark is called a partial discharge. The discharge is referred to as "partial" since the spark is only in the air pocket or void. The rest of the insulation is intact and can support the applied voltage. (A complete discharge is really a phase–ground breakdown, which would trigger the ground fault relay.) For a more complete analysis of partial discharge phenomena, see Reference 1.15.

Since the breakdown strength of typical solid groundwall insulation materials is in excess of 300 kV/mm, there is practically no possibility of the groundwall insulation itself experiencing electrical breakdown. Rather, PD will occur only when there are gas-filled voids within the groundwall. These discharges are harmful to the groundwall, because repeat PD will eventually degrade the solid insulation by breaking the chemical bonds. The spark contains electrons and ions that bombard the insulation surfaces of the void. This bombardment can rupture the chemical bonds, especially in organic materials such as asphalt, polyester, and epoxy, all common groundwall insulation materials. Over the years, the constant impact from the electrons and ions will erode a hole through the groundwall, giving rise to a ground fault. The consequence of PD is discussed in greater detail in Chapter 8. The important conclusion is that air pockets within the groundwall of high voltage coils can lead to PD and eventual failure.

1.4.5 Groundwall Stress Relief Coatings

The stress relief coatings are important insulation system components in stator windings operating at 6 kV or above. These coatings are present to prevent partial discharges from occurring on the surface of the stator bars or coils. They prevent PD from occurring in any air gap that might be present between the coil/bar surface and the stator core, or in the end-winding near the end of the stator core.

Slot Semiconductive Coating. The reason PD may occur between the coil and the core is similar to the reason PD can occur in air pockets within the groundwall. Since coils and bars are fabricated outside of the stator core, they must be thinner in the narrow dimension than the width of the core's steel slots, otherwise, the coils/bars cannot be inserted into the slot. Thus, an air gap between the coil/bar surface and the core is inevitable.* Figure 1.14a shows the gap that can occur in the slot, adjacent to the coil surface, since the coil is undersized. An equivalent circuit, only slightly different from the groundwall case, is shown in Figure 1.14b. As for the groundwall voids, a significant percentage of the copper voltage will appear across the air gap. If the electric stress in the air gap exceeds 3 kV/mm, PD will occur, at least in an air-cooled machine. This PD will eventually erode a hole through the groundwall, causing failure. Discharges on the coil/bar surface are sometimes referred to as slot discharge, since they can be seen in the slot. Under practical conditions, most stators rated 6 kV or more will experience this PD on the coil/bar surface. In addition to deteriorating the insulation, surface PD in air-cooled machines creates ozone. The ozone combines with nitrogen to create acids that pose a health hazard, and which can weaken rubber materials and corrode metal; for example in heat exchangers.

To prevent PD on the coil or bar surfaces, manufacturers have long been coating the coil/bar in the slot area with a partly conductive coating. The coating is usually a carbon-black-loaded paint or tape. This coating, often called a semiconductive coating (although this has nothing to do with semiconductors in the transistor sense), is likely to be in contact with the grounded stator core at many places along the length of the slot. With a sufficiently low resistance (schematically shown as Rs in Figure 1.14b), this coating is essentially at ground potential because of the contact with the core. Thus, the voltage across any air gap is zero. PD cannot occur in the gap, because the electric stress will never exceed 3 kV/mm. The result is that semiconductive coatings with surface resistance from 0.1 to 10 kilohms per square prevent surface discharges in the slot. Note that the coating cannot be highly conductive, since this will short out the stator core laminations.

Semiconductive coatings on coils in the slot are not normally needed for stators rated at 6 kV or less. Clearly, this is because it is unlikely that the critical threshold of 3 kV/mm (electric breakdown strength of air) will occur at this low operating voltage, even if a substantial gap occurs between the coil and the core.

---

*As discussed in Chapter 3, one stator manufacturing process called global VPI, may fill in the gap with epoxy or polyester between the coil surface and the core, thus, in theory, eliminating the need for stress relief coatings. However, because of thermal cycling considerations (Section 8.2), most machine manufacturers still use a semiconductive coating with global VPI stators.

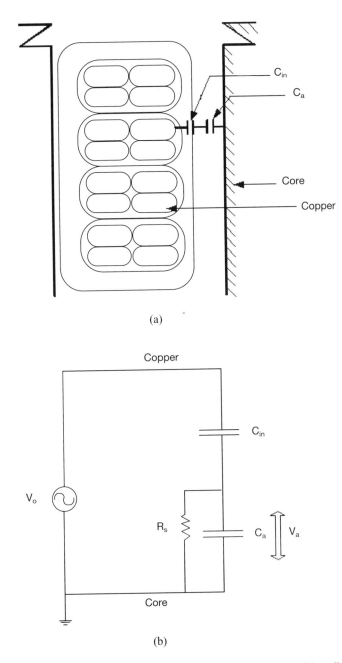

(a)

(b)

Figure 1.14. (a) Cross section of a coil in a slot where PD can occur at the surface of the coil. (b) The equivalent electrical circuit.

Silicon Carbide Coating. The low-resistance semiconductive slot coating usually extends only a few centimeters beyond each end of the slot,* otherwise the grounded surface would be brought too close to the connection of one coil/bar to another. Decades of experience shows that even a fully insulated connection between adjacent coils will be a weak spot that may prompt a future insulation failure if a ground potential is nearby. In addition, as described in Chapters 3 and 4, it is often difficult to avoid air pockets in the insulation occurring in the end-winding during manufacture. If a low-resistance semiconductive coating grounded the endwinding surface, such air pockets are likely to initiate PD as described above, eventually leading to failure.

The slot semiconductive coating cannot end abruptly a few centimeters outside the slot since the thin coating would give rise to a high, localized electric field. This field would exceed 3 kV/mm, and PD would occur at the end of the coating. Such PD would eventually destroy the insulation in the vicinity, leading to failure. The thin edge of the coating creates a very non-uniform electric field at the end of the slot coating since the electric stress depends strongly on the inverse of the radius. The smaller the radius (i.e. the thinner the semiconductive coating) the larger the electric field. For example, a needlepoint at voltage V with a radius r, and a distance d between the needle and a flat ground plane, will create a maximum electric field at the needle tip of approximately [1.16]:

$$E = \frac{2V}{r \ln(4d/r)} \qquad (1.8)$$

where ln is the natural logarithm. This shows that the radius is critical to the maximum electric field.

Thus, just as for high-voltage cables, the end of the semiconductive coating must be "terminated." In the early days of high-voltage rotating machines, the electric field just outside of the slot was made more uniform by embedding concentric floating metal foils of specific lengths within the volume of the groundwall, in the area where the semiconductive coating ends. This is the same principle used in high-voltage condenser bushings in transformers [1.14]. However, when a unique material called silicon carbide became available, it superceded the bushing approach.

Silicon carbide is a special material that has an interesting property: as the electric stress increases in this material, its resistance decreases. That is, it is not "ohmic." In the past, silicon carbide was used in high-voltage surge arrestors to divert high-voltage surges from lightning to ground (i.e., it has a low resistance state), while being fully insulating during normal operating voltage of a transmission line. When applied to stator coils and bars, the silicon carbide has a very low resistance in the high-stress region at the end of the slot semiconductive coating, and gradually increases its resistance further along the end-winding from the core. This varying resistance makes the electric field at the end of the semiconductive coating more uniform. Usually, the stress is reduced to below the critical 3 kV/mm (in air) that would initiate PD.

Silicon carbide is usually mixed into a paint base, or incorporated into a tape that is applied to the coil/bar surface. The length of the silicon carbide surface coating depends on the

---

*Some manufacturers of very high voltage stator windings may use a high resistance (usually hundreds of kilohms or more per square) over the entire end-winding to control the electrical stress more closely, reducing the probability of flashover during an AC hipot test (Section 12.4). Since the coating is high resistance, the further a point is away from the core, the higher the point will be above ground potential. At the connection to the next coil or bar, the voltage of the coating is assumed to be at the same potential as the voltage of the copper underneath the insulation.

voltage rating, but 5 to 10 cm is usual. In Figure 1.6b, the black tape in the middle of the coil is the slot semiconductive coating, whereas the short gray areas at the end of the semiconductive coating are the silicon carbide coatings (not visible due to a surface paint). The silicon carbide coating is also called the gradient coating. Further information on stress control coatings can be found in References 1.17 and 1.18.

1.4.6   Mechanical Support in the Slot

The coils and bars are subject to large magnetically induced forces during normal motor or generator operation, as described in Equation 1.3. These forces are at twice the AC line frequency. Furthermore, if a phase–phase fault occurs, the transient fault current can be many times the rated current, which, due to the I^2 component in Equation 1.3, causes a magnetic force many times larger than that occurring during normal operation. The coils must be restrained from moving under these steady-state and transient mechanical forces. If the coils become loose, then any relative movement that occurs between conductors or coils in the slot or the end-winding will lead to abrasion of the insulation or fatigue cracking of the insulation. Both are obviously undesirable, since they can lead to shorts.

Within the stator slot, the restraint of the coil or bar is accomplished by several means. Often, more than one method is used. As shown in the slot cross sections in Figures 1.8 and 1.9, fundamentally, both random-wound and form-wound slots are filled as much as possible with conductors and insulation, and then the slot is closed off with a nonconductive and usually nonmagnetic wedge. Filling the slot as much as possible reduces the probability that the turns coils/bars will become loose. The slot wedge is usually made from an insulating material and is critical to restricting movement. Originally, wedges were of a simple "flat" design, held in place by dovetail grooves in the stator slot (Figure 1.9). Some insulation filler strips (also known as sticks, depth packing, or side packing) are also inserted to take up any extra space in the slot, either under the wedge, or on the side of the slots. In machines rated at more than 6 kV, often the filler strips have been filled with carbon black to make them semiconductive. This ensures that the semiconductive coating on the coils/bars is grounded to the core via the filler strips, preventing slot discharge.

For large machines, in this context those rated at approximately 20 MW, many manufacturers use a more sophisticated wedge that has two parts (Figures 1.9b, 1.15). The two-part wedge has a tapered slider that is pushed between the main wedge and the coils/bars. This creates a slight spring action in the radial direction that can keep pressure on the coils, even if the slot contents shrink slightly over time.

As an alternative to the two-part wedge (or sometimes as an addition), large machines may use "ripple springs." The ripple spring is a laminated material sometimes filled with carbon black to make it partly conductive (Figure 1.16). The waves in the ripple spring are normally flattened during installation. If the slot contents shrink over time, the ripple spring expands, taking up the new space and holding the coils/bars tight. Ripple springs under a wedge are usually nonconductive. Side ripple springs (between the side of a coil and the core) are usually semiconductive.

Other technologies have also been developed to keep coils and bars tight within the slot in large machines. Coils and bars have been manufactured with a compliant semiconductive silicone rubber compound bonded to the surface of the coils/bars, outside of the semiconductive slot coating [1.19]. Since the silicone rubber is compliant, the coils are made to a zero clearance fit and gradually forced into the slots with hydraulic jacks. If the slot contents shrink over time, the silicone rubber expands to fill the new space. At least some of the silicone rubber must be loaded with carbon black to ensure the semiconductive coating is grounded. Oth-

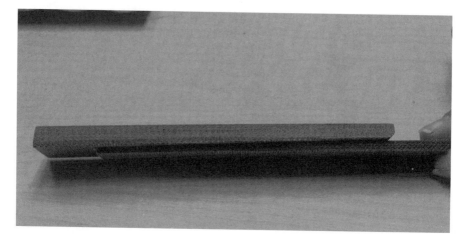

Figure 1.15. Side view of a two-part wedge, where a beveled slider is placed under the wedge to increase radial pressure on the coils.

Figure 1.16. Photograph of semiconductive ripple spring material lying on top of nonconductive ripple spring material.

er designs use a U-shaped hollow tube that is placed in the slot. After wedging (and perhaps after some thermal cycling to compact the slot contents), the hollow tube is filled with an insulating material under pressure, to take up any remaining space.

Another way to ensure that the coils/bars do not become loose is to glue the coils/bars into the slot. This is very common for motor stators, and a few manufacturers have used it on high-speed generators rated up to about 200 MVA [1.20, 1.21]. For random-wound stators, the glue (usually a varnish) is applied by simply dipping the entire stator in a tank of liquid varnish. Trickle impregnation and other variations are also sometimes used to introduce the glue into the slots. For form-wound machines (and now even some critical random-wound stators), a process called global vacuum pressure impregnation (VPI) is used. The latter uses epoxy or polyester as the glue. These processes are described in detail in Chapters 3 and 4. The result is coils or bars that do not move in the slot.

1.4.7 Mechanical Support in the End-Winding

The principle purpose of the end-winding is to allow safe electrical connections between coils/bars that are in series and to make safe connections to other parallels. As discussed above, for machines rated at 1000 V or more, these connections must be made well away from the grounded stator core, to prevent eventual problems with insulation failure at the insulated connection points. Generally, the higher the voltage rating, the longer will be the "creepage distance" between the core and the connections. For geometric reasons, high-speed machines also tend to have long end-windings. On large two-pole generators, the end-windings can extend up to 2 m beyond the core.

These end-windings must be supported to prevent movement. If not suppressed, end-winding vibration will occur because the stator winding currents in each coil/bar will create a magnetic field that will interact with the fields produced by adjacent bars/coils. The resulting force vectors can be quite complicated [1.22]. However, both radial and circumferential vibrating forces occur at twice AC line frequency (Figure 1.17). Since the end-winding coils and bars in large machines are essentially cantilevered beams, the vibration forces can lead to fatigue cracking of the insulation and sometimes even the conductors, at the core end or at the connections. In addition, if the coils/bars are free to vibrate relative to one another, this relative movement can lead to insulation abrasion in the endwinding. If the machine is expected to see many system disturbances, or a motor is to be frequently started, the high inrush currents will create even larger transient forces, which must be accommodated.

End-windings can be supported against the vibratory magnetic forces in many ways. In random-wound machines, the coil end-windings are essentially right up against the core. With the short extensions and low currents involved, the normal "glues" that impregnate such windings, together with modest lashings (Figure 1.5) around the coil end-windings, are often adequate to support the end-winding.

For the form-wound machines, which always have a significant end-winding structure, there is almost always a "support ring" or "bracing ring" of some sort (Figures 1.6a, 1.7). The ring can be made of steel, which is usually insulated, polyester– or epoxy–glass laminate, or fiberglass rope (which is impregnated with stage B epoxy or is dry and impregnated during a global VPI process). This ring is placed either inside the coil/bar layers (i.e., closer to the axis than the coils/bars), or radially outside of the coils or bars. Each coil/bar is lashed to the ring. Large machines expecting many starts and stops may have more than one support ring and large generators usually have both inner and outer end-winding support rings. The ring tends to prevent coil/bar movement in the radial direction. Insulating blocks a few centimeters in

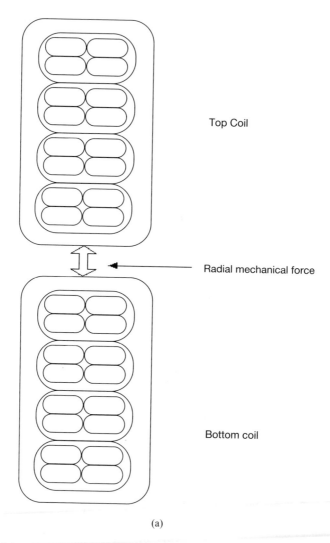

Top Coil

Radial mechanical force

Bottom coil

(a)

Figure 1.17. Schematics showing the magnetically induced mechanical forces occurring (a) between the top and bottom layers of coils in the end-winding.

length, placed between adjacent coils, provide circumferential support. One or more rows of such blocking may be present in each end-winding (Figure 1.6a).

Support against the radial magnetic forces comes mainly from insulating blocking that is placed between the bottom (i.e., coils/bars furthest from the shaft) and top layers of coils/bars in the endwinding. The hoop strength of the support ring also helps to ensure that the coils/bars do not move radially.

Most of the endwinding blocking materials, as well as any lashing that may be used to bind coils/bars to one another and to the bullring, are made from insulating material that will bond to other components. For stators that are entirely impregnated, for example, in the glob-

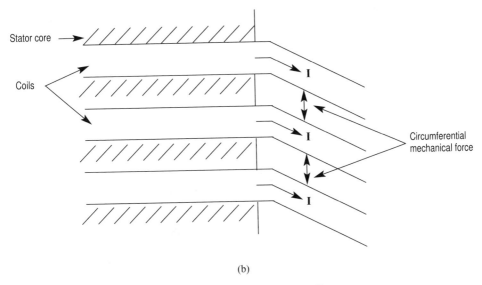

Stator core

Coils

I

I

I

Circumferential
mechanical force

(b)

Figure 1.17. (b) between adjacent coils.

al VPI process (Section 3.10.4), the endwindings are less likely to suffer from relative movement.

Another consideration in end-winding design, especially for large two- and four-pole generators, is the growth of the coils in the slot and the end-winding as a result of high operating
temperatures. As a stator goes from no load to full load, the copper conductors heat and, due
to the coefficient of thermal expansion, the bars grow in length. The end-winding support
system must be able to compensate for this growth, otherwise the support system and even
the bars can become distorted. Accomplishing this tends to be somewhat of an art. See Reference 1.23 for more information on one manufacturer's method of allowing for axial expansion in large turbogenerators.

1.4.8 Transposition Insulation

The final stator winding insulation component considered in this chapter is the transposition
insulation. Transpositions occur only in form-wound machines. Multiturn coils may have
what may be called an internal transposition of the copper conductors (often called an inverted or twisted turn). Roebel bars (half-turn coils) always have a transposition. Usually, extra
insulation is needed in the vicinity of the transposition to ensure that strand shorts do not occur.

The need for transpositions of the copper strands will be briefly explained first for Roebel
bars. The reader is encouraged to refer to reference 1.2 for more comprehensive explana-
tions. Magnetic flux is higher near the rotor side of the bar in a generator than at the bottom
of the bar (i.e., furthest from the rotor). Consequently, if the strands in a Roebel coil were always in the same position within the bar, throughout the length of the bar, the strands closer
to the rotor would have a greater induced voltage on them than strands furthest from the rotor. In a Roebel bar arrangement, all the strands are normally braised together at each end of
the bar. If strands are connected together that have a potential difference induced by the dif-

ferent flux levels, then, by Ohm's Law, an axial current will be forced to flow up and down the bar. Since the resistance of the copper is low, the circulating current flow will be substantial. This will increase the I^2R loss, which reduces efficiency and increases the temperature of the copper.

In a Roebel transposition, each copper strand is placed into every possible position in the bar as the strand moves along the length of the bar. That is, a strand that is initially in the top-left position (Figures 1.18a and b), after several centimeters along the bar will be shifted one position lower on the left. The same strand will then be forced another position lower (i.e., two strand positions from the top) a few centimeters further along the bar. About half way

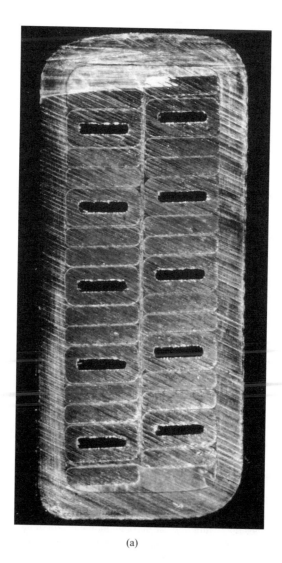

(a)

Figure 1.18. (a) Photo of cross-section of a water-cooled Roebel bar with transpositions.

along the bar, the strand will be in the bottom position on the left. Then the strand is shifted from the left side of the conductor stack to the right side, and gradually moves up to the top right position in the other half of the bar. Eventually it reaches its original position. This is called a 360° transposition. If this 360° transposition is done in the stator core, where the magnetic flux is the highest, then a single strand will have been in each radial position for the same distance along the slot. The total induced voltage on this strand will then be the same as the induced voltage on all the other strands that were transposed. Thus, one can safely connect all the strands together at both ends of the bar and not give rise to axial circulating currents.

There are many different ways to accomplish the movement of the strands into all radial positions. Beside the approach describe above, another popular way is to have the strands move back and forth between the left and right (Figure 1.18b) as it moves from top to bottom. Also, some manufacturers prefer 540° or even 720° transpositions. Transposing can also occur in the end-winding, since stray magnetic fields are present there. The purpose the transposition is to improve stator winding efficiency and reduce operating temperature. The mechanical process of shifting strands from one position to another is called Roebelling, after the inventor of the original equipment (actually developed for wire rope).

As discussed above, the copper strands each have their own insulation layer. Since the difference in magnetic flux between the top and bottom positions is not great, the potential difference between the strands is only a few volts. Thus, normal strand insulation is sufficient to insulate the strands from one another. Although all bar manufacturers ensure that strand shorts are not present in new bars, such shorts can occur as a result of in-service aging. Some strand shorts are easily tolerated, especially if they are close to one another. However, in the vicinity where a copper strand is shifted from one position to another, an air pocket can occur. Extra insulation (often a putty) is needed in these areas to eliminate the air gap.

Most multiturn coils do not use Roebel transpositions, that is, shifting of strand positions continuously along the slot, because the distance between the top strand (i.e. the strand closest to the rotor) and the bottom strand in a turn is usually much shorter than occurs in a Roebel bar. Thus the difference in magnetic flux between the top and bottom strand is much smaller, and the potential difference is much smaller. However, to improve efficiency, many

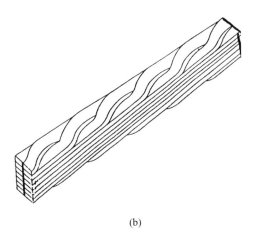

(b)

Figure 1.18. (b) Side view showing one way of transposing insulated strands in stator bar.

coils have an inverted turn in the knuckle region of the coil (the end which is opposite to the connection end). What effectively happens here is the strand bundle in the turn is rotated through 180°. Thus the strand that was in the bottom position in one leg of the coil is in the top position of the other leg of the coil. This approximately balances out the induced voltage on each strand in the turn. Alternatively, some manufacturers accomplish a transposition over several coils by more laborious connection arrangements outside of the coil.

1.5 ROTOR WINDING INSULATION SYSTEM COMPONENTS

Synchronous machine rotor windings have turn-to-turn insulation and slot (ground) insulation. The rotor windings in synchronous motors and generators are subject to relatively low DC voltage, unlike the high AC voltage on the stator. Even in very large turbogenerators, the rotor voltage is usually less than 600 Vdc. This low voltage implies that the turn and ground insulation can be relatively thin. Since the voltage is DC, there is no need for strand insulation,* because the skin effect is not present under DC voltage. If the DC is provided by static excitation systems, that is, DC created from thyristor operation, some high spike-type voltages may be superimposed on the DC. This is an additional stress that can cause partial discharge [1.24].

The other stresses on the rotor insulation include temperature and centrifugal forces. As with the current through the stator winding, the DC current through the rotor winding copper conductors creates considerable I^2R losses, leading to copper heating. The adjacent turn and ground insulation must be able to withstand these high temperatures.

The primary mechanical force on the rotor is not due to magnetic field interaction (although these are present to some degree), but rather due to rotational forces. The centrifugal force is enormous and unidirectional (radially out). The weight of the copper conductors will tend to compress and/or distort any insulation. Thus, the rotor winding insulation must have very high compression strength and be supported to prevent distortion. An additional mechanical (or more properly thermomechanical) stress occurs whenever the machine is turned on and off. When the rotor winding is excited, the current through the conductors causes the copper temperature to rise, and the copper will experience axial growth due to the thermal coefficient of expansion. This growth leads to differential movement of the winding components, and the copper or the insulation can be abraded. Therefore the rotor insulation components should have good abrasion resistance. This is usually obtained by using a slippery coating on the slot insulation.

In a synchronous machine rotor, the winding potential is floating with respect to the rotor body. That is, neither the positive nor the negative DC supply is electrically connected to the rotor body (assumed to be at ground potential). Thus, a single ground fault in the rotor winding has no harmful consequences for the rotor. However, if a second ground fault occurs some distance from the first fault, a large closed loop will be formed in the rotor body through which the rotor DC can flow. This can lead to very high currents flowing in the rotor body, causing localized heating at the fault points and possible melting of the rotor itself. A second ground fault must therefore be avoided at all costs. Most synchronous machine rotors come equipped with a ground fault alarm to warn operators of the first ground fault, or even remove the machine from service if the first ground fault occurs. However, since a rotor winding can operate with a single ground fault, machine owners who are willing to take the

---

*However, due to a problem called copper dusting (Section 9.1.3), some manufacturers now apply strand insulation to prevent copper strands rubbing against adjacent copper strands.

risk of rotor damage, can install a second rotor ground fault relay, which would only remove the machine from service if a second ground fault occurs [1.25].

Rotor turn insulation faults in synchronous machines are much less important than similar turn faults in the stator winding. Since only DC is flowing in the rotor, there is no equivalent to the "transfomer effect," discussed in Section 1.4.2, which induce a large current to flow around the shorted turn. The main impact of a rotor turn short is to reduce the magnetic field strength from the affected rotor pole. This creates an unsymmetrical magnetic field around the rotor periphery, which tends to increase rotor vibration. Vibration can also occur due to nonuniform heating of the rotor surface, since unfaulted turns in slots will have a higher current and, consequently, a higher local temperature, than slots containing shorted turns. Thermal expansion of the rotor body will be different around the periphery, bowing the rotor and increasing vibration (Section 13.10.2). In addition, if too many shorted turns occur, then much more excitation current will be needed to create the magnetic field for the required output. Although shorted rotor turns in themselves may not be hazardous to the machine, if the number of shorts are increasing over time, then it may be an indicator that there is a greater risk that a rotor ground fault may occur.

Wound induction motor rotors experience much the same thermal and mechanical environment as synchronous machine rotors. However, instead of a constant DC current, this type of motor rotor winding sees an initial 50/60 Hz inrush current during motor starting, followed by a gradual reduction in current and current frequency as the motor comes up to speed. The critical time for the rotor winding then, both in terms of heating and in voltage, is during starting. The open-circuit induced voltage is less than about 1000 V, so the insulation can be very thin. As for synchronous rotors, a single ground fault or multiple turn faults do not necessarily render the motor inoperable.

The following describes some of the main characteristics of the general physical construction and unique insulation system requirements for each rotor type.

1.5.1 Salient Pole Rotor

Hydrogenerators and four or more pole synchronous motors have salient pole rotor windings. Each field pole is constructed separately and the rotor winding made by mounting the completed poles on the rotor rim or directly to an integral solid steel body (Figure 1.19). The poles are then electrically connected to the DC supply in such a way as to create alternating north and south poles around the rim. Each field pole consists of a laminated steel core or solid steel body, which looks rectangular when viewed from the rotor axis. Around the periphery of each pole are the copper windings. Figure 1.19 is a photo of a single field pole from a large motor. There are two basic types of salient pole designs.

The older type of field pole design, and still the design used on motors and generators rated at less than a few megawatts, is called the multilayer wire-wound type. In this design, magnet wire is wrapped around the pole (Figure 1.20). The magnet wire usually has a rectangular cross section, and many hundreds of turns are wound on the pole, several magnet wire layers deep. The turn insulation is the magnet wire insulation. Looking from the axial direction, the laminations are shaped to have a pole tip (which is the part of the rotor pole closest to the stator) to support the winding against the centrifugal force. Insulating washers and strips are placed between the magnet wire and the laminations to act as the ground insulation.

For larger machines, the "strip-on-edge" design is favored since it can be made to better withstand rotational forces. In this case, a thin copper strip is formed into a "picture frame" shape, so that the "frame" can be slid over the pole. Laminated insulating separators act as

Figure 1.19. Photo of a pole on a salient pole rotor. The V block is between two adjacent poles.

turn insulation to insulate each copper "frame" from one another. On some copper "frames," especially those near the pole face, an insulating tape may be applied to the copper to increase the creepage distance. The tape and separators form the turn insulation, and the copper "picture frames" are connected in series to make the coil. As with the multilayer design, the winding is isolated from the grounded pole body by insulating washers and strips (Figure 1.21). Often, the entire pole may be dipped in an insulating liquid to bond the various components together. For solid pole rotors, the pole tips are usually bolted on (Figure 1.19).

1.5.2 Round Rotors

Round rotors, also called cylindrical rotors, are most often found in two- and four-pole turbo-generators. The rotor body is usually made from a single steel alloy forging. Axial slots are cut into the forging. The copper turns, made from copper strips up to a few centimeters wide, are placed in these slots. There are usually five to 10 strips (turns) in a slot forming a coil (Figure 1.22). Rather than directly insulating the copper turns with a tape or film insulation, the turns in large rotors are kept separated from each other by insulating sections. Insulating strips or L-shaped channels line the slot, to act as the ground insulation. Since the components are not bonded to one another, the components can slide relative to one another as the current, and thus the temperature, cycles. The copper coils are kept in the slot by means of wedges. Since the rotational forces are so enormous (in overspeed conditions, the acceleration can reach in excess of 5000 G), only metal wedges can be used. The wedges are made from copper alloys (in older machines), nonmagnetic steel, magnetic steel, or aluminum alloy. As shown in Figure 1.22, air or hydrogen passages are often cut in the copper turns and wedges, to allow the gas to directly cool the rotor winding. As will be seen in Chapter 9, these passages, together with the use of insulating strips, allow short creepage paths to devel-

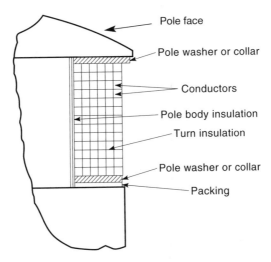

Figure 1.20. Cross section of a multilayer salient pole.

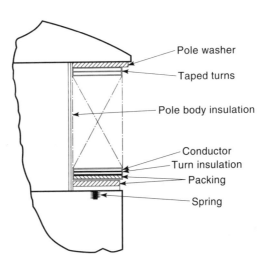

Figure 1.21. Cross section of a "strip-on-edge" salient pole.

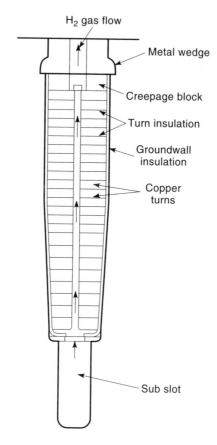

Figure 1.22. Cross section of a round rotor slot.

op if pollution is present. Also, these relatively small channels are easily blocked by foreign material or turn insualtion migration, causing local hotspots in the rotor.

One of the most challenging design aspects of a round rotor is the end-winding. After the copper turns exit from the slot, they go through a 90° bend and are directed around the rotor circumference until they reach the appropriate slot, at which point they go through another 90° bend. This rotor end-winding area must be supported to prevent the centrifugal forces from breaking off the copper conductors. Furthermore, space must be left and slip planes inserted to allow the rotor winding to expand axially as it heats up when current is circulated. Also, blocking must be installed between coils in the end-winding to prevent distortion from thermal expansion forces. The copper can grow several mm along the slot in a large turbo rotor that may be 6 m long.

The end-winding is supported against rotational forces by retaining rings at each end of the rotor body. In most round rotors for machines larger than about 10 MW, the retaining rings are made from stainless steel. Considerable technology is used to make these rings and shrink them onto the rotor body. An insulating sleeve that allows the components to slip with

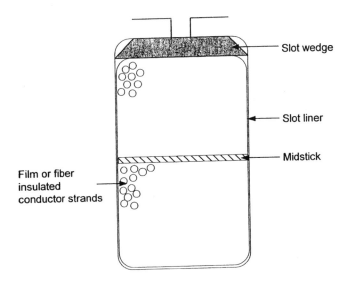

Figure 1.23. Cross section of a small wound-rotor slot.

respect to one another is needed between the copper endwindings and the retaining ring. In open ventilated air-cooled machines, as well as older machines, the copper turns were often taped in the end-winding region with an insulating tape. However, it is now common to use insulating strips, sheets, and blocks.

1.5.3 Induction Motor Wound Rotors

These rotors see a 50 or 60 Hz current during motor starting, which gradually reduces to a near DC current once the rotor is up to rated speed. The rotor consists of a laminated steel core with slots around the periphery. Usually, the slots are partially closed off at the rotor surface by the steel, so that the coils cannot easily be ejected into the air gap, even if the rotor wedges are not present. Two basic types of rotor windings have been used.

Random rotor windings are made from magnet wire inserted in the slots (Figure 1.23). The turn insulation is the magnet wire insulation. Once the turns are inserted, an insulating wedge closes off the slot. An insulating slot liner serves as the ground insulation. It is very important to completely fill the slot to ensure that the turns do not move relative to one another under the rotational forces. All space is filled with insulating strips and the entire rotor is impregnated with a liquid insulation that is cured, using a dipping, trickle impregnation, or global VPI process (Chapter 3). The interturn and ground voltages are low, so the insulation is thin. In the end-windings outside of the slot, the overhang region is only several centimeters to 30 cm long, depending on the rotor size and speed, thus end-winding movement can be restrained by impregnated cords tying together coils separated by impregnated felt pads. Insulating fiberglass bands are then applied over the end-winding area to support it against centrifugal forces.

For larger rotors, copper bars are used for the winding conductors. These bars are rectangular in shape, with the long dimension oriented in the radial direction (Figure 1.24). Usually, either two or four bars are inserted in each slot by dropping through the top, or pushed

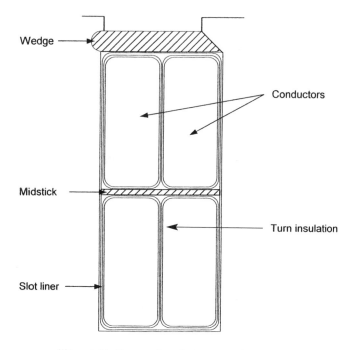

Figure 1.24. Cross section of a large wound stator slot.

through the slot from the end of the slot. Prior to insertion, the bars are preinsulated with a few layers of insulating tape. The ground insulation is a slot liner. As with random-wound rotors, an insulating band restrains the end-windings and the rotor is impregnated to prevent relative movement.

REFERENCES

1.1 S. J. Chapman, *Electric Machinery Fundamentals,* 3rd ed., McGraw-Hill, 1999.

1.2 M. G. Say, Alternating Current Machines, Pitman Publishing, 4th ed., 1976.

1.3 D. V. Richardson, *Rotating Electric Machinery and Transformer Technology,* Prentice-Hall, 1982.

1.4 IEC Standard 60034, "Rotating Electric Machines—Part 5, Classification of Degrees of Protection Provided by Enclosures; Part 6, Methods of Cooling."

1.5 NEMA Standard MG 1-1998, "Motors and Generators."

1.6 B. E. B. Gott, "Application of Air-Cooled Generators to Modern Power Plants," *IEEE Electric Machines and Drives Conference,* Seattle, May 1999, pp. 317–319.

1.7 A. P. Kopp et al., "Comparison of Maximum Rated Air-Cooled Turbogenerators with Modern Hydrogen Cooled Generators," CIGRE Paper 11-202, September 1992.

1.8 B. J. Moore, R. H. Rehder, and R. E. Draper, "Utilizing Reduced Build Concepts in the Development of Insulation Systems for Large Motors," In *Proceedings of IEEE Electrical Insulation Conference,* Cincinnati, October 1999, pp. 347–352.

1.9   R. H. Engelmann and W. H. Middendorf, *Handbook of Electric Motors,* Marcel Dekker, 1995.

1.10   P. Clark, N. Glew, and R. Regan, "Solutions for MV Motors and Generators in Hazardous Locations," In *Proceedings of IEEE Petroleum and Chemical Industry Conference,* Philadelphia, Sept 1996, pp. 49–60.

1.11   B. K. Gupta et al., "Turn Insulation Capability of Large AC Motors, Part 3—Insulation Coordination," *IEEE Trans EC,* Dec. 1987, pp. 674–679.

1.12   J. F. Calvert, "Forces in Turbine Generator Stator Windings," *Trans of AIEE,* 1931, pp. 178–196.

1.13   IEEE Standard 100-1996: "IEEE Standard Dictionary of Electrical and Electronics Terms."

1.14   E. Kuffel, J. Kuffel, and W. S. Zaengl, *High Voltage Engineering,* Butterworth-Heinemann, 1998.

1.15   R. Bartnikas, "Corona Measurement and Interpretation," *ASTM, STP 669,* 1980.

1.16   J. H. Mason, "Breakdown of Solids in Divergent Fields," *Proc IEEE, Part C,* 1955, p. 254.

1.17   F. T. Emery, "The Application of Conductive and Stress Grading Tapes to VPI'd High Voltage Stator Coils," *IEEE Electrical Insulation Magazine,* July 1996, pp. 15–22.

1.18   J. Thienpont and T. H. Sie, "Suppression of Surface Discharges in the Stator Windings of High Voltage Machines," *CIGRE,* June 1964.

1.19   D. S. Hyndman et al., "Twintone Bar/Slot Interface to Maximize the Benefits of Modern High Voltage Stator Ground Insulation," In *Proceedings of IEEE Electrical Insulation Conference,* Chicago, October 1981.

1.20   K. Nyland and R. Schuler, "Insulation Systems for Synchronous Machines," In *Proceedings of International Conference on Synchronous Machines,* Zurich, August 1991, pp. 182–188.

1.21   N. Didzun and F. Stobbe, "Indirect Cooled Generator Stator Windings with Micalastic Post Impregnation Technique," presented at ISOTEC Workshop 93, Section 4, Zurich, November 23–24, 1993.

1.22   M. R. Patel and J. M. Butler, "Endwinding Vibrations in Large Synchronous Generators," *IEEE Trans PAS,* May 1983, pp. 1371–1377.

1.23   J. M. Butler and H. J. Lehman, "Performance of New Stator Winding Support System for Large Generators," In *Proceedings of American Power Conference,* p. 461, 1984.

1.24   S. Nagano et al., In "Evaluation of Field Insulation of Turbine Generators with Higher Thyristor Excitation," *IEEE Trans EC,* Dec. 1986, pp. 115–121.

1.25   J. M. Hodge et al., "Generator Monitoring Systems in the United Kingdom," CIGRE Paper 11-08, September 1982.

1.26   S. Hirabayashi, T. Kawakami, T. Tani, K. Shibayama, and S. Matsuda, "A new Corona Suppression Method for High Voltage Generation Insulations," In *Proceedings of the 12th E/EIC,* pp. 139–142, 1975.

CHAPTER 2

EVALUATING INSULATION MATERIALS AND SYSTEMS

Owners of motors and generators expect a minimum rotor and stator winding life. For most industrial and utility applications, at least a 20 to 30 year life is achieved before a rewind is needed. Decades ago, the insulation system design to accomplish a specified life was largely achieved by trial and error. If an older winding design failed prematurely, then extra insulation was added or other corrective measures were taken in a new design. Generally, to prevent premature failure, "safety margins" were added to the design; that is, the copper may have been given a greater cross section than strictly needed to make sure the operating temperature was low, the groundwall insulation made thicker to avoid electrical breakdown, etc. The common result was an insulation system that would greatly outperform the specified life. Indeed, many rotor and stator windings made in the first half of the 1900s are still operating today as a result of the conservative design methods used then.

In the past few decades, it has been recognized that the safety margins in the design were excessive and can greatly increase the cost of the rotor and stator. For example, Draper has indicated that a 20% reduction in the groundwall insulation thickness in a large SCI motor stator will allow the width and the depth of the stator slot to be reduced, for the same output power. The stator bore can then be smaller, and the distance from the bore to the back of the stator core can be reduced, while maintaining the same core mechanical stiffness. Depending on the speed of the motor, the weight of the steel needed for the stator is reduced from 13–33%, the copper is reduced 5–64%,* and the mass of the insulation is reduced 12–57% [2.1]. These are significant reductions. Since the price of a motor or generator is very dependent on the mass of the steel, copper, and insulation, the stator cost is significantly reduced by increasing the design electric stress level, i.e., by reducing the groundwall thickness [2.1].

The recognition that manufacturing costs can be reduced by eliminating unnecessary design conservatism, together with the very competitive global market for motors and genera-

---

*This is due to shorter circuit ring busses and less copper cross section in the coil, since the heat from I^2R losses can be transmitted through the thinner groundwall insulation more easily.

Electrical Insulation for Rotating Machines. By Stone, Boulter, Culbert, and Dhirani
ISBN 0-471-44506-1 © 2004 Institute of Electrical and Electronics Engineers

tors, has led to more scientific methods in the design of insulation systems in rotor and generator windings. The main tool that manufacturers now use to design the insulation system is called an accelerated aging test. The basic concept is to make a model of the insulation system, and then subject the model to higher than normal stresses (temperature, voltage, mechanical force, etc.); that is, the stress is elevated to increase (accelerate) the rate of deterioration. The model will then fail in a much shorter time than would occur in normal service. Usually the higher the stress applied, the shorter the life. By testing insulation system models at a variety of stress levels, one can sometimes extrapolate the results from high stress to the expected operating stress, and estimate the service life. Subjecting the different insulation system designs to all the accelerated aging tests will identify the design that achieves the desired life with the least cost.

The preceding is very simplistic. As described in Chapters 8 to 10, the rotor and stator insulation systems may fail from more than two dozen failure processes. Each failure process is driven by one or more stresses like voltage, temperature, pollution, and mechanical force. That is, one stress alone does not define the life of a winding. The result is that to qualify an insulation system design for service, the system and its constituent parts must be subjected to a wide battery of accelerated aging tests that will establish its probable life over most of the failure processes discussed in Chapters 8 to 10. Another problem with accelerated aging tests is that the results of the tests tend to have a great deal of scatter. This requires much repeat testing and statistical analysis.

This chapter describes the practical aspects and limitations of accelerated aging tests used to determine the capability of winding insulation systems. A large number of accelerated aging tests have been developed to model various failure processes, and many of these have been standardized by international organizations such as IEC and IEEE as well as national standardization organizations. Special attention is paid to these standardized test procedures since users can refer to such standards in an effort to compare the offerings from different manufacturers.

In addition to aging tests, many tests have been developed to evaluate the specific properties or capabilities of insulation materials and systems. For example, tests are needed to determine melting temperature, tensile strength, resistance to heat shock, cut-through resistance, electrical breakdown strength, chemical resistance, etc., in which aging is not a direct factor. Summaries of such tests, which can help ensure that the insulation can withstand known operating conditions, are discussed in Section 2.8.

Before proceeding, a distinction needs to be made between insulation materials and insulation systems. An insulation system is composed of insulating materials. The insulation system usually contains conductors and one or more insulating materials of specific thickness and shape. To determine the properties of each insulating material; aging, and other tests are often made on a slab of insulation material itself. Many of the tests discussed below, in fact, evaluate the material; for example, the temperature at which the insulation material degrades. However, the life of a winding is determined by the capabilities of the system, not its individual constituents. Thus, material tests alone are not sufficient to determine the expected life of a complete insulation system.

2.1 AGING STRESSES

There are many different stresses that can affect the rate of insulation degradation in stator and rotor windings. Some were introduced in Chapter 1. Broadly speaking, there are thermal, electrical, ambient, and mechanical stresses, the so-called TEAM stresses [2.2]. Each of these will be described below.

Before dealing with the individual stresses, it is important to consider that the stresses can be constant, or they can be present for only a brief time, i.e., they are transient. Constant stresses include the operating temperature, the 50 or 60 Hz AC voltage, and the 100 or 120 Hz magnetically induced mechanical stresses. In general, if failure is caused by a constant stress, the time to failure is proportional to the number of operating hours for the motor or generator. Transient stresses include those such as motor starting, out-of-phase synchronization of generators, lightning strikes, etc. If deterioration is primarily due to these transients, then the time to failure is proportional to the number of transients the machine experiences.

2.1.1 Thermal Stress

This is probably the most recognized cause of gradual insulation deterioration and ultimate failure, hence, a winding insulation system must be evaluated for its capability under thermal stress. The operating temperature of a winding causes thermal stress. This temperature results from I^2R, eddy current and stray load losses in the copper conductors, plus additional heating due to core losses, windage, etc. As will be described in detail in Sections 8.1, 9.1.1, and 9.2.1, in modern insulations the high temperature causes a chemical reaction (oxidation in air-cooled machines) when it operates above a threshold temperature. The oxidation process makes all types of insulation brittle and/or tends to cause delamination in form-wound coil groundwalls. Delamination is the separation of the groundwall tape layers due to loss of bonding strength and/or impregnating compound.

To a first approximation, the oxidation process is a first-order chemical reaction in which the rate of the reaction is governed by the Arrhenius rate law. As first proposed by Dakin [2.3], the life of the insulation (L, in hours) is related to the temperature (T, in °K) by

$$L = Ae^{B/T} \tag{2.1}$$

where A and B are assumed to be constants. Saying that the life of the winding will decrease by 50% for every 10°C rise in temperature often approximates this equation. Equation 2.1 is an approximation for two reasons. First, it is only valid at relatively high operating temperatures. Below a threshold, which is different for each insulation material, no thermal aging will occur. Second, more than one chemical reaction usually occurs at a time. Thus, a simple first-order reaction rate model is not strictly valid. However, because Equation 2.1 is firmly entrenched in standards, there is little desire to make the model more accurate.

Clearly, the higher the temperature, the shorter is the expected life of the insulation and, thus, the winding. Equation 2.1 is the basis of all accelerated aging tests which are used to estimate the thermal life of a winding, and is also used to define the insulation thermal classes, e.g., A, B, F, and H. This will be discussed further in Section 2.3.

A variation of thermal stress is thermomechanical stress. It is primarily applicable to very large machines. As described in Sections 8.2 and 9.1.2, changing machine load will cause the winding temperatures to change. If the winding temperature quickly goes from room temperature to its operating temperature, the copper conductors will expand axially. In contrast, modern insulations have a lower coefficient of thermal expansion than copper and, in a transient situation, are much cooler just after a load increase. The result is a shear stress between the conductors and the insulation as the copper grows more quickly than the groundwall insulation. In stator windings, after many thermal (i.e., load) cycles, the bond between the insulation and the copper may break. No simple relationship has been developed to relate the number of cycles to failure as a function of temperature. However, the higher the temperature difference between the insulation and the copper, the fewer the number of cycles to failure.

Temperature has an effect on many other failure processes that are not strictly thermal. For example, in stators with electric stress relief coating problems (Sections 8.5 and 8.6), the higher the temperature, the faster the deterioration rate. In some cases, operating windings at high temperatures can be beneficial. High temperatures tend to prevent moisture from settling on windings, and thus reducing the risk of electrical tracking failures (Section 8.8). In addition, if the groundwall of stator coils/bars is delaminated either from poor manufacturing (Section 8.3) or thermal deterioration (Section 8.1), then operation at high temperature will swell the insulation somewhat, reducing the size of any air pockets in the insulation and decreasing the partial discharge activity [2.4]. In addition, some older types of form-wound groundwall insulations swell on heating, reducing the likelihood of abrasion if the coils or bars are loose at low temperatures (Section 8.4).

2.1.2   Electric Stress

Power frequency electric stress has little impact on the aging of the electrical insulation in stator windings rated at less than about 1000 V. The thickness of the insulation in such low-voltage stator windings, as well as all rotor windings, is primarily determined by mechanical considerations. That is, the insulation must be thick enough to withstand the rigors of coil winding and the mechanical forces that are impressed on the winding in service.

In stator windings rated at above about 1000 V, the thickness of the insulation is primarily determined by the electric stress; that is, the rated power frequency voltage divided by the insulation thickness. Power frequency voltage can contribute to the aging of the insulation if partial discharges (PD) are present. As discussed in Section 1.4, partial discharges are small electric sparks that occur within air pockets in the insulation or on the surface of coil insulation. These sparks contain electrons and ions that bombard the solid insulation. Organic materials such as films, polyesters, asphalts, and epoxies degrade under this bombardment due to breaking (scission) of certain chemical bonds such as the carbon–hydrogen bond. With enough time, the PD will erode a hole through the organic parts of the groundwall, leading to failure.

If partial discharges are present, then the effect of the stress level (E in kV/mm) on the life of the insulation (L in hours) is most often represented by the inverse power model [2.5]:

$$L = cE^{-n} \tag{2.2}$$

where c is a constant and n is called the power law constant. This model is based on work done by Eyring, among others [2.6]. As with thermal aging, below a threshold electrical stress, there is effectively no aging. This threshold is the discharge extinction voltage (DEV), where E is the DEV divided by the insulation thickness. Sometimes E in Equation 2.2 is replaced by $E - E_0$, where E_0 is the threshold stress below which aging does not occur. If electric stress versus time to failure is plotted on log-log graph paper, the slope of the line according to Equation (2.2) would be n.

The power law "constant" is usually reported to range from 9 to 12 for machine insulation systems [2.7–2.9]. If one assumes n to be 10, then a two-times increase in electric stress will reduce the life by about 1000 times. Thus, the electric stress (voltage) has a very powerful influence on service life if PD is occurring.

Although *n* is referred to as a constant, some have reported that *n* may change with stress level [2.10]. Thus, sometimes an exponential model is used to model the influence of stress on life. For example:

$$L = ae^{-bE} \qquad (2.3)$$

where *a* and *b* are constants. However, this model is rarely used in rotating machine insulation applications.

As with the Arrhenius model for thermal stress, the model in Equation 2.2 in principle enables calculation of the service life based on accelerated aging tests done at high electric stress. As will be seen in Section 2.4, standardized methods have been developed to perform electric stress endurance tests assuming a PD failure process. Such tests are the basis for determining the groundwall thickness of the insulation applied in form-wound stator coils.

Another important way that electric stress can age the insulation occurs when many repetitive voltage surges are impressed across the turn insulation in random-wound stators or synchronous machine rotors. Inverter-fed drives (IFDs) which use electronic switching devices and pulse width modulation (PWM) technologies can create tens of thousands of fast-rise-time surges per second. As described in Section 8.7, these surges can impose relatively high voltages across the first few turns in a stator winding. In random-wound machines in which the insulation is thin and air pockets are plentiful, partial discharges have been detected on machines rated as low as 440 V during IFD operation [2.11, 2.12]. These discharges gradually erode organic film insulation, leading to failure. Even in the absence of PD, some have suggested that the stress is high enough in some situations that space charge injection occurs [2.13, 2.14]. This process involves the emission of electrons from surface imperfections on the copper into the insulating film with every voltage surge. This repetitive injection on each surge breaks some chemical bonds in the film insulation. Eventually, enough bonds are broken that puncture can occur. On rotor windings in which the DC is obtained from a "static excitation system" that uses thyristors or other electronic switches, the voltage surges created have been reported to cause aging by a PD mechanism [2.15] (see Sections 9.1.5 and 9.2.6).

Thus, the power frequency AC stress can age the insulation in form-wound stators. The voltage surges caused by electronic switching devices can also age the insulation in random-wound stators and very large synchronous rotors.

2.1.3 Ambient Stress (Factors)

Ambient stress refers to a collection of factors, which come from the environment surrounding the motor or generator, that can lead to failure. Some of these factors are:

- Moisture condensed on the windings
- Oil from the bearings or oil seal system in hydrogen-cooled machines
- High humidity
- Aggressive chemicals
- Abrasive particles in the cooling air or hydrogen
- Particles from brake shoe wear (if fitted) or carbon brush wear (if fitted) within the machine
- Dirt and debris brought into the machine from the environment, such as insects, fly ash, coal dust, and powders that are by-products of associated industrial processes (cement, pulp, chemical residues, etc.)
- Radiation

Each of these can affect the rotor and stator insulation in different ways. In some cases, these "factors of influence" in themselves do not cause aging but, when combined with another stress, can lead to aging. For example, moisture and/or oil, combined with partly conductive dirt, carbon brush particles, etc. can make a partly conductive film over the insulation, in which the electric stress then causes surface currents and electrical tracking (Sections 8.8, 9.1.4, 9.2.3). Oil/moisture/dirt combinations can collect in the rotor and stator ventilation passages and between coils in the end-winding to block cooling airflow, which increases the risk of thermal deterioration (Sections 8.1, 9.1.1, 9.2.1). Oil can also be a lubricant that facilitates relative movement between coils and the slot in rotor and stator windings, leading to insulation abrasion (Section 8.4). Low humidity in the cooling air or hydrogen reduces the breakdown voltage of the gas, leading to greater partial discharge activity in stator end-windings. Similarly, chemicals such as acids and ozone can decompose the insulation, reducing its mechanical strength (Section 8.10). With these factors, it is often not possible to relate the level of a factor directly to the rate of deterioration. As will be seen in Section 2.2, these factors are usually either present or not present in an accelerated aging test.

Radiation due to nuclear reactions is somewhat different from the other factors. Radiation itself can lead to chemical bond scission and, thus, insulation embrittlement. The higher the radiation level, the faster the aging. The process is much like thermal deterioration, with the exception that the surface of the insulation ages more quickly than the interior. This is the reverse of the thermal deterioration situation. Since the process is similar to thermal deterioration, the chemical reaction rate law (with radiation intensity replacing temperature) in Equation 2.1 is often used. Modern insulting materials used in motors tend to have a threshold (about 10^7 rads in total gamma dose over their service life) below which radiation aging does not occur, and some insulation systems have been qualified for a total integrated radiation dose of 200 Megarads [2.40]. Of course, only motors operating in a nuclear plant or a nuclear powered ship are likely to experience radiation-induced aging.

2.1.4   Mechanical Stress

There are three main sources of mechanical stress. On a rotor, the insulation system is exposed to high centrifugal force. This is a nonvibrating force that tends to crush or distort the insulation. For the most part, the insulation either has, or has not, the capability to endure such forces. Various short-time mechanical tests can evaluate this. There is little aging involved, although some materials may "cold-flow," that is, slowly creep away from the high stress areas, eventually leading to a fault.

The second common mechanical stress is caused by the power frequency current, which gives rise to a magnetic force oscillating at twice the power frequency. Equation 1.3 shows the relationship between the mechanical force and the current flowing in a stator coil/bar. If the coils are loose in the stator slot, the force causes the coils to vibrate, and the groundwall insulation is abraded (Section 8.4). A similar magnetic force occurs in the end-winding. If the coils/bars are free to vibrate relative to one another or against blocking or support rings, the insulation may again abrade.

Unlike thermal and electrical stresses, there are no well-accepted models to describe the relationship between vibration amplitude and life. Although models do describe the amount of abrasion that may occur [2.16], they are not practical and none have become the basis for standard accelerated aging tests under vibration. Similarly, a form-wound stator coil or bar in the end-winding can be modeled as a laminated cantilevered beam and, in principle, a relationship between the number of cycles to failure and the vibration amplitude (referred to as the SN curve) can be established [2.16]. However, none are widely accepted.

Transients cause the third important mechanical stress: switching on of motors or out-of-phase synchronization of synchronous machines. Both give rise to a large transient power frequency current that may be five times, or more, greater than normal operating current in the stator. The result is that the magnetically induced mechanical force is 25 or more times stronger than normal service. The "DC" component (Equation 1.3) of this transient force tends to bend the coils/bars in the stator end-winding. If the force cannot be withstood, the coil/bar insulation cracks. If many transients occur, such as frequent motor starting, then the end-winding may gradually loosen over time, allowing relative movement between the end-winding components, and insulation abrasion under normal power frequency current, as described above. No model exists to relate the transient level to the number of transients that can be withstood. Instead, manufacturers calculate the forces that could occur under various transient current situations, and determine if a single transient can be withstood. Aging is usually not considered.

2.1.5 Multiple Stresses

Many of the failure processes to be described in Chapters 8 to 10 do not, in fact, depend on a single stress causing gradual deterioration of the winding insulation. As pointed out above, when discussing ambient stresses/factors, two or more stresses/factors often need to interact to result in deterioration. In addition to those described in the previous section, other examples include:

- Thermal deterioration in form-wound stators that creates delamination, allowing partial discharges that ultimately erode a hole through the insulation
- Fatigue cracking of copper conductors in direct-water-cooled stator bars caused by end-winding vibration, with the leaking water delaminating the insulation (Section 8.13). This enables PD to occur, leading to a ground fault.
- Form-wound coil/bar semiconductive coating deterioration caused by poor manufacture and/or high-temperature operation, which leads to PD, creating ozone that chemically attacks the insulation

The key feature is that if two or more of these factors/stresses are present, the failure process is much faster than if only a single stress/factor were present if, indeed, the single stress/factor would ever lead to failure [2.41].

Aging models that allow machine manufacturers to predict the capability of an insulation under such multistress situations tend to be very complex and, to date, somewhat impractical. However, Section 2.2 will address this to some extent.

2.2 PRINCIPLES OF ACCELERATED AGING TESTS

The purpose of accelerated aging tests is to increase one or more stresses above normal levels, to speed up a specific failure process that may occur in service. This will accelerate the process and result in a much shorter time to failure. A shorter time to failure is necessary since one can not realistically wait, say, 20 years to see if a new design or material is acceptable.

Before specific aging tests for each type of stress are described, it is useful to discuss a number of issues that are common to all accelerated aging tests on insulation material and systems.

2.2.1 Candidate and Reference Materials/Systems

There are two basic possible outcomes from accelerated aging tests. One outcome can be an equation that enables prediction of the life of the insulation under operating stress. For reasons that will be clear later, such predictions are very dubious. No standards are based on this principle.

The second outcome is based on comparing the life measured under an accelerated stress of a "candidate" system (that is, with a new insulation material or system that is to be evaluated) with a "proven" material or system tested under the same conditions. The proven material or system will be one that has given good service life in actual operation. The proven material/system is often referred to in the accelerated aging test standards as the "reference" material/system. If the candidate system performs as well or better than the reference system in the same aging test, then the candidate system is expected to do as well or better than the reference system in service. The comparison of results from different materials/systems is the usual way aging tests are evaluated in standards.

2.2.2 Statistical Variation

If 10 specimens of an insulation system are subjected to an accelerated aging test using high voltage, there will normally be as much as a ten-to-one difference in the life of the insulation. That is, if the first failure occurred after 100 hours, the tenth sample may survive for 1000 hours. This occurs even if the specimens are all identical and the voltage exactly the same for all tests. This huge variation in outcome is typical of aging tests. Times to failure under thermal and mechanical accelerated aging tests also vary widely. This parallels the situation for human beings: healthy individuals at 20 years of age may live from 0 to 80 more years.

When outcomes from testing are extremely variable, statistical methods are needed to analyze the results. Statistical analysis helps determine if there are real differences between a candidate and reference insulation system, or whether a result is just due to normal variation. For example, if two insulation models using different materials are tested at high temperature, and one model endures twice as long as the other, is the longer-surviving model better, or is this just normal statistical variation? Furthermore, if the lifetime is measured on a specimen tested at a moderate temperature, and another specimen is tested at a high temperature, can one predict the life at a third (possibly lower) temperature?

Statistical methods have been developed to help answer such questions. One aspect of statistical methods is to quantify the normal amount of variation in an outcome by using statistical distributions, also known as probability distributions. Another aspect is to derive an equation based on aging tests at a few stress levels that can be used to predict the outcome of a test at a different stress level. This is called regression, more colloquially known as curve fitting. Both aspects are the subject of many textbooks on statistics. The following will briefly review these two topics and discuss the relevant terminology and methods for insulation testing. The reader is encouraged to refer to any standard textbook on the subject, for example, Reference 2.17.

Statistical Distributions. Most engineers and scientists are familiar with the normal probability distribution, sometimes called the Gaussian distribution. This distribution allows the calculation of the mean (μ) and standard deviation (σ) of a collection of test outcomes, as well as confidence intervals. As long as two or more outcomes have been measured, the mean and standard deviation can summarize all the collected data in just two numbers, assuming that the normal distribution fairly represents the distribution of the outcomes. The mean is the most probable outcome. The standard deviation is a measure of the amount of variation possible. For example, there is little variation in the mass of American 25 cent coins, whereas there is tremendous variation in the mass of people, even if they are the same age and of the same sex. The variation is an inherent property of whatever is being measured, just like the mean is.

It is not possible to know the actual mean and standard deviation, since this would require an infinite amount of data to be collected. However, we can estimate what the mean and standard deviation are for a collection of data. For normally distributed data, the mean is estimated by calculating the average ($\bar{x}$):

$$\bar{x} = \frac{\Sigma x_i}{N} \tag{2.1}$$

where x_i are the individual results and N is the number of specimens tested. The standard deviation is estimated by s:

$$s = \left[\frac{\Sigma(x_i - \bar{x})}{(N-1)} \right]^{1/2} \tag{2.2}$$

$\bar{x}$ and s are estimates of the true values of the mean and standard deviation. The more specimens tested, the more closely will $\bar{x}$ and s reflect the actual value of the mean and standard deviation. Since $\bar{x}$ and s are not the true values of μ and σ, statisticians have developed methods to calculate confidence intervals, i.e., the bounds surrounding $\bar{x}$ and s, within which μ and σ are likely to occur with a set (usually high) probability. For example, the breakdown voltage of an insulation may have an average of 9.2 kV with a 95% confidence interval of 9.0 to 9.4 kV. This means that the lower and upper confidence limits are 9.0 and 9.4, respectively. The 95% interval indicates that if the same number of specimens (say, 10 insulation specimens) were repeatedly tested, and the $\bar{x}$ and s for each group of 10 specimens calculated, $\bar{x}$ and s would be within the lower and upper confidence bounds 95% of the time. Thus, the confidence limits indicate the uncertainty in the $\bar{x}$ and s estimates. Methods to calculate the confidence intervals for normal data are described in any statistics text [2.17]. However, in general, the width of the confidence interval (i.e., the uncertainty) decreases with the square root of N, the number of specimens tested.

Although the normal distribution is widely applicable to a large number of physical properties, this distribution does not seem to fit insulation lifetime data or physical strength data such as electrical breakdown strength or mechanical tensile strength. For lifetime and strength data, the lognormal and Weibull probability distributions are more commonly used.

Censoring. Before discussing the distributions most commonly used in insulation aging tests, let us consider another important statistical phenomenon relevant to such tests—censoring. In many aging tests, most of the specimens fail in a reasonable time. However, a few of the specimens tend to "live" forever. This can be very frustrating to the testing personnel. The equations described above to estimate the mean and standard deviation depend on all the specimens being tested to failure. That is, data must be complete to be applicable. However, methods have been developed to estimate the distribution parameters (such as μ and σ) when some of the specimens have still not failed. When an experiment is performed and the parameters are estimated before some of the specimens have failed, this is called a censored experiment. There are different types of censoring depending on when the experiment is stopped—after a fixed number of specimens or after a fixed amount of time (or voltage or tensile strength). More details on censoring are described in most advanced statistical methods books, for example, Reference 2.18.

For the most popular probability distributions used in aging tests, methods have been developed to estimate parameters using censored data. As seems reasonable, the higher the percentage of specimens that are tested to failure, the narrower will be the confidence intervals on the parameters.

Lognormal Distribution. The lognormal distribution is a simple transformation of the normal distribution. This distribution is almost exclusively used to analyze the results of accelerated aging tests in which temperature is the accelerated stress, and is a critical element in the thermal classification of insulation. The lognormal distribution is also sometimes used for breakdown voltage and tensile strength tests. IEEE 930-1987 [2.5] gives the general principles for calculating the log mean and log standard deviation for the lognormal distribution, as well as for calculating the parameter confidence intervals. The application of the lognormal distribution to thermal aging tests is described in IEEE 101 [2.19] and IEC 60216 Part 3 [2.21].

Instead of directly using the failure times or voltages as the x_i, the logarithm of the voltages, times, etc. are first taken. Thus, using the terminology in Equations 2.1 and 2.2:

$$x_i = \log y_i \qquad (2.3)$$

where y_i are the individual outcomes (failure times, breakdown voltages, etc.) from an experiment. Either the natural logarithm or the logarithm to base 10 can be used, as long as one is consistent. Once the transformation is done, estimates of the log mean and log standard deviations are calculated using Equations 2.1 and 2.2. The parameter estimates can then be converted to normal engineering units by taking the inverse log. Confidence intervals on the parameter estimates are easily calculated using the Student's t and χ^2 (chi square) tables, found in all statistics texts.

If two types of insulation are subjected to a thermal accelerated aging test, then it is common to ask if the two types of insulation behave the same, or if one insulation type has superior thermal performance, i.e., lasts longer at the same temperature, or lasts the same amount of time at a higher temperature. If the confidence intervals on the mean for each type of insulation system overlap, then it cannot be proved that there is a significant difference between the two insulation types. However, if the confidence intervals do not overlap, then one insulation type (the one with the higher mean time to failure) is better. A rigorous analysis of whether the insulations are different can be done using the t test and the F test, which allow one to say if the two insulations are different with a (high) level of confidence, given the natural variation in the test results [2.17]. IEEE 101 and IEC 60216 [2.21] give many examples of using statistical methods to determine superior insulting materials under thermal aging tests.

Analysis with the lognormal distribution is difficult if the data is censored. No simple equations or tables have been developed to help one calculate the log mean and log standard deviation if all the specimens under test have not failed. Thus, virtually all thermal aging tests assume that all the specimens will be tested to failure. However, some tables found in [2.20] can be used in some censoring situations. In addition, some very powerful computer programs can calculate lognormal parameters and confidence intervals of censored data. Although many packages are now available, the SAS Institute SAS/STAT personal computer software has been shown to be effective [2.23].

Weibull Distribution. The Weibull distribution has been widely used for voltage endurance test data, i.e., tests in which a constant AC voltage is applied to insulation specimens and the time to failure is measured. This distribution is also employed in experiments in which the breakdown voltage of insulation is measured. The Weibull distribution is a member of the class of distributions called extreme value distributions [2.18]. In principle, the distribution is most applicable where complete breakdown is caused by the failure of a "weak link." Since insulation will puncture at the point where the insulation is the thinnest, or the biggest defect is present, the weak link model seems appropriate.

The cumulative distribution function of the two parameter Weibull distribution function is given by

$$F(x) = 1 - \exp - (t/\alpha)^\beta \qquad (2.4)$$

where $F(x)$ is the probability of failure if the specimen survives time t. The α parameter is called the scale parameter and has the same units as t. The bigger α is, the longer the time to failure, since α corresponds to the time to failure for 63.2% of the specimens under test. β is the "shape" parameter. It is a measure of the spread of the data. The larger β is, the smaller is the range in failure times (or voltages). Thus, it is proportional to the inverse of the standard deviation. In general, β is on the order of 1 to 2 for voltage endurance tests, and tends to be around 10 for tests in which the breakdown voltage is measured.

There is an alternate form of the Weibull distribution called the three parameter Weibull distribution [2.18]. This form allows for the existence of a minimum possible time to failure or minimum failure voltage. However, this form has rarely been used for rotating machine insulation systems.

Although simple equations are not available to calculate estimates of α and β, a simple computer program exists to calculate the estimates using the method of maximum likelihood [2.5]. Simple graphs have also been produced to allow the calculation of approximate confidence intervals on the parameter estimates [2.5]. Many commercial computer programs are also available to calculate the parameters and confidence intervals, for example the SAS/STAT program described above. As discussed above for the lognormal distribution, one can easily tell if two insulation samples are significantly different by determining if the confidence intervals for α overlap. If there is overlap, then one cannot be sure there is a statistical difference, or if the scale parameter for the two different insulations are due to natural variation.

One of the reasons for the popularity of the Weibull distribution is that censored data are easily analyzed. IEEE 930 [2.5] gives a program and charts for this. Similarly, most modern statistical programs can easily accommodate censored data, including calculating confidence intervals and performing hypothesis tests to determine if two insulation systems are significantly different.

Regression Analysis. In addition to allowing the determination whether two types of insulation are significantly different (a comparison test), statistical methods are also available to enable the prediction of insulation system behavior at a temperature or voltage different from that used in testing. The calculation procedures are called regression analysis, also known as curve fitting. This subject is a standard part of any fundamental statistics textbook [2.17], and most of the principles discussed in such texts are applicable to insulation accelerated aging tests.

As an example, tests may be performed on an insulation system at two or more temperatures, usually at temperatures higher than expected in service. For a Class F insulation material, it is expected that the material will survive, on average, for 20,000 hours at 155°C (see Section 2.3). This corresponds to about 3 years. This is too long a time to wait for test results; thus, higher temperatures are used for the testing, perhaps 175°C, 195°C, and 205°C. These will give significantly shorter times to failure (recall Equation 2.1 and the strong influence temperature has on life). Using the times to failure for several specimens at each of these three temperatures, regression analysis can be used to define an equation to relate failure time to temperature. Further, the equation can be used to predict the time to failure at a lower temperature, for example 155°C. Figure 2.1 shows an example plot of life versus temperature for an insulation material, taken from IEEE 1.

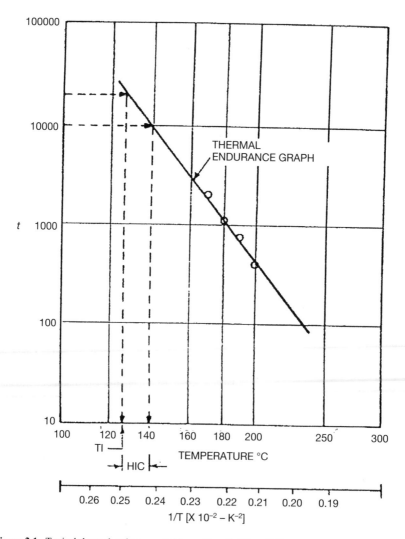

Figure 2.1. Typical thermal endurance plot for a Class B (Class 130) material. Note the thermal index corresponds to a life of 20,000 hours. (Copyright © 1989 IEEE. All rights reserved.)

The regression procedures are well established for normal distributed data, or data that can be transformed into the normal distribution, such as thermal life data represented by the lognormal distribution. Any statistics textbook details the procedures. Examples of the procedure applied to thermal aging data are in [2.19 and 2.21]. Note that these two standards assume that Equation 2.1 is valid (i.e., there is only one chemical reaction causing the thermal deterioration) and that all specimens are tested to failure. Procedures for other situations have not been generally accepted.

Regression for Weibull distributed data is, in principle, the same as for lognormal. However there are a number of complications. Since the assumption underlying linear regression is that the distribution of times to failure are normal (or can be transformed to normal), then the usual procedures for linear regression should not be directly applied to Weibull data [2.22]. From a practical point of view, most people ignore this restriction and perform a conventional regression, perhaps assuming that the data fits the lognormal distribution. Alternatively, rigorous regression methods assuming the Weibull distribution are described in [2.18]. Since most engineers find the latter method difficult to implement, various statistical programs, for example, the SAS/STAT package, can perform the calculations, derive the equation, and calculate confidence limits on the derived equation.

Number of Specimens. Since the results from aging tests have large statistical variation, the question arises about how many specimens need to be tested to determine if a new insulation material or system is equivalent to or superior to a reference material/system. The more specimens tested and the greater the difference between the two types of insulation, the easier it is to detect if a significant difference is present or not. Unfortunately, it takes time and money to produce the test results. Consequently, there is tremendous pressure to test as few specimens as possible.

The variation in times to failure from voltage or thermal aging tests is much larger than from experiments in which tensile strength or breakdown voltage are being measured. The result is that more specimens need to be tested in constant-stress aging tests than in progressive stress tests. Based on the normal variance expected in thermal aging tests, many key IEEE and IEC standards (Section 2.3) recommend the number of specimens to be tested at each temperature. Typically, this is five specimens. There is less agreement for voltage endurance tests [2.24]. To detect an improvement in life of two times between a new and a reference insulation material, 10 specimens of each type should be tested. In reality, usually fewer specimens are tested, leaving the analysis of the results open to question.

2.2.3 Failure Indicators

When the aging stress is voltage, in most cases it is obvious how to determine the lifetime of an insulation specimen: measure the aging time it takes to puncture the insulation. Electrical aging deteriorates the insulation such that a hole is eroded through the insulation and the high-voltage conductor shorts to ground, which trips out the power supply. Thus, time to failure (or breakdown voltage) is directly measured as the time that voltage is applied before the power supply automatically turns off. Similarly, this is often the case for some mechanical properties such as tensile strength. The end of life occurs when the insulation material parts.

For most other aging tests, how to determine the end of life is much less clear. For example, take a thermal endurance test. When an insulation material is subjected to thermal aging, the insulation may lose mechanical strength, become more brittle, lose some mass (i.e., lose weight) and/or have a lower electrical breakdown strength. However, it is unlikely to spontaneously "fail," unless the insulation catches fire. The question then occurs of how we determine the "life" of the insulation in such a test.

The solution is to use a diagnostic test that is applied after certain intervals of aging. The diagnostic test can be destructive; for example, a tensile strength or electrical breakdown strength test. The diagnostic test can also be nondestructive; for example, insulation resistance, dissipation factor, or partial discharge tests can be used. In both cases, unaged specimens are measured according to the diagnostic test and typical values are established. The specimens are aged at the desired stress for a fixed time interval and removed from the aging test. Then the diagnostic test is applied. If there is a defined change in the diagnostic, then the specimen has "failed," and is removed from further aging testing.

For thermal aging tests, the most common diagnostic tests are the electrical breakdown strength and the tensile strength. Failure is often deemed to have occurred if either of these strengths is reduced to 50% of the original (unaged) breakdown or tensile strength. An example of a nondestructive diagnostic test is power factor tip-up or partial discharge magnitude, when applied to a thermal cycling test [2.25]. If the tip-up increases or the PD magnitude increases by, say, two times over the duration of the aging test, then the insulation specimen is deemed to have failed. The problem with using diagnostic tests is that variability in failure time comes not only from the inherent variability in the aging process, but the diagnostic itself adds some variability. Recall that electrical breakdown strength itself is not fixed, but is a variable. Consequently, all other things being equal, where life is measured by means of a diagnostic test, more specimens will be required to overcome the additional uncertainty in the data.

2.3 THERMAL ENDURANCE TESTS

2.3.1 Basic Principles

In Section 2.1.1, the causes of thermal stresses were introduced and Equation 2.1 was given as a theoretical means of establishing the relationship between the life of insulation and its exposure to heat. Organic materials may degrade by several chemical mechanisms when subjected to heating. The most common experience is gradual oxidation of these materials, although polymer depolymerization (chain scission) and cross linking may also occur. Increasing the temperature generally accelerates these chemical reaction rates, although the presence of other materials and other factors of influence (the TEAM stresses introduced earlier in this Chapter) may either accelerate or decelerate the reaction rate.

Organic materials have an activation energy requirement. When the molecular energy of thermal motion is below a level characteristic of each material, little or no chemical change takes place. Increasing the temperature adds to the molecular energy, and when energy is sufficient to overcome the stability of chemical bonds within a material, the reactions start. Initially, near the activation energy, only a few molecules of the material will have sufficient energy to overcome the bonding energy and so only a slow aging of the material will occur. As the temperature is raised, more molecules will attain the necessary energy and the aging rate will increase.

Thermal aging is most straightforward when single insulating materials are evaluated. Combinations with other organic materials, as in some composites, may change the aging rate. Thermal aging of insulating materials is well established. Results of these tests are generally available from the suppliers of the materials and are used as a screening test for insulation engineers when selecting the components for an insulation system for a particular application. However, material thermal aging data is not acceptable for establishing the thermal endurance of an insulation system.

When designing an electrical insulation system, the engineer must take into account the other material components of the application, including also the inorganic and metallic materials in the rotating machine. They must design for the duty cycle and for the particular stresses of the machine environment. Thus, insulation system tests are generally performed by the equipment builder and are usually done on models or sections of the machine with all of the associated parts. Some larger suppliers of materials perform insulation system tests for high-volume applications, using their own materials when available and adding materials from other suppliers to complete the insulation system.

Thermal aging tests for materials use endpoints that are arbitrarily chosen and which may not be proper for particular applications. For example, testing may continue until a diagnostic endpoint is reached, such as loss of 5% of initial sample weight or 50% loss of flexural strength. Thermal aging tests for insulation systems usually run until electrical failures are obtained during periodic elevated voltage exposures.

2.3.2 Thermal Identification and Classification

An electrical insulating material is a substance in which the electrical conductivity is very small (approaching zero) and provides electric isolation. An electrical insulation system is an insulating material or a suitable combination of insulating materials specifically designed to perform the functions needed in electric equipment. A simple combination of insulating materials, perhaps containing a number of them without associated equipment parts, may be tested to evaluate any interaction between them.

There are slight differences between IEEE and IEC definitions for the thermal identification and classification of insulating materials and insulation systems. For this book, the IEEE definitions will be used. Historically, the term *thermal classification* has been used in reference to both insulation systems and to electric equipment. Thermal classification should be used in combination with the words *system* or *equipment* to clearly denote to which the term applies.

Electrical insulating materials are thermally rated by test to establish a thermal endurance relationship. This is an expression of aging time to the selected failure criterion as a function of test temperature in an aging test. A thermal endurance graph (Figure 2.1) is a graphical expression of the thermal endurance relationship in which time to failure is plotted against the reciprocal of the absolute test temperature. From this work, a material may be assigned a temperature index (TI) or a relative temperature index (RTI). The former is the number that corresponds to the temperature in °C, derived mathematically or graphically from the thermal endurance relationship at a specified time (often 20,000 hours for utility and industrial machines). The latter is the temperature index of a new or candidate insulating material, which corresponds to the accepted temperature index of a reference material for which considerable test and service experience has been obtained. Table 2.1 shows the accepted temperature indices of materials according to IEC 60085. The older classification system was according to a letter designation (e.g., A, B, ...). Presently, a numerical designation is preferred (105, 130, ...). A Class F or Class 155 material should have an average life of 20,000 hours (about 3 years) when operating at 155°C.

As noted above, in current practice, insulation systems by themselves are not given a temperature class. Technical committees may prepare standards for specific types of rotating machinery in which test procedures are established to enable the temperature class of their insulation systems to be established [2.42, 2.43]. These tests will usually require a reference

Table 2.1. Thermal Classification of Rotating Machine Insulation Materials (from IEC 60085)

Numerical Classification	Letter Classification	Temperature (°C)
105	A	105
130	B	130
155	F	155
180	H	180

insulation system, defined as an electrical insulation system assigned a system temperature rating, based on service experience or testing, that is used to qualify a new or modified electrical insulation system.

2.3.3 Insulating Material Thermal Aging Tests

In IEEE practice, the basic standard for all insulating materials and insulation systems tests is the latest version of ANSI/IEEE Standard No. 1—Recommended Practice for Temperature Limits in the Rating of Electrical Equipment and for the Evaluation of Electrical Insulations. Subsidiary standards for materials are IEEE No. 98—Standard for the Preparation of Test Procedures for the Thermal Evaluation of Solid Electrical Insulating Materials and IEEE No. 101—Statistical Analysis of Thermal Life Test Data. The comparable International Electrotechnical Commission general standard is IEC publication No. 60085—Thermal Evaluation and Classification of Electrical Insulation. Specifically for materials thermal evaluation there is an IEC Standard, the IEC 60216 Series—Guide for the Determination of Thermal Endurance Properties of Electrical Insulating Materials:

Part 1: General Procedures for the Determination of Thermal Properties, Temperature Indices and Thermal Endurance Profiles

Part 2: List of Materials and Available Tests

Part 3: Statistical Methods

Part 4: Instructions for Calculating the Thermal Endurance Profile

Part 5: Determination of Relative Thermal Endurance Index (RTE) of an Insulating Material

There is a lot of commonality between these IEEE and IEC standards and the responsible technical committees contain some members in common who have worked to make them compatible in the areas that both cover.

2.3.4 Insulation Systems Thermal Aging Tests

For insulation systems, there are a number of standards that supplement IEEE Standard No. 1. Some of these are general standards that apply to the thermal evaluation of all insulation systems, whereas a growing number address thermal aging for specific types of equipment. The general standard includes IEEE No. 99: Recommended Practice for the Preparation of Test Procedures for the Thermal Evaluation of Insulation Systems for Electric Equipment. The IEC general standard for insulation systems is Publication No. 60505—Evaluation and Qualification of Electrical Insulation Systems [2.2]. Several related specific standards are IEC Publications

No. 60610—Principal Aspects of Functional Evaluation of Electrical Insulation Systems: Aging Mechanisms and Diagnostic Procedures

No. 60611—Guide for the Preparation of Test Procedures for Evaluating the Thermal Endurance of Electrical Insulation Systems

No. 62114—Thermal Classification

For thermal evaluation and classification of rotating machines and their component parts, there is a growing list of IEEE and IEC standards. In IEEE there are

No. 117—Standard Test Procedure for Evaluation of Systems of Insulating Materials for Random-Wound AC Electric Machinery

No. 275—Recommended Practice for Thermal Evaluation of Insulation Systems for Alternating-Current Electric Machinery Employing Form-Wound Pre-insulated Stator Coils for Machines Rated 6900 V and Below

No. 304—Standard Test Procedure for Evaluation and Classification of Insulation Systems for Direct-Current Machines

No. 429—Recommended Practice for Thermal Evaluation of Sealed Insulation Systems for AC Electric Machinery Employing Form-Wound Pre-insulated Stator Coils for Machines Rated 6900 V and Below

No. 1107—Recommended Practice for Thermal Evaluation of Sealed Insulation Systems for AC Electric Machinery Employing Random-Wound Stator Coils.

A related IEC standard is IEC 61857-2—Procedures for Thermal Evaluation, Part 21: Specific Requirements for General Purpose Model Wire-Wound Applications.

IEEE 275 is an example of an accelerated aging test to determine the thermal classification of a form-wound stator coil insulation system. This standard describes the nature of coil specimens (called formettes) that are intended to model a coil insulation system. Formettes are specially manufactured coils with turn and ground insulation. They are typically only 0.3 m in length but have closely modeled the diamond shape. Instead of the copper turns being connected in series as in a normal coil, each turn in the formette is separately accessible, to facilitate turn-to-turn testing. To evaluate a candidate insulation system, typically, many formettes are manufactured, with two formettes reserved for destructive testing in the unaged condition. Each formette contains three or six coil sides. Sometimes, six bars are installed without being connected to each other during periodic hipot tests. The test standard recommends that four different test temperatures be used, with the heat being applied in an oven. Thus, one quarter of the formettes are used at each temperature. For a candidate system expected to be identified as Class F, typical test temperatures are 160°C for a 49-day cycle, 180°C for 4 days, 200°C for 2 days, and 220°C for 1 day. After each thermal aging cycle, the formettes are subjected to mechanical stress exposure, followed by moisture exposure, and ending with voltage exposure at a level that is based on the rated line-to-line voltage. Typically, the entire test is designed so that 10 cycles of heat aging and diagnostic exposures are required before most of the specimens fail during voltage application.

According to IEEE 275, for a candidate insulation system to be identified as class F, the thermal endurance graph for this system must result in failure times that are the same or longer than a reference insulation system already shown to have satisfactory life in service, and tested in exactly the same manner as the candidate system. The weakness in classifying a candidate system to a reference system is that few reference systems are actually operated at anywhere near the "design" maximum operating temperature in service.

IEEE 117 is a similar standardized thermal aging test for random-wound stator windings, using a "motorette" consisting of model coils installed in a simulated slot. To determine the life of an insulation system, IEEE 117 relies on usually 10 or more thermal aging cycles, with each cycle followed by shaking on a vibration table and then exposure to 100% humidity with condensation in a special chamber. After the prescribed time to allow the insulation to become saturated with moisture, an overvoltage is applied. Coil breakdown ends the test for that particular specimen, whereas those that survive are returned to the several different circulating air ovens, held at several different temperatures, for another thermal aging period. For thermal aging acceleration, it is desirable to have the lowest aging temperature (and the longest thermal exposure) set at 5 to 15°C above the maximum service temperature for which the insulation system is being qualified. The next two or three aging temperatures should each be at least 20°C above the lowest aging temperature. As with IEEE 275, the results of this test procedure on a candidate system must be compared to a similar test on a reference system.

Within IEC 60034 Part 18, Sections 1, 21, and 31, there is a comparable set of standards for several of the rotating machinery types. Usually, only the details of the nature of the test models and the diagnostic tests differ amongst the various standards.

Underwriters Laboratories in the United States has also produced UL-1446 and related documents that are based on IEEE standards. UL-1446 describes thermal testing methods for candidate insulation materials and insulation systems for use in small- and medium-sized random-wound machines as well as form-wound machines.

2.3.5   Future Trends

Technical committees within IEEE and IEC are continuing to work on both revisions to existing standards and on new standards for insulating materials and insulation systems. Both IEEE and IEC have policies requiring the review, reaffirmation, revision or withdrawal of existing standards on a periodic basis. New standards are written in response to perceived needs in the industry and must complete a thorough review of the need for and content of them in order to make them consensus standards, which will be generally accepted and used.

Recent committee work has focused on such matters as how to evaluate materials and systems when only slight changes have been made in their compositions or where the application of accepted materials and systems have been extended to different duty levels, machine sizes or applications. There is also a drive to design machines to operate at near their 'thermal class', since this will considerably reduce machine cost, because less copper would be required for the same output.

2.4   ELECTRICAL ENDURANCE TESTS

Since the main purpose of electrical insulation is to prevent electrical shorts between conductors at different potentials, it is not surprising that there are means to determine the ability of an insulation to withstand voltage. From a basic point of view, each insulation material has a short-term breakdown strength. That is, if the voltage is rapidly increased across an insulation material, at some point there will be sufficient voltage to puncture the insulation. The breakdown strength is the breakdown voltage divided by the insulation thickness. Each insulating material has its characteristic breakdown strength. Gases typically have a breakdown strength of around 3 kV/mm at atmospheric pressure, whereas most homogeneous solid insulating materials have an intrinsic breakdown strength ranging from 100 kV/mm to 1

MV/mm. Elaborate methods have been developed for measuring the short-term (voltage is applied for less than a few minutes) breakdown strength of insulation. Some of these are referred to in Section 2.8.

No practical rotating machine insulation system is subjected to electrical stresses anywhere near the short-term breakdown strength. In fact, even high-voltage stator windings have an average design electrical stress that is usually less than 3 kV/mm rms (although there are some modern designs that exceed this [2.1]). This stress is about 100 times less than the intrinsic capability of modern epoxy–mica insulation systems. The reason for the much lower design stress is because the insulation will rapidly age at higher stresses, in part due to partial discharge activity that inevitably will be present because of small pockets of gas in the insulation (Section 1.4.4). Thus knowledge of the short-term breakdown strength of the insulation is not sufficient to allow design of an insulation system that will yield long life.

To predict the behavior of an insulation system under voltage stress in stator windings rated at 1000 V or more and determine the best insulation system design with respect to voltage stress, voltage endurance tests have been developed. In a voltage endurance test, a voltage is applied to an insulation system that is higher than expected in operation, and the time to failure is measured. Sometimes, the insulation system is simultaneously subjected to high temperature. The endurance test voltage is usually orders of magnitude less than the short-term breakdown voltage. Usually, voltage endurance tests for rotating machines applications are only applied to insulation systems, and not to the materials. Although there are some voltage endurance test methods that have been standardized, the number and pervasiveness of these standards are far less than for thermal endurance testing. Voltage endurance tests are normally only applied to stator winding insulation systems, since it is usually only the stator winding insulation that may gradually age under voltage stress. However, as discussed in Sections 1.4.2 and 8.7, voltage surges may sometimes age the insulation, and such surges from static excitation systems may deteriorate synchronous machine rotors (Sections 9.1.5 and 9.2.6). Endurance tests under voltage surges for rotor or low-voltage stator windings have yet to be standardized, but this will likely occur in the early 2000s.

Most voltage endurance tests for stator windings rated at 1000 V or more are proprietary test methods that are unique to each manufacturer of stator coils. These will be discussed first. Then IEEE and IEC test methods will be presented.

2.4.1   Proprietary Tests for Form-Wound Coils

Machine manufacturers determine the thickness of the groundwall insulation in form-wound stator coils and bars on the basis of voltage endurance test methods developed in-house. Each manufacturer has different approaches. References 2.8, 2.9, and 2.26–29 present some of the test methods that have been published. By performing the voltage endurance test on a range of insulation system designs, the manufacturer can establish which design gives a satisfactory result under the accelerated conditions for the least cost.

Most of the test procedures use relatively short specimens that simulate the bar or coil in the slot. Typically, the specimens may be from 0.5 m to 1 m in length to simulate the slot section of a coil/bar, although some manufacturers sometimes use full-length coils or bars [2.27]. The coil/bar specimens may have a thinner insulation thickness than normal, enabling testing at lower voltages. There is always a semiconductive layer applied to give a ground plane. Because the test stress is high, stress relief coatings are needed at each end of the ground plane region (Section 1.4.5).

The 50/60 Hz power frequency voltage is the most common means of energizing the copper conductors of the specimens. However, many manufacturers use 400 Hz to 5000 Hz AC power supplies to energize the specimens [2.9, 2.29]. By using a frequency higher than the power frequency, the insulation seems to fail faster. In part, this "frequency acceleration" effect is probably due to the fact that the number of partial discharges per second increases proportionally with the number of AC cycles applied. Thus, many researchers believe that the voltage endurance life is inversely proportional to frequency. The frequency acceleration effect remains controversial, since different researchers have found different acceleration factors [2.9].

Most proprietary voltage endurance tests are done at a temperature that is higher than room temperature. Usually, temperatures in the range of 120°C to 180°C are reported. Heating is sometimes accomplished by putting the specimens in a large oven, circulating current through the conductors, circulating hot water or oil in the hollow conductors (where possible), or by applying heated plates to the outside of the specimens.

Clearly, the test procedures are highly variable, and often make comparison of test results between manufacturers very difficult.

2.4.2   Standardized Test Methods for Form-Wound Coils

The only standard, detailed voltage endurance test procedure for form-wound bars or coils that is widely applied is IEEE 1043 [2.30]. This method is based on a test developed by a utility in the 1950s [2.7]. Key features of the test are:

- Usually, full length coils or bars are tested.
- The specimens have the normal end-winding, with the usual stress relief coatings, since many problems in service originate from this area.
- Only one method of heat application is permitted: strip heaters applied to the surface of the specimen, in the slot region (Figure 2.2).
- Only power frequency voltage (50 or 60 Hz) is permitted.

The test procedure does not specify a test voltage or the test temperature. Some utilities, however, have published the temperatures and voltages that they employ [2.31, 2.32]. These suggest a 13.8 kV rated bar or coil, test voltage either 30 kV or 35 kV, and a minimum acceptable time to failure of either 400 hours or 250 hours, respectively. The 35 kV test is considered more onerous. The test temperatures are typically in the 110 to 120°C range and should be the maximum actual operating temperature for a winding. There have been recent

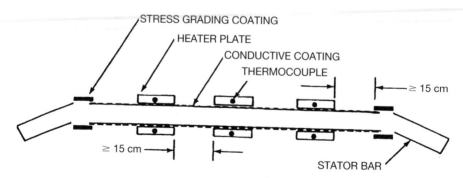

Figure 2.2. Application of heater plates to a stator bar for the IEEE 1043 voltage endurance test.

efforts to make an IEEE standard based on IEEE 1043, which specifies the test voltages, temperatures, and minimum times to failure. This standard is confined to hydrogenerator stator coils and bars [2.33].

The IEEE 1043 test method is now used by many purchasers of windings as one of the vendor selection criteria. The procedure has also been used as a quality control test.

IEC has developed procedures to explain the general principles for voltage endurance testing [2.34]. However, no standards have been produced by IEC that give specific and detailed test procedures for rotating machine insulation systems.

The voltage endurance tests are good for establishing the effect of voltage on life for the slot portion of a coil or bar. However, such tests do not enable determination of the voltage endurance capabilities in the end-winding, where electrical tracking and PD between coils may occur (Sections 8.8 and 8.11).

2.5 THERMAL CYCLING TESTS

Thermal cycling occurs when a motor or generator rapidly changes load, causing the current and, thus, the conductor temperature, to rapidly change between a low temperature and a high temperature. As the conductor temperature rises and falls, the conductors expand and contract due to the thermal coefficient of expansion; this can gradually fatigue crack the insulation or abrade it away. The specifics of the deterioration processes are in Sections 8.2, 9.1.2 and 9.2.2.

Thermal cycling in round rotor windings of large turbine generators is of great concern. However, no standardized thermal cycling endurance procedures for complete rotor insulation systems have been developed. Indeed, not even proprietary methods have been published. Instead, machine manufacturers rely on material property tests such as abrasion resistance or fatigue strength (Section 2.8).

Thermal cycling tests that mimic the associated failure mechanism in form-wound stators have been standardized, both by IEEE and IEC. These tests are normally intended to be used on full-size coils or bars. The tests are most applicable to coils or bars in stators in which the load is rapidly changed, usually hydrogenerators, pump-storage generators, or gas turbine generators. As discussed in Section 8.2, motors also experience rapid load changes, but due to the shorter stator core length in most motors, insulation aging is less likely.

2.5.1 IEEE Thermal Cycling Test

IEEE 1310 is a test method developed in the early 1990s to allow manufacturers and users to assess the relative capability of stator winding insulation systems to resist the thermal cycling failure process [2.25]. The test only evaluates the ability to resist failure of the bond between the groundwall insulation and the copper conductors. It does not enable evaluation of the slot mechanical support system (wedges, side packing, ripple springs, etc.), nor is IEEE 1310 relevant for global VPI stators in which the coils/bars are bonded to the stator core. IEEE 1310 is a recommended test procedure, intended to ensure that results from different testing organizations can be compared. It does not specify a pass/fail criterion.

The method uses full-size bars or coils, which are "floating" on supports, allowing free axial movement of the bar/coil. Thermal cycling is achieved by circulating DC or AC current through each bar/coil, sufficient to raise the temperature of the copper to the thermal class temperature for the insulation (155°C for Class F systems). The temperature is raised to this upper limit in about 45 minutes. When the high temperature limit is reached, the current is shut off and cool air is blown over the coil/bar to reduce the copper temperature to 40°C in about 45 minutes. This temperature swing from low to high to low temperature constitutes one thermal cycle. The test proceeds until 500 thermal cycles are completed. At least 4 bars (or 2 coils) are needed to establish a clear trend. While the thermal cycling is occurring, high voltage is *not* applied across the insulation. The test is destructive (the coils/bars can not be installed in a stator after undergoing the test). The test is one of the most expensive of all the aging tests described in this chapter, and is only used if the application warrants it.

Like the thermal aging tests in Section 2.3.4, the thermal cycling test does not result in failure by itself. Diagnostic tests are needed to establish when deterioration has occurred. The best diagnostic test is the voltage endurance test. But this is also the most expensive, since the voltage endurance test requires even more test specimens. With a voltage endurance test (Section 2.4.2) as a diagnostic test, the times to failure of several "virgin" bars, which have not been subjected to thermal cycling, are established. The voltage endurance test is then done on the coils/bars that have been subjected to 500 thermal cycles. If there has been a significant decrease in voltage endurance life, then, clearly, the thermal cycling has deteriorated the insulation. Since IEEE 1310 provides only a test procedure, the test user must define what constitutes a significant decrease in voltage endurance life due to thermal cycling.

Other less expensive diagnostic tests are available that are not destructive. The most useful are the partial discharge test and the power factor tip-up test. These are described in Sections 12.10 and 12.9, respectively. These diagnostic tests are sensitive to the presence of delamination, i.e., voids within the groundwall insulation that may occur as a result of the thermal cycling. Either test is done on the bars or coils prior to any thermal cycles. These are the baseline measurements. Then, after 50, 100, 250, and 500 thermal cycles, the coils and bars are subjected to repeat partial discharge or tip-up tests. If there is a significant increase in the measurement as compared to the baseline measurement, then deterioration has occurred. The IEEE document does not provide guidance on what constitutes a significant increase, but a doubling of the PD magnitude or a 0.5% increase in tip-up may be considered significant.

Many generator owners are using this test as a vendor qualification test, to decide which manufacturers make windings that are resistant to thermal cycling. In addition, some manufacturers use the test to optimize their insulation system design, especially with respect to the choice of bonding resin formulations.

2.5.2 IEC Thermal Cycling Test

Early in 2000, IEC published two thermal cycling test procedures [2.35]. One of the procedures is very similar to the IEEE 1310 document; that is, it describes a method for evaluating the bonding between the groundwall and the copper, using floating bars/coils. The only real difference is that the IEC version is less restrictive on the test parameters.

The other IEC procedure is much more comprehensive, in that the bars or coils are inserted in simulated slots, with all the normal slot contents, including wedges, fillers, ripple springs, and side packing. In addition, realistic end-winding bracing is applied. Thus this procedure not only evaluates the ability of the coil/bar itself to withstand thermal cycling, but also assesses the entire contents of the slot. As an alternate to circulating AC or DC current to produce the thermal cycles, hot or cold liquid circulated through hollow conductors is also permitted in the procedure.

This method is primarily intended for use by manufacturers who are evaluating the optimum design for wedging, side packing, and/or groundwall insulation. The results of any new insulation material or new wedging approach are compared to a reference system that has seen successful service. The test is very expensive, costing many tens of thousands of dollars to perform.

2.6 MULTIFACTOR STRESS TESTING

Section 2.1.5 discussed the notion that some failure processes in rotor and stator windings do not occur as a result of a single stress or factor but, in fact, depend on two or more stresses/factors. Thus, to reasonably duplicate some failure processes in an accelerated aging test, more than one stress needs to be applied at the same time. In the three types of tests described above (thermal, thermal cycling, and voltage endurance), there was only one stress that was "accelerated," that is, the stress was applied at a level higher than would occur in normal service. Although another stress or factor may be present (for example, thermal stress in the voltage endurance test), this additional stress is not an "accelerated" stress. If the applied temperature is above the rated temperature, then it could be considered "accelerated."

A multifactor accelerated aging test is one in which two or more stresses/factors are simultaneously or sequentially applied at higher levels than would occur in service. The aim is both to realistically simulate a real failure process, and to have the failure occur in a time much shorter than would be expected in service. As with all aging tests, the purpose is to find the insulation system design that performs satisfactorily in comparison to a reference system, at the lowest cost.

In practice, multifactor aging tests are not used to any great extent to simulate rotor and stator winding failure. No specific test procedures have been standardized. What has been standardized, and about which several papers have been presented, are the general procedures that should be employed in a multifactor aging test on insulation systems. Paloniemi introduced the concept of "equalized aging," that is, all the relevant stresses in a failure process should be accelerated to such an extant that the aging rate due to each stress is about the same [2.36]. This principle has been incorporated in an IEC guide for developing multifactor aging tests for insulation systems [2.2]. A similar IEEE document has also been developed [2.37]. A general guide to the principles involved in multifactor tests in which both thermal and electrical stresses are accelerated is given in IEC 60034–18–33 [2.38].

2.7 NUCLEAR ENVIRONMENTAL QUALIFICATION TESTS

Stator windings in certain motors used in nuclear power plant applications have to be qualified for operation in "harsh or mild radiation environments," which can either be inside or outside a radiation containment area. Since the insulation systems must be shown to be able to withstand such conditions, a brief description of what is in reality a multi-factor tests are presented. Such tests are sometimes called EQ tests, or environmental qualification tests. The particular environments are discussed in more detail as follows.

(a) *Loss of Coolant Accident (LOCA).* Normally, motors exposed to this type of accident are subjected to mild radiation fields under normal operating conditions, and during a LOCA they see higher radiation fields together with high pressure and temperature transients. The motor may also be subjected to chemical spraying during the LOCA.

(b) *Main Steam Line Break (MSLB).* Motors exposed to this type of environment can be either inside or outside containment areas and have to be qualified for operation during or after a high-pressure steam line break that may also be associated with a seismic event. Such an event results in the motor being subjected to a high-temperature steam environment.

(c) *Radiation Only Environments.* Such environments may contain nuclear radiation fields that exist during normal motor operation and/or after a LOCA, but not the high-temperature and pressure conditions from a LOCA or MSLB.

(d) *Mild Environments.* In this environment, stator windings only see normal aging conditions, but may have to be qualified to be functional after a LOCA and/or seismic event.

Environmental qualification must be based on the assumption that the motor will have to be capable of operating after being subjected to a LOCA, MSLB, seismic event, or a combination of these, at the end of the plant life. At this time, these windings will have been subjected to their maximum normal thermal, mechanical, electrical, chemical, and, if appropriate, radiation aging. Where LOCA and MSLB qualification is required and the motor has an open-type enclosure or is totally enclosed but the steam pressure is very high, a sealed winding insulation system (IEEE 429) is required.

IEEE Standards 323 and 344 give some guidance on qualification techniques and the use of complete motors or formettes (stator winding models) for qualification testing. The various options for the EQ of a stator winding insulation system are as follows.

2.7.1 Environmental Qualification (EQ) by Testing

(a) *Test Specimens.* Complete winding or formette specimens must first be constructed of materials that are considered to be resistant to both normal and accident conditions. Detailed coil and winding manufacturing procedures and methods of verifying all of the winding materials used have to be prepared before specimen manufacture. This is required to ensure that if the insulation system is environmentally qualified, it can be exactly replicated. Specimens of all materials used should be kept and insulation resistance measured often to provide baseline data for future material verification.

Once constructed, the test specimens must then be subjected to accelerated aging that simulates normal aging plus accident radiation and seismic conditions (if applicable). After such aging, those specimens that represent stator winding insulation systems that have to be qualified to survive a LOCA or MSLB are placed in a sealed test chamber and subjected to high-pressure steam and temperatures that simulate such conditions. Insulation systems to be qualified for LOCA conditions are also subjected to a chemical or demineralized water spray, if applicable, used to control pressure buildup in the reactor building.

(b) *EQ Test Sequence.* The test sequence that is normally used is as presented in Table 2.2.

(c) *EQ Test Report.* After EQ testing is complete, a report containing details of all the testing performed and the results must be prepared. This is used to justify the use of any insulation system that is qualified for normal plus accident conditions that have been enveloped by the test program.

2.7.2 Environmental Qualification by Analysis

This is usually only used for stator winding insulation systems to be used in Radiation Only and Mild Environment applications for standby motors that do not see any significant thermal aging. The insulating and bracing materials used are types known to have a radiation damage threshold above the total radiation dose the motor will see in its lifetime. This allows environmental qualification for such applications to be performed by analysis. Examples of

Table 2.2. EQ Test Sequence

Activity or Test	Type of Qualification			
	LOCA	MSLB	Radiation Only	Mild Environment
Test specimen incoming inspection	X	X	X	X
Winding baseline functional testing [normally IR, PI (if form-wound), winding resistance, surge, and hipot tests]	X	X	X	X
Normal radiation aging[a]	X		X	
Functional testing [normally IR, PI (if form-wound), winding resistance, and reduced voltage hipot and surge tests]	X		X	
Thermal aging in an oven	X	X	X	X
Repeat functional testing	X	X	X	X
Vibration aging and, if applicable, mechanical cycling[b]	X	X	X	X
Repeat functional testing	X	X	X	X
Seismic simulation test[c]	X	X	X	X
Repeat functional testing	X	X	X	X
Accident radiation aging	X		X	
Repeat functional testing	X		X	
Simulated LOCA or MSLB	X	X		
Repeat functional testing	X	X		
Visual Inspection	X	X	X	X

Notes: [a]Sometimes, the test specimen is subjected to normal plus accident radiation aging at this time. [b]Mechanical cycling can only be performed if the specimen is a complete motor. [c]Only performed if the insulation system has to be seismically qualified. [d]For a particular test program, the normal and accident conditions should envelope the worst conditions the insulation system is likely to see.

materials with high radiation resistance are Nomex[153], mica, woven glass, epoxy resin, and polyimide magnet wire insulation.

2.7.3   Environmental Qualification by a Combination of Testing and Analysis

This is best illustrated by the following example. Suppose an insulation system is to be qualified for use in a motor that is located in a radiation environment, is only operated for testing, and has to operate after a seismic event. By constructing the insulation system with materials having a radiation damage limit above the total integrated radiation dose the motor is subjected to, radiation aging qualification can be performed by analysis. On the other hand, seismic qualification may be most easily obtained by making a formette model of the insulation system and subjecting it to a vibration (shaker) table test that simulates the worst postulated seismic event. Thus, a combination of analysis and testing is to environmentally qualify such a stator winding insulation system

2.8   MATERIAL PROPERTY TESTS

The majority of this chapter has been concerned with accelerated aging tests on insulation systems. Such tests simulate a specific failure process in the rotor or stator winding. However, there is a large collection of tests intended to measure specific insulation properties, usually on materials, but sometimes on systems. The idea is that with knowledge of a failure mechanism, it is often possible to identify one or more insulation material properties that can be measured on a candidate insulation material to give an indication of how well the insulation will resist a specific failure mechanism.

For example, the slot liner (or ground) insulation in a turbine generator rotor is subject to abrasion during load cycling (Section 9.1.2). Rather than simulate the entire failure process

Table 2.3. ASTM Test Standards Relevant to Rotating Machine Insulation Material

D 149-81	Test Method for Dielectric Breakdown Voltage and Dielectric Strength of Solid Electrical Insulating Materials at Commercial Power Frequencies
D 150-81	Test Methods for AC Loss Characteristics and Permittivity (Dielectric Constant) of Solid Electrical Insulating Materials
D 257-78(1983)	Test Methods for DC Resistance or Conductance of Insulating Materials
D 495-84	Test Method for High-Voltage, Low-Current, Dry Arc Resistance of Solid Electrical Insulation
D 651-84	Test Method for Tensile Strength of Molded Electrical Insulating Materials
D 1676-83	Methods for Testing Film-Insulated Magnet Wire
D 1868-81	Method for Detection and Measurement of Partial Discharge (Corona) Pulses in Evaluation of Insulation Systems
D 2132-85	Test Method for Dust-and-Fog Tracking and Erosion Resistance of Electrical Insulating Materials
D 2275-80	Test Method for Voltage Endurance of Solid Electrical Insulating Materials Subjected to Partial Discharges (Corona) on the Surface
D 2303-85	Test Methods for Liquid-Containment, Inclined-Plane Tracking, and Erosion of Insulating Materials
D 2304-85	Method for Thermal Evaluation of Rigid Electrical Insulating Materials
D 2307-86	Test Method for Relative Thermal Endurance of Film-Insulated Round Magnet Wire
D 2519-83	Test Method for Bond Strength of Electrical Insulating Varnishes by the Helical Coil Test
D 3145-83	Test Method for Thermal Endurance of Electrical Insulating Varnishes by the Helical Coil Method
D 3151-79	Test Method for Thermal Failure Under Electric Stress of Solid Electrical Insulating Materials
D 3251-84	Test Method for Thermal-Aging Characteristics of Electrical Insulating Varnishes Applied Over Film-Insulated Magnet Wire
D 3377-85	Test Method for Weight Loss of Solventless Varnishes
D 3386-84	Test Method for Coefficient of Linear Thermal Expansion of Electrical Insulating Materials
D 3426-75(1980)	Test Method for Dielectric Breakdown Voltage and Dielectric Strength of Solid Electrical Insulating Materials Using Impulse Waves
D 3638-85	Test Method for Comparative Tracking Index of Electrical Insulating Materials
D 3755-79	Test Method for Dielectric Breakdown Voltage and Dielectric Strength of Solid Electrical Insulating Materials Under Direct-Voltage Stress
D 3850-84	Test Method for Rapid Thermal Degradation of Solid Electrical Insulating Materials by Thermogravimetric Method

in an accelerated aging test, the designer knows that a slot liner that is very resistant to abrasion would result in longer winding life. Thus, the designer selects a material that performs well on a material test that is called the "abrasion resistance" test.

Similarly, the end-winding of a high-voltage stator winding can suffer from a failure process called electrical tracking (Section 8.8). Rather than try to duplicate and accelerate the entire deterioration process, the designer recognizes that a single property called "tracking resistance" can be measured on slabs of candidate groundwall insulation materials. The material that has the highest tracking resistance will perform the best in a rotating machine that may be subject to a polluted environment.

There are many dozens of tests that have been developed to measure a wide range of insulation properties. Reference 2.39 describes many of these tests and how many common rotating machine materials behave under them. In addition, standards organizations such as ASTM and IEC have published standard test procedures to measure many of the properties that are relevant to rotating machine insulation systems. Table 2.3 contains a list of common material tests applicable to rotating machine insulation systems from ASTM. Also, many of these tests are described in the following chapters and their relevance to materials is described in Chapters 3 to 6.

REFERENCES

2.1   B. J. Moore, R. H. Rehder, and R. E. Draper, "Utilizing Reduced Build Concepts in the Development of Insulation Systems for Large Motors," in *Proceedings of IEEE Electrical Insulation Conference,* Cincinnati, Oct. 1999, pp. 347–352.

2.2   IEC Standard 60505, "Evaluation and Qualification of Electrical Insulation Systems."

2.3   T. W. Dakin, "Electrical Insulation Deterioration Treated as a Chemical Rate Phenomenon," *AIEE Trans., Part 1, 67,* 113–122 (1948).

2.4   M. Kurtz, J. F. Lyles, and G. C. Stone, "Application of Partial Discharge Testing to Hydro Generator Maintenance," *IEEE Trans PAS,* Aug. 1984, 2588–2597.

2.5   IEEE Standard 930, "Guide for the Statistical Analysis of Electrical Insulation Voltage Endurance Test Data."

2.6   H. S. Endicott, B. D. Hatch, and R. G. Sohmen, "Application of the Erying Model to Capacitor Aging Data," *IEEE Trans. CP-12,* March 1965, 34–40.

2.7   A. W. W. Cameron and M. Kurtz, "A Utility's Functional Evaluation Tests for High Voltage Stator Insulation," *AIEE Trans. Vol. 78, Part III,* June 1959, 174–184.

2.8   Wichmann, "Accelerated Voltage Endurance Testing of Micaceous Insulation Systems for Large Turbogenerators Under Combined Stresses," *IEEE Trans. PAS,* Jan. 1977, 255–260.

2.9   D. R. Johnston et al., "Frequency Acceleration of Voltage Endurance," *IEEE Trans. EI,* June 1979, 121–126.

2.10 IEC 1251, "Electrical Insulating Materials—AC Voltage Endurance Evaluation-Introduction," 1993.

2.11 E. Persson, "Transient Effects in Applications of PWM Inverters to Induction Motors," *IEEE Trans. IA,* Sept. 1992, 1095–1102.

2.12 G. C. Stone, S. R. Campbell, and S. Tetreault, "Inverter Fed Drives: Which Motor Stators are at Risk," *IEEE Trans. IA,* Sept. 2000, 17–22.

2.13 G. C. Stone, R. G. VanHeeswijk, and R. Bartnikas, "Investigation of the Effect of Repetitive Voltage Surges on Epoxy Insulation," *IEEE Trans. EC,* Dec 1992, 754–760.

2.14 J. P. Bellomo et al., "Influence of the Risetime on the Dielectric Behavior of Stator Winding Insulation Materials," in *Proceedings of IEEE CEIDP,* Oct. 1996, pp. 472–476.

2.15   S. Nagano et al., "Evaluation of Field Insulation of Turbine Generators with Higher Thyristor Excitation," *IEEE Trans. EC,* Dec. 1986, 115–121.

2.16   G. C. Stone et al., "Motor and Generator Insulation Life Estimation," EPRI Report TR-100185, Jan. 1992.

2.17   G. W. Snedecor and W. G. Cochran, *Statistical Methods,* Iowa State University Press, 1967.

2.18   J. F. Lawless, *Statistical Models and Methods for Lifetime Data,* Wiley, 1982.

2.19   IEEE 101–1987, "IEEE Guide for the Statistical Analysis of Thermal Life Test Data."

2.20   D. Gladstein, W. Nelson, and J. Schmee, "Confidence Limits for Parameters of a Normal Distribution from Singly Censored Samples Using Maximum Likelihood," Technometrics, May 1985.

2.21   IEC 60216, "Guide for the Determination of Thermal Endurance Properties of Electrical Insulating Materials; Part 3—Instructions for Calculating Thermal Endurance Characteristics."

2.22   J. F. Lawless and G. C. Stone, "The Application of Weibull Statistics to Insulation Aging Tests," *IEEE Trans. EI-14,* October 1979, 233–238.

2.23   G. C. Stone, R. G. vanHeeswijk, and R. Bartnikas, "Electrical Aging and Electroluminescence in Epoxy Under Repetitive Voltage Surges," *IEEE Trans. DEI-27,* April 1992.

2.24   G. C. Stone, "The Statistics of Aging Models and Practical Reality," *IEEE Trans. EI,* Oct. 1993, pp. 716–728.

2.25   IEEE 1310, "Recommended Practice for Thermal Cycle Testing of Form Wound Stator Bars and Coils for Large Generators."

2.26   Kelen, "The Functional Testing of HV Generator Stator Insulation," CIGRE Paper 15-03, Paris, August 1976.

2.27   R. Schuler and G. Liptak, "Long Term Functional Test on Insulation Systems for High Voltage Rotating Machines," CIGRE 15-05, Paris, August 1976.

2.28   J. Carlier et al., "Ageing Under Voltage of the Insulation of Rotating Machines: Influence of Frequency and Temperature," CIGRE 15-06, Paris, August 1976.

2.29   Wichmann, "Two Decades of Experience and Progress in Epoxy Insulation Systems for Large Rotating Machines," *IEEE Trans. PAS,* Jan. 1983, 74–82.

2.30   IEEE 1043–1996, "IEEE Recommended Practice for Voltage Endurance Testing of Form Wound Bars and Coils."

2.31   B. E. Ward, G. C. Stone, and M. Kurtz, "Quality Control Test for High Voltage Stator Insulation," *IEEE Electrical Insulation Magazine,* Sept. 1987, 12–17.

2.32   S. E. Cherukupalli and G. L. Halldorson, "Using the Voltage Endurance Test to Ensure the Quality of Insulation in a Generator," *Hydro Review,* October 1997, 48–57.

2.33   IEEE 1553-2002, "Standard for Voltage Endurance Testing of Form-Wound Bars and Coils for Hydrogenerators."

2.34   IEC 60034, Part 18, Section 32, "Rotating Electrical Machines—Functional Evaluation of Insulation Systems: Test Procedures for Form-Wound Windings—Electrical Evaluation of Insulation Systems Used in Machines up to and Including 50 MVA and 15 kV."

2.35   IEC 60034, Part 18, Section 34, "Rotating Electrical Machines: Functional Evaluation of Insulation Systems: Test Procedures for Form-Wound Windings—Evaluation of Thermomechanical Endurance of Insulation Systems.

2.36   P. Paloniemi, "Theory of Equalization of Thermal Aging Processes of Electrical Insulating Materials in Thermal Endurance Tests, Parts 1, 2 and 3," *IEEE Trans. EI,* Feb. 1981, 1–30.

2.37   IEEE 1064, "IEEE Guide for Multifactor Stress Functional Testing of Electrical Insulation Systems."

2.38   IEC 60034, Part 18, Section 33, "Rotating Electrical Machines: Functional Evaluation of Insulation Systems: Test Procedures for Form-Wound Windings—Multifactor Functional Evaluation—Endurance Under Combined Thermal and Electrical Stresses of Insulation Systems Used in Machines up to and Including 50 MVA and 15 kV."

2.39   W. T. Shugg, *Handbook of Electrical and Electronic Insulating Materials,* 2nd ed., IEEE Press, 1995.

2.40   P. Holzman, in "EPRI Motor Insulation System Seminar and Workshop—Material Testing for Steam Tolerance," September 19–21, 2000, Norfolk, VA.

2.41   E. A. Boulter, "Multi-Factor Insulation System Functional Tests," paper presented at IEEE Inernational Symposium on Electrical Insulation, 76CH1088, June 1976.

2.42   E. A. Boulter, "Functional Evaluation of High Voltage Turbine-Generator Insulation Using a Generette Core Model," paper presented at AIEE CP59–1083, Fall General Meeting, October 1959.

2.43   J. C. Botts, E. A. Boulter, and R. J. Flaherty, "Aging in Rotating Machinery Insulating Materials," paper presented at IEEE CP59-PWR, Winter Power Meeting, January 1970.

CHAPTER 3

HISTORICAL DEVELOPMENT OF INSULATION MATERIALS AND SYSTEMS

The selection of electrical insulation systems for rotating machines has always been dependent on the materials available, their cost, the technical needs of the motor or generator application, and the relative costs of the several manufacturing processes available at the time. In the early years of the industry, there was a near total reliance on naturally occurring materials and much trial and error experimentation to find systems that met minimum design criteria. Thus, operating temperatures, as well as mechanical and electrical stresses, were kept low to accommodate the limitations of these materials.

Service experience with the early machines generally evolved into the widespread use of conservative designs to ensure long life. In an increasingly competitive environment and with growing knowledge of the capabilities and limitations of the existing natural materials, combinations were developed that enabled systems that could be operated at higher TEAM (thermal, electrical, ambient, and mechanical) stresses (Section 2.1).

From the beginning of the industry until the advent of the mainframe computer, engineers used slide rules and handbook tables to design rotating machines. This laborious process led to a range of machine ratings within the same frame size and a distribution of insulation stresses from low to fairly high over the rating range. Some machines seemed to never wear out and could be operated at power levels well above their nameplate ratings with safety. In fact, many machines installed in the early part of the 1900s are still in service. Today, with sophisticated finite element design software and manufacturing processes, windings are designed to operate at higher stresses than in the past. The result is that today windings are less expensive than those of the past, but operation beyond the original design life is much less likely, and operation beyond nameplate ratings is usually not possible.

This chapter describes the chronological development of insulation materials and systems, together with some of the key innovations in manufacturing. More details on the systems in widespread use today are in Chapters 4 to 6. For the most part, it was the innovations

Electrical Insulation for Rotating Machines. By Stone, Boulter, Culbert, and Dhirani
ISBN 0-471-44506-1 © 2004 Institute of Electrical and Electronics Engineers

in materials and manufacturing for form-wound stator groundwall insulation that paced the development of machine insulation systems. Thus, the following concentrates on stator groundwall insulation. The reader should also refer to Shugg [3.1], which has an excellent review of the development of insulating materials for all applications.

3.1 NATURAL MATERIALS

The first insulation systems used materials that were then common in industry for other uses. These included natural fibers of cellulose, silk, flax, cotton, wool, and, later, asbestos. The fibers were used both as individual strands for applications such as wire servings, in groups of strands for support ties, and in combined forms as in nonwoven papers and woven cloths. Natural resins derived from trees, plants, and insects, and petroleum deposits, were used in combination with the fiber forms to make insulating materials. These included refined petroleum oils, waxes, asphalts, and natural resins such as pitch, shellac, rosin, linseed oil, etc. Solids such as sand, mica, asbestos, quartz, and other minerals were often used as fillers in ground or powdered form. Among the earliest materials used as stator groundwall insulation was varnished cambric, which is still manufactured and used today in some electrical applications. This material was originally composed of a fine grade of cotton cloth impregnated with a natural resin of the drying oil variety, which oxidizes and changes to a hard, tough, elastic substance when exposed in a thin film to air. Modifications in the choice of oil and blending were used to modify the properties for specific uses.

These resins were applied from solvent solutions and had good shelf life when sealed to prevent oxidation and solvent evaporation. Tower ovens, with good air circulation, were used to speed up evaporation and oxidation to produce continuous rolls of the material that could then be cut to desired shapes for use as insulation pieces or to be built up into thicker sections or wrapped onto conductors as groundwall insulation.

During World War 1, asphaltic resins (also called bitumen) were combined with mica splittings for the first time to make improved groundwall insulation for turbine generator stator coils. The mica splittings were supported by a fine grade of cellulose paper on both sides. Although practices differed between manufacturers, a common method that was in use until the 1980s was to impregnate the supported mica splitting sheet with a drying-oil-modified asphaltic varnish solution in petroleum-based solvents such as toluene. Tapes, slit from rolls of this sheet, were applied by hand to stator coils and bars and then covered with two layers of cotton tape. The wet coils were then transferred to a large, steam-heated coil tank. After a suitable time period to warm up a batch of many insulated coils, a vacuum was applied to the tank and the remaining solvent and air was evacuated over a period of several hours. This left the insulation in a dry and porous state. While still under vacuum, the entire batch of coils was then flooded with more melted hot asphalt. The tank was then pressurized, usually with dry nitrogen gas, to about 550 kPa (80 psig) and held under pressure for a number of hours, to force the liquid asphalt between the mica layers. Then the nitrogen pressure was used to push the asphalt back into a heated storage tank to await the next batch of coils. After venting the residual nitrogen gas, the hot coils were removed and allowed to cool to room temperature, which caused the asphalt-impregnated and covered coils to harden. Next, the surface layer of sacrifice cotton tape was stripped off, leaving a smooth, solid, impregnated cotton insulation surface. This vacuum, impregnation, and pressurizing process was called the VPI process. As discussed later, the VPI process, used with other resins, is still a cornerstone of groundwall insulation manufacturing today.

These mica splittings and asphalt, together with the coil manufacturing process, led to some variations in the uniformity and thickness of the groundwall insulation. To help overcome this limitation, a final hot-press cycle was frequently used for the straight-slot sections of the coils. The pressing led to some redistribution of the impregnating materials and produced a coil with a small enough insulation thickness variation to be inserted into the stator slots.

For lower-voltage, form-wound coils, a finish made with drying-oil-based paints was applied to the coil surface to give some protection against contaminates in service, such as lubricating oils and cleaning materials. These paints were usually pigmented to give a characteristic color for manufacturer identification. For machines of about 6600 volts and higher, it was customary to armor the coils by applying a ferrous asbestos tape to help control destructive slot electrical discharges between the stator core and the coil surface (Section 1.4.5). This material has a natural low conductivity, and when the asbestos in the slot sections of coils was impregnated with a varnish containing added carbon black to obtain the desired range of resistivity, partial discharge (slot discharge) in service was prevented. Some manufacturers carried out the final hot press after the armoring process was completed.

Asphaltic-micafolium-insulated coils (as the groundwall was called) varied in their heat resistance properties. The incorporation of drying oils in asphalts, during formulation in large cooking kettles, led to varnishes or resins that would cure to a weakly thermoset condition; that is, they would greatly soften on heating but would not completely melt. Natural asphalts are thermoplastic materials that will repeatedly melt on heating. Natural asphalts, in contact with mica splittings tapes containing drying-oil-modified asphaltic varnishes, will slowly harden during service heating, but will not become true thermoset materials in the modern sense.

All asphaltic mica coil insulations will relax somewhat after winding into a stator core and being subjected to service heating. The resulting puffing or expansion causes a tight fit in the slots, but tends to create small voids (delaminations) in the groundwall that would lead to partial discharges in service (Section 1.4.4). To avoid these discharges, the average electric stress (Equation 1.4) was controlled to be less than the partial discharge inception voltage of the voids in the insulation. Thus, the design stress in asphaltic mica groundwall insulation was generally designed to be below 2 kV/mm. In contrast, some designs now employ a stress as high as 5 kV/mm.

The extent of use and types of drying oil formulations varied widely between generator manufacturers, and the resulting coils thus had different thermal properties. At some suppliers, a drying-oil-modified varnish was brushed between each layer of half-lapped tape while other suppliers relied on the tank liquids to complete the impregnating process. Over time, generator ratings grew, so that by about 1940 the slot length of the largest machines approached or exceeded 2.5 meters in length. When such machines were placed in service to meet wartime production power demands and frequently subjected to rated loads or beyond while still relatively new, a new failure phenomenon developed. The insulation relaxation described above led to well-bonded tape layers next to the copper strands and to surface tape layers that often were locked into the cooling ducts of the stator core (Section 8.2). Since the copper windings heat up faster and to a higher temperature than the stator core, the insulation is subjected to shear and may fail by differentially migrating with each start and stop or major load cycle. The resulting condition has been called "tape separation" or "girth cracks" and led to many failures of generators during the 1940 to 1960 period. As will be seen later, this history of both materials and insulation methods led to the many modern variations of stator groundwall insulation systems now in use.

Micafolium insulation systems were being manufactured at the same time as the asphaltic mica systems. Micafolium was first used for sheet wrapping of high-voltage coils and the making of shaped insulating parts. A common early construction consisted of clear muscovite mica splittings bonded with natural shellac to a backing of kraft paper. Systems in the 1950s and 1960s used a glass cloth and even a B-stage epoxy varnish (see Section 3.4.2) to make sheet products. Micafoliums are generally supplied in rolls about 1 m wide and are usually not slit into tapes. Such insulation systems had many of the properties and limitations of asphaltic micas systems.

3.2 EARLY SYNTHETICS

The history of synthetic products for insulation started with the work of Dr. L. H. Baekeland in 1908. This led to the development of a workable and reproducible process for the production of phenol–formaldehyde resins that were used to make many types of electrical products. Although "Bakelite" was formerly a trade name, in contemporary usage it describes the broad variety of condensation products of phenols and cresols with formaldehydes that grew out of Baekeland's work ("phenolics"). Other early synthetic materials for insulation use were introduced throughout the 1920s and 1930s. Alkyd resins were used in 1926 for electrical bases and in the same year aniline–formaldehyde formulations were used for terminal boards.

Alkyds, made with saturated long-chain fatty acids and alcohols, are similar to the naturally occurring drying oil resins. During the late 1920s and 1930s, these new alkyds were used to both replace and to blend with the natural resins in many applications. New equipment finishes and insulating varnishes in solvent solutions found use in many rotating machines. The varnishes were used for dipping and coating entire windings as well as to make improved versions of insulating materials like varnished cambric. Most machines made from the late 1920s through the 1950s utilized these improved synthetic versions of the old natural materials, which still have a small place in the insulation industry and in protective paints for equipment. The volatile organic chemicals (VOCs), used to solvate these paints and varnishes for application, became a liability with widespread use because of their contribution to air pollution. When limits were established for VOC emissions, either their removal from the air venting from drying and baking operations or the switch to low-VOC materials or modern solventless resins became necessary.

Other pre-World War II materials now used in insulation include polyvinyl chloride (introduced in 1927), urea–formaldehyde (1929), acrylic (1936), polystyrene and nylon (1938), and melamine–formaldehyde (1939).

There was a great increase in the types of synthetic polymers and resins introduced during the 1940s and 1950s. Polyesters and polyethylenes date from 1942, fluorocarbons and silicones from 1943, and epoxies from 1947. In the 1950s polyurethane, polypropylene, and polycarbonate were introduced. Although the early versions of these materials often lacked the sophistication and properties enhancements of current offerings, their arrival on the scene led to an explosion of new applications in electrical insulation. Polyesters, derived from experience with alkyd chemistry, which used both saturated and unsaturated polyfunctional carboxylic acids and polyhydric alcohols, became very common after World War II and began to penetrate the electrical insulation market.

An important film insulation is made from polyethylene glycol terephthalate or PET, the reaction product of the saturated aromatic terephthalic acid and ethylene glycol. This linear, thermoplastic polymer has a high degree of crystallization and exhibits a high melting point (265°C) and great stability. Films and fibers of this polymer are manufactured in Russia as Lavasan, in England as Terylene, and in United States as Mylar and Dacron.

When the dibasic acid or its anhydride is unsaturated (contains double bonds between carbons, or C=C bonds), esterification with polyhydric alcohols gives rise to polyesters that are also unsaturated. With the double bonds present, these polyesters may, in the presence of catalysts, enter into a copolymerization reaction with unsaturated monomeric materials or with more complex compounds containing double bonds. This reaction allows the conversion of liquid starting materials into solid substances with no volatile by-products and without the use of volatile solvents. A typical polyester of this type may be formed from maleic acid or its anhydride and ethylene glycol. When the resulting unsaturated glycol maleate is copolymerized with styrene monomer, in the presence of a small percentage of the initiator benzoyl peroxide, the cured product has very good electrical insulation and mechanical properties. This composition, and similar formulations using other unsaturated organic acids or anhydrides and monomers, has been widely used for impregnating the windings of electric motors. As will be shown in Section 3.4.1, this chemistry has also been used for high-voltage generator insulation.

When natural materials and the early synthetics were the only choices, service experience was the major factor in determining insulation thermal classification. It was recognized as early as 1913 in an AIEE paper by Steinmetz and Lamme on "Temperature and Electrical Insulation" [3.2] that insulation deteriorates with time when exposed to heat. Materials were placed into three thermal classes on the basis of their generic composition. Class A materials could be used to temperatures of 90°C and included the natural organic fibers such as paper and cotton and the natural oils, resins, and gums described earlier. Class B materials were for use at temperatures to 125°C and included the heat-resistant minerals of mica, asbestos, and silicas, usually used in combination with the natural impregnating and bonding materials. When these inorganic materials were used without dependence on organic binders to maintain their forms, they were classified as Class C— 150°C to incandescence. The above thermal classifications do differ than those used today (see Table 2.1).

In 1948, Dakin published an AIEE paper, "Electrical Insulation Deterioration Treated as a Chemical Rate Phenomenon" [3.3]. In this paper, he showed that the thermal deterioration of organic insulation was the result of internal chemical change during aging. The relationship he described was taken from studies of organic chemical reaction rates experienced in the laboratory and is known as the Arrhenius chemical deterioration rate equation (Equation 2.1). When plotting the log of insulation life, as measured at several elevated temperatures, against the reciprocal of the absolute temperature (°K), a straight line will be produced (Figure 2.1), except where second- and higher-order chemical reactions become a part of the aging experience. If the straight portion of the line is extended to lower temperatures, it can be used to predict the insulation system life at the lower temperatures. Thus, after relatively short aging periods at several temperatures above expected service temperature, a life prediction, based on an arbitrarily chosen end-of-life diagnostic test, can be made.

In Section 2.3, the concept of thermal endurance tests was introduced and their use to establish an insulation system thermal classification was discussed. This topic is reviewed here because of its relationship to the history of the development of insulating materials and insulation systems.

The variety of new synthetic materials that became available to the industry after World War II forced the development of new means to quickly evaluate them as a substitute for extended service experience. Instead of the costly building of complete machines and operating them at various service stress levels, a new method of accelerated functional testing was developed. During the 1950s, several organizations developed models of parts of electric motors and small generators that reproduced the essential combinations of insulating materials and associated electrical and supporting parts [3.4, 3.5]. These inexpensive models evolved into the "motorettes" that we know today, which are the basic models used in IEEE No. 117-1974 (R1991) Standard Test Procedure for Evaluation of Systems of Insulating Materials for Random-Wound AC Electric Machinery. Large numbers of these models were built to evaluate many different types and combinations of the new synthetic materials. Section 2.3.4 describes the now-standardized aging test procedure in more detail.

In the course of carrying out motorette tests on many new synthetic materials and on variations of them of the same generic type, it was found that systems of these materials had significantly different test lives and could not be thermally classified by chemical type, as had been common at that time. As a result, today insulation systems are characterized by their performance in accepted aging life tests, such as IEEE No. 117 and 275 as well as IEC 60034, Part 18, Sections 21 and 31, and are thermally classified for specific applications on the basis of such tests. Experience has shown that these comparative functional tests are good indicators of winding life in service. As the new materials and systems accumulate successful service experience, they become the reference systems for comparison with still newer materials and systems.

3.3   PLASTIC FILMS AND NONWOVENS

In 1950, insulation engineers started to investigate the proliferation of new materials made with synthetic plastic films and, later, polymer-fiber-based nonwovens, being offered by suppliers as slot, turn, and phase insulation for random-wound induction motors. Their need for a realistic and economical way to screen these many new materials were the direct cause of the development of the motorette tests described above. The new materials offered the potentials of lower costs, better performance, and ease of manufacture. As a result, extruded and calendered films of nylons and polyesters and solution-cast films of other materials were tested. Although some plastic films were used in production motors, a particularly successful insulation product was a thin laminate of an electrical grade PET polyester film sandwiched between layers of calendered, nonwoven polyester mat or paper. These composites have been widely used in low-voltage small and medium motor construction and, by the IEEE 117 test method, have been accepted as insulation systems for both Class 130°C (B) and Class 155°C (F) when used with the appropriate varnish.

Indeed, the motorette accelerated thermal aging tests were widely used to evaluate insulation systems containing films and nonwovens that are treated with many types of varnish, both solvent-based and solventless. In addition to modified natural varnishes, the synthetic varnishes using phenolic, polyester, silicone, epoxy, and other chemistries were used for coil treatments on motorettes.

The success of the polyester mat, polyester film, and polyester mat laminates, bonded with thermosetting adhesives, for use in fractional motors up to the largest rotating electrical equipment as slot, phase, and end-winding insulation and as ground and wrapper insulation in dry-type transformers is well established. Many variations in thickness (4 to 1 ratio) surface treatment and bonding resins are available from a number of suppliers. The polyester film contributes high dielectric, tear, tensile, and burst strengths. The high-density polyester mat, laminating adhesive, and surface resin treatments enhance the thermal stability. The polyester mat may be partially or fully saturated with resins from the laminating adhesives and the topcoats.

Many material variations in this type of laminate have been developed to enable them to be economically used for service in machines with insulation systems classes from Class B,

Class F, and Class 180 (H) up to Class 220 (C). To achieve these service temperature levels with the most economical construction for each class, the polyester mat may be replaced with similar structures made with glass fibers, mica paper, cotton paper, wood-pulp kraft paper, vulcanized paper, ceramic paper, liquid crystal polymer paper, or aramid fiber paper. The films can be replaced with ether–imide, amide–imide, or solely imide polymers and biaxial extruded liquid crystal polymer. In addition to heat resistance, some constructions are optimized for use in hermetic motors or in coils subjected to high levels of nuclear radiation.

3.4 LIQUID SYNTHETIC RESINS

One of the major developments was the replacement of solvent-borne natural and synthetic resins with solventless synthetic resins. These materials are normally thermosetting under the action of heat, catalysts, hardeners, or radiation. In addition to improved thermal stability and physical properties, the elimination of solvents makes their application more environmentally friendly and less likely to form voids within the groundwall. For groundwall insulation in form-wound stators, there are two families of resins that are very important—the polyesters and the epoxies. Other solventless resins that are sometimes used for special applications include silicones, acrylates, imides, and blends of phenolics with other resins.

3.4.1 Polyesters

Polyester resins evolved out of alkyd resins and became available starting in 1942. The first wartime uses were for military needs, including boats and pontoons, where they were combined with glass fibers to make strong laminates. During the decades following the end of the war, extensive development of this resin family took place and now most of their uses are nonelectrical. Their advent was, however, very important in the development of modern coil insulations.

In Section 3.1, the problems of "tape separation" or "girth cracks" developing during service with high-voltage asphalt mica flake groundwall insulation were introduced. The softening temperature of the asphaltic resins used by various manufacturers was dependent on the percentage of drying oils used in their preparation. Insulation, containing the most weakly thermoset formulations, could only withstand a small amount of shearing loads from thermal cycling (Section 8.2) before the layers of insulating mica-splitting tape began to separate. Both the generator slot length and the service temperature were important factors in determining the shear loads within the insulation. Other factors included the tightness of fit of the coils in their slots, the duty cycle of the generator, and the degree of cumulative heat exposure in air before cyclic duty was started.

When drying-oil-modified asphaltic resins are heated in air, they will gradually harden by slow oxidation, becoming stronger and with a higher softening temperature. However, the introduction of hydrogen-cooled generators in the 1930s removed the oxygen from contact with the insulation and nearly eliminated the in-service oxidation or drying process that had helped avoid the tape separation phenomenon. Insulation engineers recognized that a new binding and filling resin that cured during coil manufacture to a thermoset, infusible state was needed. The wartime advances in polyester resin chemistry provided the materials to begin the development of a new generation of improved generator insulation.

Not long after the end of the war, Westinghouse Electric Corporation engineers began the laboratory work needed to turn the new polyester chemistry into a workable generator and motor coil high-voltage insulation system. The system they developed was trade named

"Thermalastic." Variations of the same basic system have been licensed to several other rotating machine manufactures and also duplicated by others as patents expired.

Westinghouse chose mixtures of organic, ethylenically unsaturated (containing C=C bonds) dicarboxylic acids, which contain reactive free carboxyl groups (–COOH), to be reacted with polyhydric alcohols having only reactive hydroxyl (–OH) groups. When substantially molar equivalents of these materials are mixed and heated in a closed reaction vessel with an esterification catalyst, which may be mineral acids, the esterification reaction takes place. Removal of the water formed in the reaction, to increase the degree of esterification, may be accomplished by azeotropic distillation when the reaction is carried out in the presence of a volatile organic liquid such as toluene, xylene, or the like. The starting materials and the degree of esterification were chosen so that the final product was a solvent-free, syrupy polyester resin at room temperature or when heated.

This resin was typically used to partially impregnate mica splittings that were laid down on a thin, pliable sheet of backing material, such as rice paper or supercalendered rope paper. Many other supporting materials, including plastic films, synthetic resin papers, woven glass cloth, glass fiber paper, etc., may be substituted on one or both sides of the sheet. Tapes cut from these sheets were wrapped onto coils to build up a few or many layers, depending on the service voltage of the coil design. The tapes were permanently flexible and, with proper storage before use, did not age, harden, or deteriorate significantly at warm or low temperatures for appreciable periods of time.

Westinghouse chose to impregnate the groundwall with polyester using a modified VPI process first used for asphalt mica coils. The wrapped coils where placed into an impregnating tank and subjected to a heat drying and vacuum cycle to remove substantially all moisture, air, and any other undesirable volatile materials from the coils. Then a low-viscosity impregnating material consisting of a liquid, unsaturated, and reactive monomer containing 1–2% of an addition-type polymerization catalyst, usually an organic peroxide, was admitted into the tank until all of the coils were covered. While the coils were submerged, compressed gas, such as air, nitrogen, or other relatively inert gas, was introduced under pressure into the tank to assist the impregnating monomer in fully penetrating the coil wrappings and to ensure that all of the voids and interstices were filled. The pressurizing time depends on the number of layers of tape that must be penetrated. This can be 10 to 15 minutes for a low-voltage design to several hours for a high-voltage coil with thicker a groundwall.

The impregnating material was selected for storage tank stability, low viscosity, and economic availability. Examples of these unsaturated reactive monomers, containing the group –C=C–, are styrene, vinyl styrene, methyl methacrylate, and a number of other monomers that can be used or mixed with the primary monomer to make up the impregnating composition. These compositions are fluids that will not readily gel or thermoset before coming in intimate contact with the polyester resin in the mica flake tape, even when they are exposed to moderate heating.

After full saturation of the mica splittings insulation, the impregnating monomer was pushed back into storage by the pressurizing gas and the wet, uncured coils were removed from the tank, allowed to drain briefly and then moved to a curing procedure. In the simplest case, the cure may be carried out in an oven, heated above the activation temperature of the catalyst chosen. Temperatures from 80°C up to about 135°C were used to carry out the final polymerization and cure, which converts the resins into a stable thermoset solid. For coils with many layers of insulating mica tape, more uniform dimensions could be obtained by transferring the coil to a heated sizing and curing press. Equipment was developed to apply molding pressure to the curved sections of coils or bars, so that the entire structure could be sized and cured simultaneously. A final armoring and painting step was applied, using materials the same as or similar to those used with asphaltic mica flake insulation, before winding the coils or half coils into the generator or motor stator core.

Many of the process steps previously used with asphaltic mica flake insulation and adaptations of similar equipment were also required with the new polyester thermoset insulation system. The fundamental similarity of the two manufacturing methods was the vacuum pressure impregnation of a low-viscosity liquid into a mica tape containing only part of the material needed to completely fill the voids created in the tape layers by the taping process.

Versatile polyester chemistry has yielded many other insulating materials that have benefited the rotating machine industry. Polyester films and film/fiber laminates were described in Section 3.3. Treating varnishes, wire enamels and servings, molded insulation pieces, and electrical laminates, based in whole or in part on polyester chemistry, have proliferated and been optimized for many different types of rotating machines and service temperature levels.

3.4.2 Epoxides (Epoxy Resins)

As mentioned above, polyesters were first introduced in 1942, whereas epoxides were not commercially available until 1947. Most of the early epoxides were solventless thermoplastic solids at room temperature and were not very suitable for use as low-viscosity impregnating resins. The solid epoxy resins could be mixed with low-viscosity monofunctional epoxides, like the glycidyl ethers, to make a viscous liquid at room temperature. However, these cured mixtures did not have the strength, heat distortion temperatures, or hot electrical properties of the high-molecular-weight solid epoxides. Epoxy resins did have several advantages over the available polyester resins available during the early and mid 1950s. Epoxides cure to a stronger polymer and tend to have improved thermal stability. Epoxy groups are unstable and react or cross-link readily with compounds that have groups with mobile hydrogen atoms, such as carboxyl, hydroxyl, and amine groups. When these resins and compounds are mixed, a reaction takes place with the polymerization (elongation) of the molecule and the formation of transverse bonds. These crosslinking reactions result in the formation of solid, stable polymers that contract very little on hardening, only 0.05 to 2%, whereas polyester compounds may shrink by as much as 10%.

The General Electric Company, which was the first to introduce the asphalt mica flake insulation system about the year 1915, continued to make changes in the materials and final properties of coils insulated with the system into the 1980s. A higher percentage of drying oils were used to make the asphaltic resins employed in both the initial tape construction and in the brushing varnish that was applied between each layer of tape as it was applied. As discussed in Section 3.1, these asphaltic resin variations were weakly thermoset materials when processed through the several shop manufacturing operations. There was, therefore, a 5 to 7 year delay, as compared to the experience of Westinghouse Electric Corporation with the asphaltic mica flake insulation system, before increasing generator ratings and the associated increased slot lengths led to the frequent occurrence of tape separation in generator stator coils.

When General Electric Company began looking for an improved thermosetting resin system in the late 1940s, several new polymer chemistries, including polyesters, were evaluated. To achieve the superior stator coil insulation system desired, epoxy chemistry was chosen. GE wanted to develop an epoxy system that did not depend on impregnating the wrapped coils with a low-viscosity resin containing a reactive hardener. Such resins often require refrigerated storage to delay advancement of viscosity. The tapes and brushing varnish for the Micapal 1™ system (Section 4.2.2) that evolved from the high-molecular-weight, early epoxides contained a solvent mixture to liquefy the resin and hardener or coreactant. Following the lead of the solvent-containing asphalt mica flake system then in current use, the epoxy development also became a so-called "Resin Rich" system in which all of the binding and filling resin was either in the tape, as applied, or in the brushing varnish used between layers as the tape was wrapped on the coils.

Epoxy resins can be cured by several means. The most common curing agents are Lewis bases, such as amines and amides; Lewis acids, such as boron trifluoride; and other materials including phenols, organic acids, and anhydrides. Alkyds have been developed since 1926 for use in paints and protective coatings and GE had much experience with them. An alkyd is the reaction product of a polyhydroxy alcohol and a polybasic acid or anhydride, the same way that polyesters are formed. In the late 1940s, these materials were widely available from various alcohols and organic acids at reasonable costs. The epoxy curing reaction with organic acids depends on the carboxyl groups on the acids. When equivalent molar ratios of hydroxyl and carboxyl groups are reacted, most of these groups are used up in the esterification reaction, which produces the alkyd or polyester resin, and such materials have little reactivity with epoxy resins. When there is an excess of the acid, there will be unused carboxyl groups left over that may then be used to react with the epoxy group to cure the system with little or no volatile by-products. These alkyds are called high-acid-number resins. By choosing a medium-length carbon chain organic acid, such as adipic acid with six carbons, to react with glycerol, a three-carbon alcohol with three hydroxyl groups—glycerol adipate resin—is formed. When five moles of adipic acid are reacted with four moles of glycerol, the resulting high-acid-number polyester will have an average molecular weight of about 800. Using this resin as a hardener or coreactant with an epoxy resin will impart a degree of toughness, without brittleness, to the epoxy formulation. GE used variations of this chemistry in making their Micapal 1 epoxy-bonded-mica high-voltage ground insulation system. GE produced this system into the 1990s.

The former VPI cycle for the asphaltic system now became a hydraulic pressure molding and curing cycle for the new epoxy system, using the same processing equipment still in use for the old system. In the asphalt mica system, there was no harm in having some of the tank asphalt incorporated in the surface layers of the groundwall insulation. However, the tank asphalt would be a contaminant in the top layers of an epoxy mica system. Therefore, a system of shape aids and sacrifice materials were evolved that excluded tank asphalt from the groundwall insulation layers. After a batch of uncured coils or bars were loaded into the empty tank, they were gradually heated, while under vacuum, to remove all of the solvents, moisture, and air present in the raw tapes. When the tank was flooded with hot asphalt and pressurized with compressed gas, usually nitrogen, the gas pressure was hydraulically transferred to the shape aids and the insulation layers underneath, compressing and molding them into a dense, relatively void-free, high-voltage insulation system. The heat of the pressurized hot liquid asphalt was transferred to the coils, curing the epoxy composition over a period of 8 to 10 hours. After venting the nitrogen and removing the load of coils to cool down to room temperature, the sacrifice materials and shape aids were removed and the individual coils were given the appropriate semiconducting slot coating, silicon carbide coating at the slot ends, and identifying colored epoxy paints.

Like Westinghouse Electric Corporation's licensing of the Thermalastic vacuum pressure impregnation (VPI) manufacturing system and materials to other motor and generator manufacturers, General Electric Company licensed the epoxy mica paper and resin-rich technology to its partners. Over time, as the early patents ran out, other manufacturers developed their own insulation systems, using both the VPI and resin-rich technologies. One variation was press curing of resin-rich, fully taped coils or bars in special presses, equipped with forms to follow the shape of the coil. It was found that a super-resin-rich tape, on the parts of coils outside of the slot sections, could be compressed and cured with only the use of oriented, heat-shrink tapes made with polyethylene terephthalate (PET) films.

Both generator manufacturers started out with processes derived from those used with asphaltic resins and evolved into the use of quite different resin-rich and VPI processes that still persist. Low-viscosity polyester systems became VPI systems, whereas the early high-molecular-weight and viscous epoxies were more easily handled by the resin-rich process. Within a few years, epoxy technology evolved to produce a variety of solventless fluid resins and hardener materials that were suitable for both press cure and VPI uses. However, the various motor and generator manufacturers have developed their process equipment and associated insulating materials for one or the other of these basic processes and will not easily change.

3.5   MICA

Mankind has used natural mica since prehistoric times. The earliest uses were in the form of sparkling dust or powder for ornamental purposes, including body paints or cosmetics. Ground mica is still used as a special filler in paints, plastics, and cosmetics, and the largest tonnages of mica today are still in these applications. Because of its transparency and its resistance to fracture and heat, mica, also called isinglass, was widely used for lantern globes and for windows in coal, wood, and kerosene stoves. In recent years, there has been a revival of the use of mica for decorative electric-lamp shades and other decorative surfaces. In addition to these ordinary uses, the Russians found mica to be useful as a covering for portholes in warships, as it could stand the shock of cannon fire better than any glass available at that time. The Russian mica became known as muscovite, the term still used to describe white or India mica.

Mineralogically, mica is the name given to a group of minerals of related composition and similar physical properties. Micas are characterized chiefly by having a perfect basal cleavage, so that they can be split readily in one direction into a great number of thin, tough, flexible laminae.

Chemically, the micas are complex silicates of aluminum with potassium, magnesium, iron, sodium, lithium, fluorine, and traces of other elements. The principal micas are muscovite, $H_2KAl_3(SiO_4)_3$; phlogopite, $[H,K(Mg,F)_3]Mg_3Al(SiO_4)_3$; and biotite, $(H,K)_2(Mg,Fe)_2(Al,Fe)_2(SiO_4)_3$. Other micas that are not so well known are lepidolite, paragonite, and zinnwaldite.

The micas have a hardness of 2–3, specific gravity of about 3, index of refraction from 1.5 to 1.7, and usable temperature limits of about 550°C (for muscovite) up to about 980°C (for some grades of Phlogopite). Muscovite and biotite occur chiefly in pegmatite dikes associated with feldspar and quartz, although they are abundant also in granites and syenites. They are chiefly obtained in India, Brazil, and the United States, although, for electrical uses, about 70% of the world supply comes from the first two countries. Phlogopite occurs in crystalline limestones, dolomites, and serpentines in Canada and Madagascar. Muscovite is also called India, white, or potassium mica. Phlogopite is called Canadian, ruby, amber, or magnesium mica. Biotite is called black or magnesium iron mica.

3.5.1   Mica Splittings

Mica is mined in chunks or "books" that are rough trimmed to exclude pieces of the associated minerals in which they are found. These books are then given further preparation with a knife, sickle, or shears to remove all edge defects, cracks, etc. before being hand-delaminated or separated into irregular thin sheets. These irregular shapes are then graded to size on the basis of area of the minimum rectangle that can be cut from each plate. Standard gradings range from 6.5 square centimeters up to 650 square centimeters.

The present mining and preparation of mica for electrical use is a very labor-intensive operation with often whole families engaged in the process. This is why most of the mica used in the developed world comes from India, Brazil, and Madagascar. During World War II, when overseas sources were either cut off or in a state of hazardous supply, muscovite was mined in the State of New Hampshire for the U.S. electrical industry. Mica splittings have been or still are used in many electrical insulation applications. These include microwave windows, electronic capacitors, transistor mounting washers, various interlayer insulations, vacuum tubes, and several rotating machine applications. Except for the latter, most of the other uses are for fabricated individual mica pieces that are cut from selected mica books or individual pieces, leaving a high-quality scrap that is preferred as a raw material for conversion into mica paper. Rotating machine use of mica splittings is generally in the form of built-up sheets or tapes with supporting backers and resinous impregnants or as molded sheets made with bonding resins. The sheet forms are then used in secondary operations to make items such as molded cones and segment insulation for DC motor armatures. The tapes are generally applied by hand as ground insulation for higher-voltage motors and generators.

3.5.2   Mica Paper

The World War II dependence on overseas supplies of mica splittings led to the development of mica papers. In the United States, a government-supported program in the General Electric Company Research Laboratory and at the GE Pittsfield, Massachusetts plant, was started with the help of fine paper manufacturer, Crane & Co. Inc., located in the same geographical area. This mica paper development during the war was to help free the United States from dependence on foreign sources of mica splittings and to assure the availability of mica products to meet military and civilian electrical equipment needs. In Europe, during the German occupation of Paris, an individual Frenchman carried on a parallel program. Both of these efforts succeeded by about the end of the war and lead to patents by both parties. The French work was supported by the Swiss electrical insulation industry, particularly by the Swiss Insulation Works, now part of Von Roll ISOLA.

In 1947, Mr. Gerard de Senerclans, representing the Swiss Insulation Works, visited the United States to promote the use of the French-developed mica paper. During his meetings with GE, it was realized that there were some overlaps in the two patents and, to overcome these potential conflicts, a cross-licensing arrangement was later worked out between the two parties.

Mica paper is currently made by several processes and with several sources of raw materials. One approach to preparing natural mica for papermaking, with minimum labor, is to first roast the mica splittings or other micaceous material to partially drive off the natural water of crystallization in the mica. This is carried out in a rotary furnace, a process called calcining, followed by quenching in an aqueous medium. After quenching, the mica is wet ground into small platelets before adding it to a liquid suspension or slurry. This slurry is admitted to the head box of a modified Fourdriner papermaking machine, where it is metered out to a porous endless belt. The combination of gravity and vacuum remove most of the water and allow the small mica platelets to bind together into a single structure, without the addition of any adhesive. At this point, the mica is weakly held together by intermolecular Van der Waals forces. The mica paper, still in pulp form, is then transferred to a steam-heated drying drum where the rest of the moisture is removed to produce a continuous dry sheet. Rolls of this pure mica product are then sent to a treater for application of resins, backing, and facing materials and solvent drying or curing as may be required. These additions enhance the strength and provide mechanical protection of the product for subsequent slitting into tapes for insulating machine windings. Many different combinations of resins, binders, and surfacing materials are used, depending on the ultimate use of the mica paper composite.

An alternate method of preparing small flakelets from natural mica by mechanical delamination was developed after the war. Several different ways of accomplishing this delamination have been developed. One example is described as the water jet process, in which ground flakes are admitted to a water-filled cylindrical tank which has a number of high-speed water jets arranged tangentially on the inner surface of the tank. These jets cause the water in the tank to spin, which, in turn, causes the larger and heavier mica clumps to migrate toward the tank wall and orient themselves parallel to it. In this position, the water jets impinge edge-on to the mica clumps and begin to delaminate them. Water bearing the finer, most delaminated, and lightest platelets, is withdrawn from near the bottom center of the tank. Grains of heavier minerals, found in the raw mica, can be collected for removal from the bottom or can be removed later from the mica slurry by settling. These mechanically delaminated and cleaned platelets are then admitted to the Fourdriner machine to form the mica paper.

Calcined mica paper tends to have smaller platelets than mechanically delaminated mica paper. Both forms of mica paper make good blotters for impregnating resins, since up to 50% of the volume of untreated paper consists of small voids. Experience has shown that the mechanically delaminated papers are somewhat easier to impregnate during vacuum pressure impregnation, although both forms are used interchangeably in this process. It is possible to mix the two forms to make a paper for special applications, although it is not often done.

Mica papers have a consistent uniform thickness, resulting in about a 300% improvement in thickness tolerance compared to flake mica. This means that less total material and less thickness are required to achieve the same insulating value. In preparing laminates of impregnated mica papers, without backer layers, the uniform controlled compressibility and mica thickness result in uniformity in the distribution of binder resin throughout the sheet. Pressed and cured sheets or plates for commutator segment insulation are ideally suited for automatic assembly and punching processes, as the plates do not flake, scale or delaminate and have a smooth surface.

Mica splittings can be fully impregnated with less binder resin, but this advantage is lost when a greater thickness of flake tapes have to be used to minimize the possibility that thin spots or areas with missing mica pieces do not compromise the insulating value. In resin-rich tape insulation, the use of alternate layers of mica paper tape and mica splittings tape has some advantages. The paper tape layers serve as a reservoir for uncured binder resin, whereas the flake layers minimize the wrinkling that is experienced with all mica paper constructions during compaction and cure, when some of the excess resin is squeezed out. Fully saturated, resin-rich mica paper tapes have been made with one-third less binder resin. This can be achieved by absorbing some of the excess resin from the impregnating step in a sacrifice blotter paper when both materials are passed through warm squeeze rolls. The resulting tape is thinner and, with a fixed insulation thickness, a higher percentage of mica can be obtained in the groundwall insulation. This method gives electrically stronger insulation with a better heat transfer capability.

After 50 years of experience with mica paper tapes and laminates, the use of mica splittings has greatly diminished. What started out as an effort to overcome wartime shortages of mica splittings has now become the engineering and economic material of choice. At first, the preferred starting material for mica paper production was mica scrap generated as a by-product of punching book mica insulation pieces for the electronics industry. This source is now insufficient, due to changes of materials in the electronics industry as it moved from vacuum tubes to solid state technology. At the end of the millennium, the most common starting material is a lower grade of mica, such as mine scrap, which is less expensive. Papers have been made with vermiculites that are derived from large deposits of a number of micaceous minerals that are hydrated silicates and that have been greatly expanded at high temperatures by driving off some of their bound water.

3.6 GLASS FIBERS

The creation and production of glass fibers dates from at least 1922. However, their application to rotating machine insulation is more recent. By 1944, the fibers were being used to reinforce polyester resins, which were commercially introduced in 1942, initially as laminates for military uses, such as small boats and pontoons. Following the war, the uses rapidly grew, including the first melamine glass cloth high-pressure laminates for arc-resistant electrical applications. By the end of the decade, glass fiber filaments were being used as magnet wire servings, as individual filament reinforcements for cellulose backing papers in mica tapes, as woven cloth for mica tape backings, and in a variety of resin laminates for electrical insulation. By the first half of the 1950s glass fiber papers appeared and began to be used as components of insulation. Many resins were used with both woven glass cloths and nonwoven mats in both flat sheets and in molded shapes for electrical applications. Glass fiber rovings and polyester resins were used to make very strong support rings for generator end-windings by 1954. The combination of 50 percent volume fraction each of glass filaments and filaments of polyethylene terephthalate (PET) were introduced as magnet wire servings by 1955. When the served wire is passed through hot ovens, the PET is melted and then coats the glass fibers, fusing them to the wire. The servings may be applied in one or several layers, depending on the desired toughness and insulating qualities needed for the strand insulation. This strand insulation is known by the Dupont trade name of Daglas or by the generic name of polyglass.

Most of the glass fibers used for electrical insulation are made from a lime–alumina–borosilicate glass that is relatively soda-free. This glass formulation was developed for electrical applications where good chemical stability and moisture resistance was desired. It is generally called "E" glass because it was initially developed for electrical uses. The raw material for glass is sand and is, of course, widely available. The fibers are drawn from the melt and the special conditions that exist as the fibers are formed modify their properties as compared to bulk glass of the same formulation. The fibers are not brittle and are very strong in tensile strength as compared to bulk glass.

As an inorganic material, glass is resistant to attack from partial discharges. When glass fibers are used to mechanically strengthen electrical insulation, the electrical breakdown strength of glass is not utilized. Glass fibers are round and are generally surrounded by organic polymers, which are subject to damage from partial discharges. Examination of glass composite insulation after a period of exposure to partial discharges reveals erosion of the polymer materials but no damage to the glass. The electrical breakdown strength of the composite insulation is only dependent on the polymer. When glass fibers are used to reinforce mica tapes for coil insulation, the mica, with its continuous, overlapping sheet form, provides the most resistance to partial discharges and significantly improves the breakdown strength.

There have been a number of attempts to use E glass in the form of thin flakes as a substitute for mica splittings in high-voltage insulation. While some success has been achieved, the glass flakes are more brittle than mica splittings and have proven to be difficult to handle. As noted above, most mica splitting applications have been replaced with relatively low-cost mica paper and have minimized the need for alternative inorganic materials.

Some work has been done with thin glass ribbons, having a width to thickness ratio similar to the mica splittings, as groundwall insulation. Using special equipment, a number of glass ribbons can be laid up in a brick wall pattern to make a tape. When impregnated with resin and prepared with a supporting and backing layer, such a tape can be applied as coil insulation. However, this construction is more expensive to make and does not provide superior benefits.

Although the chemical inertness of glass is well established, the large ratio of surface area to volume of glass fiber filaments subjects them to the possibility of chemical attack. Freshly formed glass fiber surfaces have a great affinity for water and can abrade each other during gathering, weaving, and other handling operations. To minimize damage, newly formed fibers are usually coated with a lubricating sizing that may be removed after the strands have been fabricated from the individual filaments. A surface treatment known as a "finish" is then applied. A number of sizes and finishes are available and the choice of finish depends on the type of resin or polymer the fiberglass is used with. Finishes have been designed to chemically bond with both the glass fiber surface and with many of the resins commonly used. The application of the finish partially replaces the water, which quickly forms from atmospheric humidity on the glass surface.

3.7 LAMINATES

Thermosetting polymers are widely used for the production of laminated electrical insulating materials that are commonly used as wedges, blocking material, filler strips, and slot liners in windings. Laminates are generally prepared by first impregnating paper, nonwoven mats, or fabric from rolls with a solution of a thermosetting resin. The solvent is driven off by continuously passing the treated material through a hot oven with good ventilation. The dried material, usually referred to as in the "B Stage" condition and called a "prepreg," is rerolled and stored until needed. Pieces of the prepreg are laid up in a mold or in a multiple platen flat-plate press. In the process of pressing at increased temperature and pressure, the resin becomes infusible and the fibrous materials are bound into a monolithic system. The IEC and the U.S. National Electrical Manufacturers Association (NEMA) have established many grades of these laminates according to the reinforcements and resins used in their construction and their thermal, electrical, and mechanical properties.

With the use of solventless resins, prepregs do not have to be dried before laminating. The reinforcing layers may even be laid up in the unimpregnated form in a closed mold, followed by injecting liquid resin under pressure, with or without vacuum assist, and cured in place. Complex shapes can thus be molded for use as machine guards, coupling covers, motor and generator end bells, and insulators.

Continuously laminated sheets are made by passing fabric or mat webs through a polyester or epoxy resin dip and combining the layers with surface layers of a suitable release film while passing through pressure rolls. The continuous lay-up is passed through a heating zone to cure the resin. The laminate thickness and resin content are controlled by the pressure rolls as the several plies are brought together. The release films act as a carrier, keeping out air during cure of the resin and imparting a smooth finish to the continuous laminate. A traveling clamping system can be used to seal the film edges to prevent air from entering the laminate after the combining process. This device is called a tenter frame. Laminating speeds as high as 30 m per minute are obtained. The cured stock is edge trimmed by rotary shears and, if flexible, is taken up on windup rolls for shipment to customers. Rigid stock, because of greater thickness and resin choice, is cut off in sheets by metal shearing equipment.

Continuously laminated stock was first developed for liners in self-sealing fuel tanks, aircraft interior liners, transparent tracing cloth, and photo template stock. Sheets from this process have been converted into rotor turn insulation and coil slot filler strips. Another continuously laminated insulation sheet process was described in Section 3.3 for the adhesive bonding of layers of film and woven or nonwoven fabrics for motor phase and slot insulation.

3.8 EVOLUTION OF WIRE AND STRAND INSULATION

Smaller, low-voltage motors and generators are usually random wound with round wire, whereas larger and medium- to high-voltage machines use rectangular wire or strands and are form wound (Section 1.3). The wire insulation in random-wound machines must be able to withstand the voltage difference between the beginning and the end of each coil, since the random winding process may bring these ends together in the same coil. The natural resins and varnishes used for the coils were not very good as thin-film magnet wire (winding wire) enamels. Instead, the natural resins were used to impregnate served wire, both round and rectangular, using materials like cotton, silk, and flax for the serving fibers. In comparison with today's wire enamel practice, the served and varnished wire insulation was thicker, stiffer, harder to wind, and had much lower thermal capability. The gradual introduction of synthetic resins and varnishes during the late 1920s through the 1940s led to the all-enamel film-based magnet wire in common use today.

These synthetic enamels are applied out of solution by means of multiple dips, with solvent drying and some curing after each dip, in special ovens. The wire enameling horizontal or vertical tower ovens are strung with the wire so that the multiple passes through the wire enamel dies and drying/curing zones can take place continuously. The wire ovens are used for both round and rectangular wire and for both enameling and applying varnish or adhesives to served wire. The enamels now available can be used for machines rated up to Class 220°C.

Serving materials have also changed. Although asbestos servings were often used in the past for high-temperature coils, the carcinogenic property of the material has led to its elimination from all insulation uses. It has been replaced with glass servings and, as discussed in Section 3.6, with the combination of glass and polyester strand servings. The PET polyester strands are fused to the wire and the glass during passage through the wire oven.

In many medium-size motors and generators, form-wound coils are divided into a number of turns, which must be well insulated from each other. In multiturn coils, the strand insulation may also be the turn insulation, a demand that in the past was usually met by separately wrapping mica-based tape on each turn (Section 1.4.2). This step can sometimes be avoided by upgrading the strand insulation through machine taping individual strands with a thin mica paper tape, supported with a PET or imide polymer film. The polymer film may also contain additives such as alumina that protect it from partial discharges or be made of a special material with natural resistance to these discharges (Section 8.7.4).

3.9 MANUFACTURE OF RANDOM-WOUND STATOR COILS

Until about the 1960s, most random-wound coils were formed, either by hand or by hand-guided wire, using a winding machine supporting a coil winding form. After removal from the form, coils were usually directly inserted into the stator or rotor slots by hand, usually with the assistance of a padded mallet or hammer. For the very large numbers of random-wound machines now manufactured, typically small motors, high-speed winding machines have been developed that directly wind the enameled wire from spools into the correct position in the stator.

Random-wound coils are usually inserted into slots that already contain the ground insulation (Figure 1.8). The insulation may be made from folded and cut sheet materials, often a laminate of a nonwoven polymeric fiber and a polymer film. These slot cell insulation pieces are inserted into all of the slots before winding is begun. Similar materials are used as phase insulation in the end regions and as a separator between the top and bottom coils in the slots of two-layer windings. The latter insulation is usually laid into position as the coil insertion progresses.

As discussed in Section 4.4.3, most random-wound stators are dipped in a varnish to bond the magnet wire turns together and aid in heat transfer. However, some stators are subjected to a more expensive process called global vacuum pressure impregnation (global VPI), in which the winding is impregnated with polyester or epoxy to seal the winding against moisture. In such stators, the end-windings and connections are taped and felt seals are inserted on the connections before impregnation to retain the epoxy or polyester resin during processing. Sometimes, felt "dams" are also inserted at the slot ends to retain the resin.

Random-wound coils may be made from wire that is overcoated with an adhesive. A baking step after winding will cause the wires to fuse to each other, replacing the varnish treatment usually used. Some consider the use of the adhesive a disadvantage since, with thermal aging, the bonding material may shrink or disappear, allowing the strands to become loose and, ultimately, abrade the insulation.

3.10 MANUFACTURE OF FORM-WOUND COILS AND BARS

3.10.1 Early Systems

In the discussion of natural insulating materials in Section 3.1, the use of varnished cambric was mentioned. This material is a fully processed insulation in that no further impregnation or cure is needed when used for form-wound coil and bar insulation. It can be applied as half-lapped tapes or as a combination of tape and sheet. Slot pressing may be done to squeeze out the air between layers of the material and, if heat is added, to achieve weak bonding in the material as the varnish softens and becomes tacky. When applied as a sheet for high-voltage coil insulation, the varnished cambric was often first laid out in sufficient quantity to complete the entire straight section of each coil side. The sheet was cut to a trapezoidal shape, with the shortest edge about equal to the slot length, and then wrapped on the straight section with the longest edge next to the copper strands. Successive wraps thus each covered a little less of the straight section, resulting in a tapering of the buildup of insulation. When the coil insulation was completed, by wrapping half-lapped layers of slit tape around the end sections of each coil, a scarf joint, with good electrical creepage strength, was achieved. The trapezoidal sheet could be applied by hand for small coils. For large coils, it was advantageous to divide the coil into two sections and to insulate the unformed half coil with the help of a machine to rotate the bar and apply tension to the sheet during wrapping. This method, developed in Europe, was named the Haefley process.

3.10.2   Asphaltic Mica Systems

The introduction of asphalt and mica splittings materials for ground insulation was also discussed in Section 3.1. Although the Haefley process was used to some extent for a version of the asphalt mica splitting groundwall (also called bitumen micafolium) in Europe, most generator manufacturers applied the new materials by hand, as tapes, and processed the resulting coil through a vacuum and pressure impregnation process. The switch to a bar construction did not occur until generator sizes grew to the point where coils were too big to handle during winding into the stator (Section 1.3.3). Another factor favoring the bar winding method is the greatly reduced distortion required to assemble the bars into the stator. When winding full coils, insulated with asphaltic binders and mica flakes, preheating coils by circulating high current in the coil or by ovens, softens the thermoplastic insulation and allows some relative movement within the insulation to accommodate coil distortion during winding.

3.10.3   Individual Coil and Bar Thermoset Systems

The post-World War II developments of thermoset ground insulation, to prevent tape separation or girth cracks as discussed in Sections 3.4.1 and 3.4.2, produced rigid insulation and limited opportunity for heat forming of coils during winding. Although full coil noses or involutes could be left uninsulated during winding to provide some flexibility, there was little advantage over the use of bar windings for fully processed ground insulation.

The first generations of thermoset insulation used synthetic resins that would soften at moderately high temperatures, yet would not melt as some of the asphaltic resins could. The softening temperature of thermoset resins is called the glass transition temperature (T_G). It is the temperature at which the resin changes from a rigid, crystalline state to a more rubbery amorphous state. The insulation engineer prefers using resins that have a glass transition temperature above the insulation service temperature. As customer demands grew for insulation systems rated for Class 130°C service to systems rated for Class 155°C service or higher, the selection of resins having higher glass transition temperatures was needed. Such insulation systems are inherently stronger and less flexible at room temperature and do not lend themselves to full coil winding.

In the last paragraph of Section 3.4.2, the derivation of the vacuum pressure Impregnation process and the resin-rich process were discussed. Both processes may be used to make fully cured coils or bars coils for winding into stators.

Although bar-wound machines impose less stress on insulation systems than required on fully insulated coil-wound machines, there is still some bar distortion required to properly nest each half coil with those previously inserted. The ends of each bar are formed in fixtures before the insulation is applied and processed. Often, some slight deviation from the ideal form or shape occurs due to such factors as variations in the copper hardness, normal shop handling, and storage while awaiting the winding operation. Within limits that vary according to the characteristics of each insulation system, expert winders can usually do some reforming of fully cured bars. Heating of the bar by circulating current or more local heating by means of heating tapes and hot air blowers, may be enough to raise the insulation temperature to the glass transition temperature of the insulation composite, allowing some reforming to be accomplished without cracking or delaminating the insulation.

The bar to bar copper strand connections required to complete a coil and the insulation of these joints is a slow, labor-intensive operation carried out by highly paid skilled winders. They also have to make the coil-to-coil and phase connections and the special arrangements for water- or gas-cooled coils. The largest conventionally cooled generators and generators with direct-cooled windings are all bar wound with fully processed insulation systems. One advantage of this winding method is that each bar can be given a high potential voltage test as soon as it is placed in its proper position and lashed in place. These windings can also be voltage tested as individual full coils when the bar-to-bar connections are completed and as a full phase group of coils before proceeding with the next phase group. A schedule of voltage levels is used, with the highest level reserved for individual bars before winding, and successively lower test voltages for each additional test, ending with the twice-rated voltage plus 1 kV as the final acceptance value (Section 15.2.2).

3.10.4   Global VPI Systems

The high cost of making form-wound stators with coils or bars that are fully impregnated and cured prior to insertion in the slots led to the development of techniques in the 1960s for making coils and bars with insulating tapes that are not fully saturated with resin and are not cured prior to insertion in the stator slots. These soft coils (also called green coils) are less expensive to make, are more easily fitted into the stator, and all of the connections can be made before final impregnation of the windings. This technique is called the Global VPI or Post-VPI method.

In this method, a stator core that has been wound with the green coils is placed in a large tank where it can be vacuum dried with some heat if necessary. Following the vacuum cycle, the wound stator is entirely flooded with a low-viscosity solventless resin that is then pressurized with a nonreactive gas to drive the resin throughout the ground insulation and the coil-to-coil and phase-connection insulation. After the unabsorbed impregnating resin is pushed back to storage, the wet stator can be removed for drainage of excess resin and placed in an oven to cure the resin. Alternatively, after resin removal and draining of the excess, the stator may be baked to cure the impregnated winding within the same VPI tank. This practice reduces the handling operations for the stator and lowers costs.

The tapes for VPI are usually mica paper tapes reinforced with thin glass cloth on one side and sometimes with a nonwoven material on the other side. They are made with 10 to 25% of the binder resin needed to fully saturate the insulation, which is sufficient to bond the layers of the tape together for tape application. The binder resin is normally of the same resin family as the final impregnating resin. In some cases, the tape binder resin is used without a curing agent, relying on the subsequent impregnation to supply the reactive material to complete the cure. The product can be used with different impregnating resins, resulting in little or no reactive change or cure in the tape binder resin during later processing. This can be tolerated because the binder resin is such a small percentage of the total impregnate. Another technique is to mix one component of a two-part curing system in the tape resin and the other part in the tank impregnating resin. The porosity of the tape layers allows the impregnating resin to flow through the entire structure, assuring that the proper amounts of curing agent is dispersed throughout the groundwall insulation build.

The resin used for the global VPI process can be either low-viscosity epoxy or polyester. Polyester is cheaper and requires less stringent manufacturing controls than does epoxy. Today, most manufacturers usually prefer epoxy because of its superior strength and chemical resistance.

Virtually all form-wound motor stators made today employ the global VPI process. As discussed in Section 4.3, since the early 1990s several manufacturers have VPI tanks large enough to impregnate stators in excess of 300 MVA.

3.11 WIRE TRANSPOSITION INSULATION

Stator bars need to have the individual strands that make up the bar transposed so that each strand will have the same current or voltage induced in it by the magnetic fields from the other bar in the slot, as well as the rotating magnetic field produced by the driven rotor (Section 1.4.8). The most common kind of transposition is the Roebel transposition. The Roebel transposition is made by making each strand follow an elongated spiral through the bar (Figure 1.18b). A strand that starts at the bottom is raised as the next strand is inserted under it and so on with the rest of the strands. When the first strand reaches the top of the bar, it is double bent edgewise so that it crosses over the narrow side of the bar and is in position to follow the spiral down the other side until the bottom of the bar is reached. Most generators are designed with a 360-degree transition so that each strand makes one full spiral through the slot portion of the bar. Some larger generators are designed with additional transpositions, such as a 480-degree transition. The additional bending needed for the extra transpositions is usually carried out in the first and last part of the bar slot section or in the end-winding, to adjust for the varying magnetic field exposure in the end regions of the assembled coils.

Roebel transpositions were first carried out by hand. All of the loose strands for one bar were first laid out on a table. Attached to the table were step-like guide forms that allow each strand to be twice bent in the proper locations by a hand-operated device. After all of the strands were formed, they were laced together in the proper order to create the transposed bar. The entire operation was called bend and lace. Repair and rewind shops still use this technique to make small quantities of new or replacement bars. Original equipment manufacturers have now largely automated the Roebel operation with special machines designed for the purpose.

The copper strands of the bar are insulated from each other by the strand insulation. The Roebel transposition bendings and lacing may damage the strand insulation, particularly where the strands cross over from one side of the bar to the other side. To assure insulation integrity, a thin piece or chip of insulation is inserted at each crossover point on the top and bottom of the bar. These pieces may be made of varnished mica splittings, nonwoven pieces such as aramid paper, or specially molded pieces of a composite. At this point in the bar assembly, the strand package is not bonded together. Bonding is achieved by placing a vertical separator (an adhesive material) between the rows of strands for subsequent consolidation in a hot bar press. The final insulation addition to the bare bar is a material that evens out the spaces between the transposition bends. The material may be a series of specially molded pieces placed between the bends or, in more recent practice, strips of molding material laid out on both the top and bottom narrow sides of the bar. Usually, the loose bar, with all of the component pieces, is wrapped with a plastic film release tape such as PET film with a silicone coating to prevent it from sticking to the bar during pressing. The bare bar hot pressing step compacts the entire assembly and bonds all strands and insulation pieces into a solid mass over the central straight section of the bar. At this point, the end sections of the bar are unbonded and usually held together with an inexpensive tape. After pressing, the integrity of the strand insulation can be checked with a voltage test between all strands. Any failures are located and repaired before the bar proceeds to the next forming operation.

3.12 INSULATING LINERS, SEPARATORS, AND SLEEVING

3.12.1 Random-Wound Stators

Rotating machines use insulating liners as ground insulation in the slots of random-wound stators, using materials such as described in Section 3.3. Similar materials, including both nonwovens and thin laminates, are also commonly used as phase separators in the end-windings. Fractional horsepower motors, for light and intermittent duty, may entirely rely on these materials and the magnet wire enamel for strand, turn, and ground insulation.

Low-voltage motor connections between coil groups and from the phases to the power supply are usually insulated with extruded plastic sleeving, generally made from polyolefin resins. These materials are a class of crystalline polyallomers prepared from at least two monomers by copolymerization. The resulting polymer chains contain polymerized segments of each of the monomers used. An example of polyolefin widely used for extruded sleeving is thermoplastic propylene–ethylene copolymer.

Stators rated up to 600 V may use varnished sleeving. These are made from woven sleeving of glass or synthetic fibers that are passed through a bath of solvated resin varnish or a solventless liquid resin formulation before heating to remove the solvent and cure the resin. Sometimes, a combination of varnished or extruded sleeving is used directly on the leads and then covered with an unimpregnated woven sleeving designed to be filled with VPI resin or other varnish dip in a subsequent step.

Another technique frequently used with global VPI stators is to place dry felts of polyester fibers in the end-windings between coils and between phases, where the felt will be impregnated during VPI and become part of the physical support of the winding as well as part of the insulation system. These felts are also used to help seal connections form moisture ingress. The felt seals are installed between each coil lead and the coil knuckle and around connections. Felts act as a sponges to absorb the resin.

3.12.2 Rotors

The slot and end-winding insulation of large-generator round rotors presents other problems. Generator rotors have comparatively low DC voltage between the heavy copper strap coils and the rotor steel. During winding of an edge-wise wound field coil, the radially inner turns are subjected to severe flexing and distortion to place them in the proper position. Some of the forces applied to the copper are transferred to the slot cell or slot armor insulation during winding. Therefore, the material for field coil ground insulation must first be able to withstand considerable winding stresses. Fields operate at 1800 rpm for four-pole 60 Hz designs and 3600 rpm for two-pole 60 Hz designs. When operating at design speed, the ends of the coils beyond the slots are held in place by the retaining ring. Insulation is applied on top of the end-windings to insulate them from the steel retaining rings. Also, the top of the slot coil insulation is subjected to high compressive loads between the top turn of copper and the alloy steel or aluminum slot wedges. The insulation must also withstand the operating service temperature without significant thermal degradation. Even the sides of the slot insulation are subjected to centrifugal force that must be withstood during operation. Slot cell insulation is often made from hot-press molded composites, containing reinforcing fibers in woven and nonwoven forms and sometimes plastic films, all impregnated and bonded together with thermoset resins. Experience has shown that the glass transition temperature of the cured resin must be above the maximum service temperature of the winding to prevent migration of the resin toward the top of the slots during operation.

REFERENCES

3.1 W. Tillar Shugg, *Handbook of Electrical and Electronic Insulating Materials,* IEEE Press, New York, 1995.

3.2 C. P. Steinmetz and B. G. Lamme, "Temperature and Electrical Insulation," *AIEE Trans., 32,* 1913, 79–89.

3.3 T. W. Dakin, "Electrical Insulation Deterioration Treated as a Chemical Rate Phenomenon," *AIEE Trans., Part 1, 67,* 1948, 113–122.

3.4 G. A. Cypher and R. Harrington, *AIEE Trans., Part III, 71,* 1952, 251.

3.5 E. L. Brancato, L. M. Johnson, and H. P. Walker, "Functional Evaluation of Motorette Insulation Systems," *Electical Manufacturing,* Mar. 1959, 146–153.

CHAPTER 4

STATOR WINDING INSULATION SYSTEMS IN CURRENT USE

In Chapter 3, the historical development of insulation materials and systems was discussed. The development of stator insulation systems, from natural materials through asphaltic mica splittings composites and on to modern polyester and epoxy mica paper insulation, was reviewed. The influence of older and established manufacturing processes and facilities on the choice of new materials was described. Four major manufacturing processes have been and are still widely used to form and consolidate insulation systems for form-wound stators. They are:

1. Vacuum pressure impregnation (VPI) of individual coils and bars
2. Global VPI of complete stators
3. Hydraulic molding of individual coils and bars using resin-rich tapes
4. Press curing of individual coils and bars, also using resin-rich tapes

There are some combinations of these methods also in use. The binder resins can be categorized as high- or low-solvent-containing and solventless, as well as by their chemical nature. Although no longer manufactured for coils in new stators, there are many machines still in service, and expected to remain in use for several more decades, that are insulated with asphaltic mica splittings as described in Chapter 3.

There are four principal drivers that govern the selection of the insulation systems currently being manufactured. They are:

1. Good service experience with earlier versions of the same basic system
2. Commercial availability of the materials to be used
3. Relative costs of the raw materials and processes in the competitive machine-sales environment
4. Design advantages or limitations each insulation system and process brings to the final generator or motor for its expected service life and economy of operation

Electrical Insulation for Rotating Machines. By Stone, Boulter, Culbert, and Dhirani
ISBN 0-471-44506-1 © 2004 Institute of Electrical and Electronics Engineers

The middle part of the 20th century was a period of rapid advances in new polymer materials that could be exploited to obtain improved stator coil insulation systems. The systems that were developed by extensive laboratory work during this period have become well established worldwide. Since the early 1980s, advances in materials that appear to have significant potential advantages for rotating machine insulation systems have greatly slowed down. The growth rate of power generation changed from as much as 7% per year in parts of the developed world during the late 1960s to negative growth a decade or so later. The resulting loss of sales led to a worldwide excess of manufacturing capacity and a wave of mergers, closed plants, and an increase in out-sourcing of machine components and materials. Most manufacturers then significantly reduced their engineering, research, and development activities in this area. Available materials provided high-quality insulation systems, when applied with care and with proper quality control of incoming materials and manufacturing procedures.

Present systems generally do not limit the machine designer in thermal rating in the same way as the older materials did. The other important parameters of machine design are the allowable voltage stress across the groundwall insulation and the thermal conductivity of the groundwall. Increases in either capability can be used by the designer to increase the rating per frame size or to make the same rating in a smaller and less expensive frame size. Thus, electric stress and thermal conductivity are presently the main drivers for the introduction of new materials and processes for stator coil insulation systems (Section 4.3).

A related driver for change is the growth of the gas turbine generator and the steam and gas turbine generator combined-cycle units. Many customers want the simplicity of air-cooled machines and the delivery of plants that can be quickly erected. Satisfaction of these needs has led to power generators, in physically large stator frame sizes, that have challenged manufacturing and transportation capabilities. The economic growth of the late 1990s, first in the United States and then in Europe and other developed nations, also increased the demand for new generation capacity. While still a very competitive market, the increase in business began to increase the needs for insulation system and manufacturing processes improvements.

The big chemical suppliers of polymers and resins moved on from insulation system components to materials for high-technology composites for the aerospace industry during the late 1970s, when potential volume and the demand for improved materials at high prices still offered them opportunities for profitable exploitation.

As a result of these background trends, in recent years manufacturers have only been making incremental changes in their stator insulation systems, often without describing them as new systems or changing the trade names. Motivations for changes are to reduce materials costs, replace a supplier's discontinued or changed product, and, in general, to replace solvent-containing resins with very low solvent or solventless resins to minimize air pollution and the costs of abatement, while, at the same time, taking advantage of new technology to assure better electrical performance. Sometimes, a supplier changes raw materials or even the details of the processes used to make what they consider to be the same product. These changes can sometimes produce different resin reactivities that have been known to upset carefully worked out process methods and to introduce variations in the physical and electrical properties of the final insulation.

The success of most major insulation systems presently in service led most rotating machine manufacturers to stop or greatly slow down insulation materials development, including major reductions in their laboratory facilities and technical staff [4.1]. They have come to rely much more on the integrated supplier of insulating materials for the development of new systems. These suppliers can combine the needs of several customers to spread the costs of new materials investigations over a number of machine manufacturers and enhance their own market success in the process. The resumed growth of the power generation market around the end of the millennium caused the larger original equipment manufacturers (OEMs) to renew the expansion of research and development in new materials and processes, including facilities to take advantage of them.

New insulating materials may require the development of new or significantly modified manufacturing processes to obtain the insulation system improvements inherent in the materials. By the 1990s, the major insulation suppliers were offering full insulation systems, including the basic processing know-how, to their customers. For smaller OEMs and most repair shops, the insulation supplier's materials, other acceptable materials, and processing specifications are all that is needed to support work. The final insulation system may use the materials supplier's trade name, e.g., VonRoll ISOLA's Samicatherm™ and Samicabond™. Larger OEMs still work with insulation suppliers to optimize both the materials and processes for new or changed insulation systems.

This chapter reviews the state of the art in random- and form-wound stator coil manufacturing, and presents the known characteristics of the major commercially available form-wound coil and bar insulation systems. A look into the future of what revolutionary changes could be in store for stator winding insulation systems is also given.

4.1 METHODS OF APPLYING FORM-WOUND STATOR COIL INSULATION

The first consideration in using modern insulation systems is the method of applying the groundwall materials to form-wound stator coils and bars. As discussed in Section 3.10, the Haefley process was widely used decades ago to apply wide sheets of insulation material to coils. Presently, however, virtually all groundwalls are fabricated by the application of relatively narrow (2–3 cm wide) tapes. When tapes were first introduced, and for many decades thereafter, they were applied by hand by skilled tradesmen.

The advent of modern glass fabric-backed mica paper based tapes and binder resins made the development of the taping machine possible. The earliest taping machines were designed to tape only the straight or slot portions of bars and coils. The tape tension, along with the rotational and traverse speed to obtain a half lap or a butt lap, could be preset. For a long time, stator bars served as test beds for the development of the rotating taping head components, as well as a production tool to lower the cost and improve the quality of the slot section groundwall insulation. These early machines could be served by one to three operators, depending on the production rate and slot length. The machine operator could also hand tape the bar end regions or leave that to separate hand tapers who would then also make the staggered lap joints (also called a scarf joint) between the hand- and machine-taped sections. Although some small shops and development facilities still use versions of these straight-bar taping machines, most OEMs have evolved to much more complicated designs that will tape the entire bar or coil, except the bar-to-bar connection area and coil-side-to-coil-side or knuckle areas.

Tape tension control started out with an adjustable friction brake on the tape supply roll. For solvent-containing tapes, from which resin may squeeze out of the supply rolls as they grow in initial diameter, the tension variation between the beginning and end of the roll became unacceptable. This led to the use of only a small braking action on the supply roll, with most of the tension being supplied by a fixed-diameter braked roll over which the tape would pass. For dry tapes, used for VPI coils, tension applied from the supply roll often continues to be used. As taping speeds increased, it was found that the combined inertia of all elements of the tape path led to a variation in actual tape tension four times per revolution of the taping head that often caused tape breakage when taping the usual rectangular coil cross section. The addition of a spring-loaded dancer roll, over which the tape passed before going around the main tension roll, substantially reduced the tension variation when using taping head speeds of up to several hundred revolutions per minute. Even higher taping speeds can be achieved by instantaneously sensing the tape tension with a torque transducer and suitable electronic controls to adjust an electromagnetic brake on the tension roll.

The various designs of mechanical friction brakes have given way to permanent magnet and electromagnetic brakes in modern taping machines. Generally, the permanent magnet brakes are used in small sizes to impart a small tension to the tape supply roll. This tension is needed to keep the roll from spinning with its own inertia when tape demand slows or stops. This input tension is also needed to make the dancer roll and main tension roll perform properly.

Early full-bar taping machines used a pantograph mechanism to control the in-and-out motion of the taping head. The up-and-down and tilt motions in these machines were controlled by the operator, as were the coordinated rotational and transverse speeds of the taping head. The pantograph mechanism settings were established by the machine operator from planning instructions derived from the coil/bar design. Correct and fully productive operation of these machines requires a training period and substantial production practice.

The latest generations of taping machines are computer controlled, including all five degrees of motion freedom, if desired (Figure 4.1). Generally, for each coil/bar design, the operator manually steers the taping head at slow speed through all of the positions necessary to properly track the coil/bar during machine taping. During this mapping process, a measuring device that sends to the computer memory the positional information needed to run the machine automatically is attached to the taping head. This recorded information can be stored on a disk or hard drive for future work on the same design.

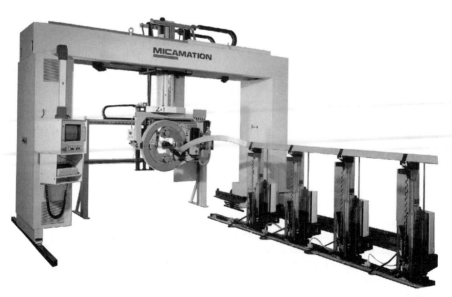

Figure 4.1. A modern stator bar taping machine that applies the tape both in the slot and end-winding portions of the bar.

Taping heads can be equipped with devices to automatically apply resin, strip out a tape interliner, or soften the tape by preheating just before it reaches the coil/bar surface. The latter device can be used to apply solvent-free, resin-rich tapes, based on improved resins.

The developments in taping technology have been one of the keys to reducing the variation in groundwall insulation quality, thus permitting design for higher insulation voltage stresses during service. The insulation improvements show up in less thickness variation and better lap control, with reductions in gaps and extra overlaps within the tape layers.

4.2 DESCRIPTION OF MAJOR TRADEMARKED FORM-WOUND STATOR INSULATION SYSTEMS

There have been many variations over time in the details of the major trademarked insulation systems. Minor changes often are not signaled by name changes, whereas major changes may be indicated by addition of a numeral or letter. The following sections attempt to discuss only the generic details of the various insulation systems, as found in technical papers used to introduce them to the industry and to customers. The exact nature of these systems is generally a proprietary trade secret.

The discussion of these trademarked stator insulation systems is limited to the groundwall insulation. It does not include the strand insulation, semiconducting slot section armoring, methods to secure the coils in the slots, measures to allow relative movement between the coils and the stator iron, the end-winding grading system, or the end-winding supports. These are all important aspects of the performance of an insulation system and they differ significantly between manufacturers. It is the proper integration of all these components that makes a good system.

4.2.1 Westinghouse Electric Co.: Thermalastic™

Chapter 3 discusses a typical 1950s era Westinghouse Thermalastic™, the first modern synthetic insulation system. The first Thermalastic insulated generator went into service in 1950 and the insulation system was introduced to industry in an AIEE paper during 1951 [4.2]. Similar insulation systems, produced under licensing agreements with National Electric Coil (Neccobond™) and Kraftwerk Union (Micalastic™), also contained solventless synthetic polyester resins and mica splittings and were introduced in the 1960's. During this period, minor changes that were made included introduction of glass cloth as a backing material for the mica, resin modifications to help VPI resin tank stability, and improvements in the partial discharge suppression treatments on generator coil surfaces.

During the 1960s Westinghouse introduced an all-epoxy Thermalastic insulation for large motors, using the individual bar/coil impregnation method. The epoxy system was used in this application for 10 years before a hybrid polyester-modified epoxy system went into commercial service on a turbine generator in August 1972. The laboratory work leading up to the use of a hybrid VPI resin was presented to the industry in 1967 [4.3]. Since 1972, all high-voltage rotating machines used different variations of the epoxy system.

Although large turbine generators continue to use the individual bar impregnation and cure method, motors and smaller generators shifted to the global VPI method in the early 1970s [4.4]. The hybrid epoxy VPI resin used for turbine generators was optimized for the previously developed processing equipment and insulation requirements. It is comprised of a modified epoxy resin, prepared in a resin cooker to create polyester linkages, and is compatible with styrene for viscosity control. The final resin cure was achieved by cross-linking through the epoxy or oxirane group. After Siemens acquired Westinghouse in the late 1990s, the Thermalastic system underwent many refinements in materials and processing while maintaining the same resin system.

4.2.2   General Electric Co. : Micapals I and II™, Epoxy Mica Mat™, Micapal HT™, and Hydromat™

The Micapal 1™ system is also described in Chapter 3. It was introduced to the industry in an IEEE Technical paper [4.5] in 1958, after several years of limited production. Micapal 1 contained approximately 50% GE Micamat™ (paper), made with calcined muscovite, and 50% muscovite splittings. The early solid bisphenol A epoxy resins and the high acid number polyester used with them to cure the system had to be dissolved in a mixed solvent to impregnate the resin-rich mica tapes. The solvents were later removed in a vacuum bake stage during the treatment cycle. This first use of the Micamat (micapaper) was a deliberate attempt to free high-voltage insulation from expensive and variable mica splittings, generally imported from India. The system remained in production into the late 1980s for small- and medium-size steam turbine generators.

Many gas turbine generators made during this period in the same shops had an advanced epoxy system to accommodate the higher operating temperatures found in peaking generator operation [4.6]. This system was then extended to medium-sized steam turbine generators as service experience with the smaller generators became available. The same needs first led to a closely related all-epoxy mica paper insulation for hydrogenerators, trade named Epoxy MicaMat in 1966 [4.7]. The gas and steam turbine generator system has been referred to as Micapal F for its outstanding Class F–H temperature resistance.

Both of these systems use mixtures of epoxy novolacs and the diglycidyl ether of bisphenol A (DGEBA) epoxy, cured with special phenolic resins, and mixed with small amounts of solvents (7–10%) to aid impregnation in making resin-rich tapes containing glass cloth and PET polyester-mat-backed mica paper. The combination cures slowly, allowing time for the vacuum bake solvent removal and the batch cure process of hydraulic molding. The cured resin has a relatively high glass transition temperature and is thermally very stable. The completed tape, slit for machine application, normally has only 2 to 4% solvent remaining by the time it is ready for application to the stator bars. The individual, fully cured bars are inserted into the stator slots and connected to each other to make complete coils and phases during the winding operation.

After a 12-year development program, General Electric announced the MICAPAL II™ insulation for large turbine generator stator windings in 1978 [4.8]. This solventless, resin-rich, second-generation epoxy mica paper insulation system has been used on most large steam turbine generators since that time. The all-mica-paper epoxy system, suitable for continuous exposure at 130°C, was developed for the electrical and mechanical conditions found in the largest generators, which are generally made with water-cooled windings that normally operate below 100°C. The special epoxy resin formulation was developed to provide the impact toughness, dimensional stability, and high voltage capability required in the largest machines.

GE has replaced the solvent-containing original Micapal F insulation system with a new solventless epoxy mica paper system, trade named Micapal HT, of similar chemistry, that is used for gas turbine and combined-cycle steam and gas turbine generators. The resin-rich system is made without any solvent and shares the same hydraulic molding production facilities used for Micapal II. The largest output volume in recent years using the new solventless system has been gas or gas and steam turbine generators.

General Electric has used partial-discharge-resistant fillers in epoxy resins or enamels in electric motors for stator strand and turn insulation since 1989 [4.9]. Adding the patented metal oxide technology to the film backers and resin in the ground insulation to the existing resin-rich epoxy mica mat hydro generator system required more than two years of engineering and laboratory effort. The program included selected utility users to evaluate individual bars during rigorous testing in their own facilities and installation of 40 bars having a 20% reduced insulation build in a planned generator rewind. This would gain early service experience in an instrumented hydrogenerator stator. With the successful conclusion of this work in 1997, GE began offering the reduced build Hydromat™ insulation system globally in both the OEM and rewind markets [4.10 and 4.11].

In 1999, GE began to offer a reduced-build strand-and-turn insulation, using similar metal oxide fillers, in the large-motor business [4.12]. These machines use the global epoxy VPI process to make the glass-fabric-supported Micamat insulation systems for machines at least up to 13.8 kV ratings. Several generations of VPI resins have been used by GE for motor manufacture. Two of these epoxy resin systems have been based on controlled reactivity chemistry. The most recent improvement [4.13] creates polyether linkages in cured diglycidyl ether bisphenol A epoxy resin and provides high reactivity at curing temperatures with excellent shelf life at room temperature.

4.2.3 Alsthom, GEC Alsthom, Alstom Power: Isotenax™, Resitherm™, Resiflex™, Resivac™, and Duritenax™

During the 1950s, Alsthom licensed the resin technology used in the GE Micapal I system to create the first Isotenax™ system. There were several differences in materials and processes between the two systems. Isotenex used only mica paper, not mica splittings. The resin-rich impregnating epoxy contained significant amounts of a solvent mixture that had to be removed after the glass-backed mica paper tape was wrapped around the stator bars. This was accomplished by a controlled vacuum bake that did not cure the resin. Next, the very loose or puffy bars were consolidated in the slot section by a warm-coil sizing and pressing operation that advanced the cure, but did not complete it. Finally, the bars were covered with release and sealing sacrifice tape layers and returned to the hydraulic molding tank. A short vacuum bake stage next removed the air in the insulation and helped in the transfer of molten asphalt into the tank, which was then pressurized with inert gas to compress and complete the resin cure. Isotenex was thus one of the earliest all-mica-paper high-voltage hydro and turbogenerator insulation systems and continued in use until at least the late 1970s.

In 1978, Alsthom Electromecanique SA introduced Isotenax N™ [4.14], made with low-solvent, resin-rich, glass-supported epoxy mica paper. The binding and impregnating resins are epoxy novolacs, selected from various epoxide resins with both short and long molecular chains to give the best mechanical and electrical properties in the finished insulation. Additional solvent is added to the mixed resins to facilitate thorough impregnation of the tape. After partial curing to the B stage, the final tape has a volatile content in the range of 6%, is still flexible, and not sticky. After taping, a batch of bars is processed in similar fashion to those made with the original Isotenex. This includes a relatively long vacuum drying followed by final curing under pressurized hot asphalt.

During the 1980s GEC Alsthom worked with solventless epoxy-resin-rich mica paper tapes that could be processed by either a shortened hydraulic molding process or by press curing. The new insulation system is trade named Duritenax™ and is based on selected novolac epoxy resins cured with a Lewis-acid-type hardener. This system has been in production since 1992 [4.15].

The UK-based operations of GEC, next GEC Alsthom, and since 2000 part of Alstom Power, used resin-rich epoxy mica paper tapes for Resitherm™, an all-rigid insulation system for solid conductor turbo generator bars for a number of years. Some distortion of bars was necessary to connect hollow conductor bars to coolant hoses, then fully cured slot insulation with permanently flexible end-winding micaeous insulation is used. This is called the Resiflex™ system and has been used for both generators and high-voltage motors.

Since the 1980s the UK operations of Alsthom have also worked with global VPI processing and an insulation system called Resivac™. Recent advances in the VPI system have used bisphenol epoxy resins with a latent Lewis acid catalyst system [4.16].

4.2.4 Siemens AG, KWU: Micalastic™

As noted earlier, Siemens began using the individual-bar VPI process with polyester resins and mica splittings as early as 1957 for hydro and steam turbine generators, with initial help from Westinghouse. This system was trade named Micalastic. Production continued with this combination of resins and processes for at least 10 years.

In the mid 1960s Siemens began investigating epoxy mica paper VPI systems [4.17], and in 1965 introduced a VPI process for individual bars and coils. About the same time, they also introduced a similar system for the production of stators up to 80 MVA using the global VPI process. In both cases, Siemens retained the Micalastic trade name. The resins were of the bisphenol epoxy type, cured with an acid anhydride and a catalyst. Except for indirect-cooled generators and direct-cooled generators rated at more than about 300 MVA or so, which still use the individual bar epoxy VPI methods, the global VPI process has been standard for all motor and turbo generator stators since 1986 [4.18]. For its large global VPI stators, this manufacturer avoids difficulties due to shear stress at the interface of the bar to the stator core (Section 8.2) by employing a slip plane. The slip plane consists of mica splittings sandwiched between two semiconductive tapes.

4.2.5 ABB Industrie AG: Micadur™, Micadur Compact™, Micapact™, and Micarex™

Brown Boveri AG started changing from resin-rich asphalt mica flake groundwall insulation about 1953, first using modified polyester resins and then switching to epoxy resins to make resin-rich tapes. The ABB Group name for bars and coils manufactured from the epoxy-resin-rich system is Micarex™. Initially, these tapes were applied by hand and later by machine taping, followed by hot-press consolidation and curing. New machine production with this system will stop with the end of turbo generator production in Sweden, although some repair licensees will continue using Micarex for some time. For some applications, such as turn insulation on multiturn windings, the latest epoxy mica paper resin-rich tapes are still in use. The backer layers have evolved over the years to include glass cloth, nonwoven synthetic papers, and plastic films.

The Micadur™ insulation system was introduced in 1955 by Brown Boveri as an individual bar VPI method. Glass-fiber-backed mica paper tape, now applied to the bars chiefly by machine taping, is dried and then thoroughly impregnated at moderate pressure in a solvent-free modified epoxy resin. Then the individual bars are pressed in molds and cured in an oven [4.19, 4.20]. The same process using similar materials still continues for the largest machines, which are too big to fit into available impregnating tanks for the global VPI of a wound stator.

The Micadur Compact™ insulation was introduced in 1965, about the same time that Siemens introduced a similar system. The Compact version uses a global VPI process that was initially for use in high-voltage motors [4.21]. As experience and confidence with the global VPI process was achieved, and some early problems with impregnating elements of the turn insulation were resolved by changes in strand and turn tapes, impregnating tanks for larger and larger machines were obtained so that only generators rated at more than about 300 MVA still use the individual bar VPI method [4.22]. The Micadur Compact process uses essentially the same groundwall insulating materials as the current individual bar VPI process. The modified epoxy resins are thought to be of the bisphenol type with an acid anhydride hardener and catalyst.

Before the merger of ASEA with Brown Boveri to form ABB in the mid 1980s, ASEA also developed the technology for individual bar VPI production, using similar materials [4.23]. The result was Micapact™, introduced in 1962 for the stator insulation of large rotating machines. It was made with glass-backed mica paper, impregnated with a special mixture of an epoxy resin, curing agent, and additives. Unlike most other VPI tapes, the glass backing and mica paper lack any impregnant or bonding resin. The adhesion between mica paper and glass was accomplished by an extremely thin layer of material, which was melted at a high temperature during formation of the tape. The tape did not contain any volatile matter, which means that the completed machine taped bar insulation was more easily evacuated and impregnated [4.24].

4.2.6 Toshiba Corporation: Tosrich™ and Tostight- I™

The Toshiba Tosrich™ insulation system for low-voltage, small-capacity generators with a relatively small number of insulation layers was based on a resin-rich mica paper tape. The solvent containing synthetic resin was impregnated into the mica tape, wound onto a coil, and cured in a mold. The residual solvent in the system, when applied to medium-capacity generators having a larger number of mica tape layers, could not be entirely eliminated during press cure in the mold and thus did not show the desired insulation characteristics in these machines. Although used successfully for many years for smaller machines, its replacement with a solventless epoxy, resin-rich mica paper tape during the 1990s allowed the improved Tosrich to be applied to medium-capacity generators; it is still gaining manufacturing and service experience [4.25].

For larger machines, the Tostight-I™ insulation system was developed. This system used either a conventional, glass-backed, calcined muscovite mica paper tape or a newer mica paper tape made with the addition of a small amount of aramid fibrid. The low-binder tape for the VPI process used an epoxy adhesive to combine the glass cloth backing with the mica paper. A solventless epoxy impregnating resin is used for the VPI process. This system has been used for a number of large-capacity water-cooled turbine generators and medium-capacity hydrogen-cooled and air-cooled generators with a relatively high operating temperature [4.25]. The use of aramid-fiber-containing mica paper has been deemphasized in recent years.

A new generation of the Tostight-I VPI insulation system was introduced in 1998 [4.25]. It has been optimized to improve heat resistance and to be environmentally friendly in materials, equipment, production methods, and disposal of waste. The mica paper has been changed to replace the aramid fibrids with short glass fibers. The new impregnating resin is principally a high-purity, heat-resistant epoxy resin, employing a complex molecular capsule, latent hardening catalyst that is activated by heat to quickly cure and produce a high-heat-resistant, mechanically and electrically strong filling material for the mica. The revised system is manufactured using new production equipment, including a fully automatic taping machine and a new vacuum pressure impregnation facility and curing oven.

The VPI tank is equipped to control vacuum and impregnation as a parameter of the coil capacitance. The new Tostight-I is intended to be usable for all types of medium and large generators [4.25].

4.2.7 Mitsubishi Electric Corporation

The groundwall insulation systems employed by Mitsubishi until the 1990s were largely based on licenses obtained from Westinghouse. During the late 1990s, Mitsubishi introduced a new global VPI insulation system for air-cooled generators up to 250 MVA. The new system supplemented an older global VPI system, used for air-cooled generators of up to 50 MVA rating. The new system uses a glass-fabric-backed mica paper tape, bonded with a very small amount of hardener-free epoxy resin as an adhesive. The global VPI resin is an epoxy anhydride [4.26, 4.27].

4.2.8 Hitachi, Ltd.: Hi-Resin™, Hi-Mold™, and Super Hi-Resin™

At Hitachi, work on synthetic resin impregnated stator insulation started in 1949 and led to the trade name SLS coil, for unsaturated polyester resin vacuum-impregnated coils. They have seen service in large AC generators since 1957. Work continued during the 1950s on the use of epoxy resins for VPI coils, which were introduced into production in 1967 as the Hi-Resin™ coils [4.28]. Development work continued with epoxy formulations, including combining a number of epoxy resins, such as aromatic, alicyclic, and fatty resins, with various types of hardeners. The objective was to obtain a resin system with low viscosity to permit easy vacuum impregnation and a longer pot life. This culminated with the introduction of the Super Hi-Resin™ coils in the 1970s. The strand insulation for these coils was glass fiber and resin so that the strand package could be molded together before the groundwall insulation was applied. A new mica paper tape, impregnated with a special mica binder, was used to obtain more flexibility and taping strength. The final surface layer was a woven glass tape. Following vacuum drying, the coils were VPI processed and, after removal from the tank, were heat cured to polymerize the resin.

Hitachi also introduced a preimpregnated or resin-rich mica paper insulation, called the Hi-Mold™ coil in 1971 [4.29]. This press-cured system uses an epoxy resin to impregnate glass-cloth-backed mica paper, which is partially cured to the B stage. The high-performance resin was selected to obtain superior electric and thermal characteristics for use in machines rated for up to Class F insulation performance. The Hi-Mold system is used for hydro and gas turbine peaking generators and for heavy duty or other unfavorable environments in synchronous and induction motors.

4.2.9 Summary Of Present-Day Insulation Systems

A review of subsections 4.2.1 through 4.2.8 shows that all of the world's larger OEMs are currently using various mixtures and types of epoxy resins and mica paper to make their stator coil groundwall insulation systems. The compositions are adjusted or tailored to accommodate the exact process used in their manufacture. The end results are comparable in terms of inherent insulation quality as related to the machine and insulation design parameters, provided that consistent quality control practices are routinely carried out. This fact is recognized by some large suppliers of rotating machines, who will, in times of extraordinary demand, out-source or purchase generators to their own design from competitors, while allowing the supplier to use their own insulation systems.

4.3 RECENT DEVELOPMENTS FOR FORM-WOUND INSULATION SYSTEMS

In the introduction to this Chapter, the recent past slowdown in polymer insulation materials development was noted. The adoption of solventless, low-viscosity, polyester and epoxy VPI resins, the use of very low-solvent and solventless epoxy resins for resin-rich tapes for hydraulic or press molding, and the advent and improvements in machine taping by the industry have produced economical and excellent coil winding products. These changes have improved the thermal stability of the winding and provided a more dense insulation with far fewer voids. Present stator insulation systems have higher dielectric strength and lower partial discharge levels, when properly made, compared to the systems in general use through the 1960s.

The two areas of insulation materials development of great interest are methods of increasing insulation system thermal conductivity and of increasing its resistance to partial discharges. Both developments have machine-uprating potential and are being actively pursued. Increasing the thermal conductivity of the groundwall insulation pays off in better heat transfer from the copper strands to the cooling systems, permitting higher current flow in the winding. Several approaches are possible. The polymer resin binders in the insulation have the lowest thermal conductivity in the system. The addition of high thermally conductive fine filler particles in the impregnating resin was an early approach. Addition of similar materials to the mica paper has also been tried, with generally poor results since the fillers tend to disrupt the tight spacing of the mica flakelets when the paper is formed. The mica paper is more thermally conductive than the resin. Thus, processing techniques in the making of a glass-cloth-reinforced mica tape, the taping operation, and subsequent consolidation all influence the thermal conductivity of the final insulation. Techniques that reduce the binder content, leaving a greater percentage of mica and glass fibers in the insulation, will improve the thermal conductivity. Recently, Toshiba introduced a new insulation system that uses tapes containing a boron nitride filler in the binding resin of the glass backing. The resulting increase in thermal conductivity will allow Toshiba to increase the MVA of its generators by as much as 15%, for the same slot dimensions and operating temperatures [4.30].

The electrical breakdown strength and the electrical endurance (Section 2.4) of the insulation are also functions of the mica type and content and degree of elimination of voids. Although the glass fiber for tape reinforcement has good thermal conductivity, the woven glass contains many small voids that must be filled with the poorer thermal conductivity resin to produce a good dielectric barrier. Although it provides essential physical support for the insulation system, this combination is not a partial-discharge-resistant insulation component. The thickness of the glass fiber cloth can be varied. Reduction in the cloth thickness leaves more room for mica, again enhancing the electrical strength and thermal conductivity of the composite insulation.

There are small differences in the thermal conductivity of mica papers made by different methods. Mica paper made with calcined mica has a wide distribution of particle sizes and includes a large percentage of fine particles. Such paper is more difficult to fully impregnate, holds more resin, and is widely used for making resin-rich tapes. During processing, some of the extra resin binder is squeezed out, helping to remove air from the groundwall insulation during hot-press curing. When resin-rich tapes are used with the hydraulic molding process, air and other volatiles are removed by the vacuum bake step before the insulation is consolidated by hydraulic pressure. Excess resin, driven out of the insulation by the molding pressure, is removed with the processing shape aids and sacrifice tapes after cure is complete. The groundwall insulation for high-voltage machines made with the VPI process is frequently made with noncalcined mica paper. These papers may be prepared by reducing mined mica to the papermaking slurry using mechanical, thermomechanical, and chemical methods. Experience has shown that when relatively large mica platelets with a small size distribution are used, impregnation is easy, tape thickness control is more difficult, and drain out of impregnated resin can be a problem. Rotating the stator during the drain and curing stages of the processing cycle can alleviate the latter difficulty.

Regardless of the methods used to delaminate raw mica, suppliers have optimized mica papers for different impregnation methods and uses by controlling the mica particle sizes and their distribution. Machine manufacturers have refined their procedures to obtain the best results from the materials available to them. The choice of mica paper may then be governed by the specific machine design and the specific use within the winding, i.e., turn, groundwall, series loop, and if it is to be applied by machine or by hand in the shop or at final assembly in the customer's plant.

A composite of about 50% calcined mica and 50% of aromatic polyamide fibers and fibrids has been available from Dupont for many years. Trade named Nomex M™, it has found a place as turn and slot cell ground insulation in some low- and medium-voltage motors and small generators. This product is substantially stronger than conventional mica papers, allowing it to be used without reinforcement. With the gain in mechanical strength has come some loss in partial discharge resistance, thus, some (but not all) manufacturers tend to use this material in applications in which such discharges are very rare or do not normally occur at destructive levels.

Japanese suppliers have developed a version of aramid-mixed calcined mica paper that contains a small amount of aramid fibrids, generally in the range of 3 to 6%. Toshiba Corporation has found that this paper is more easily impregnated and has better heat resistance in service, since the aramid reduces mica particle separation at elevated temperatures [4.25]. A disadvantage of this product is the noxious fumes produced when the insulation is burned off to allow the reuse of the conductors. The company has overcome this disadvantage by replacing the aramid fibrids with a small quantity of short glass fibers, which are similar in specific gravity and other physical properties to mica. The glass-mixed mica tape has far better impregnation ability than conventional mica tapes. When resin is impregnated into the built-up tape layers, it must pass through the small clearances between mica particles. For a VPI tape, the unimpregnated dry tape adhesive binder can be adjusted, as can the aspect ratio of mica particles. Such changes could not drastically improve the impregnation performance, as did the addition of the short glass fibers, which opened up the spaces between some of the mica platelets. The finished insulation characteristics of VPI bars made with the new glass-mixed mica tape were superior in several electrical parameters, including dissipation factor versus temperature and better stability during thermal aging, compared to similar bars made with the aramid-mixed mica tape.

Many coil/bar manufacturers use a resin-rich tape for the turn insulation of multiturn coil windings. These tapes have often replaced partially saturated VPI tapes, because industry experience has shown that complete filling of the innermost tape layers during vacuum pressure impregnation is very difficult, particularly with the relatively thick groundwall in high-voltage bars or coils. There is also a trend to replace enameled and served strand insulation with a very thin mica paper insulation, based on 0.07 to 0.08 mm mica paper. The strand and turn tape must be strong and flexible, since during application it must withstand the sharp corners of the copper strands or the group of strands in the turn group. Single and multiturn coils and bars are usually consolidated and cured by a hot-pressing operation, followed by electrical checks for strand and turn insulation integrity. Alternatively, some manufacturers VPI fully taped individual coils and then obtain proper form and cure using hot pressing.

To save space, the glass cloth backings in these thin mica tapes are frequently replaced with thermoplastic films that often are only one quarter of the thickness of the glass cloth used previously—down to about 0.0065 mm. Initially, PET films (polyethylene terephthalate) were used, since there is a long history of successful use of this polymer in rotating machine insulation. Newer polyester films, made with polyethylene naphthalate (PEN), with better thermal stability, became commercially available in the late 1990s and are replacing PET films in some applications. The Dupont product, trade named Kaladex™, has a glass transition temperature (T_G) that is 42°C higher than PET film. The film has a higher modulus, which gives it a 25% greater stiffness and an average electric strength also 25% higher than PET film. These properties help make PEN film a better supporting layer for mica paper and it may be further improved by incorporation of a few percent of reinforcing fillers in the film. The film-backed mica paper tapes yield a higher percentage of mica in the insulation and thus give better thermal conductivity. Films may also be filled with thermally conductive and/or partial-discharge-resistant fillers. Metallic oxides, such as submicron aluminum oxide, have given good results in films and, in combination with epoxy resin and mica paper [4.31], may lead to improved groundwall insulation. When these fillers are used with enamels, they can be used for partial-discharge-resistant wire strand and turn insulation [4.8].

Some North American utilities have restricted the use of films in high-voltage generator windings, based on their failure experiences during the 1970s and 1980s. Many of these failures were likely caused by poor processing of coils, rather than by the presence of the film alone. Significant improvements in the vacuum and resin filling processes, aided by careful studies of the early failures and improved instrumentation to measure impregnation, have largely eliminated problems with films and are again allowing the inherent advantages of film backers to be obtained.

The electrical stress level on the insulation is greatest at the corners of strands and turn packages, and partial discharges are more likely to occur there as a result of aging or some manufacturing deficiency. Most unfilled films are not resistant to the partial discharges that occur in the minute voids that may be present in new insulation or be produced during thermal aging. During the early 1990s, Dupont introduced a partial-discharge-resistant polyimide film, based on a GE patent, trade named Kapton CR™. When combined with uncalcined mica, the result is improved performance with increased field strength and excellent dielectric and long-term thermal stability.

4.4 RANDOM-WOUND STATOR INSULATION SYSTEMS

4.4.1 Magnet Wire Insulation

Magnet wire, i.e., copper or aluminum conductors with insulating enamels or films bonded to the conductors, is formed into coils, usually by coiling machines. The insulation materials used over the years are discussed in Section 3.8. The most common magnet wire for random-wound stators in use today is a round copper wire insulated with a polyamide–imide insulation (Class 220°C) or polyester with a polyamide–imide overcoat. The insulation thickness is usually from 0.05 mm to about 0.1 mm. The most common standard covering magnet wires is the U.S. NEMA MW1000. The insulation on the magnet wire serves as the coil turn insulation.

With the introduction of inverter-fed drives (IFDs), even motors rated as low as 440 V have been observed to have the white powder associated with partial discharge deterioration on the turns connected to the phase terminals (Section 8.7). Thus, new magnet wires have been introduced that contain metal oxides to impart PD resistance to the normal organic insulation [4.9]. These filler materials are expected to increase the life of windings subject to voltage surges from IFDs.

4.4.2 Phase and Ground Insulation

The cross section of a random-wound stator in the stator slot is shown in Figure 1.8. As with form-wound machines, there are typically two coils, often from different phases, in the same slot. Thus "phase" insulation is often used to separate the two coils. The most common phase insulations are "papers" made from the synthetic material "aramid," sold under the Dupont trade name Nomex™. Nomex has a 220°C thermal classification, is resistant to chemical attack, and has excellent tear resistance. Depending on the voltage class, the paper may be from 0.1 mm to 0.5 mm thick.

The same aramid material is used as a slot liner, to provide extra ground insulation between the coils and the stator slot. Similarly, this material is often used between coils in different phases in the end-winding.

Another material that is commonly used for random-wound stator ground and phase insulation, as well as for wedges, is Dacron™/Mylar™/Dacron or DMD, which tends to have better mechanical strength than Nomex.

For sealed random-wound stators, the endwindings are insulated with a Dacron tape to retain the varnish and then the stator is global VPI'd. Such stators are often required in harsh nuclear station environments (see Section 2.7), or in chemically harsh applications.

4.4.3 Varnish Treatment and Impregnation

Most random-wound stators are coated with a varnish or resin after the coils have been inserted in the slot. This coating imparts resistance to moisture and contamination (which lead to electrical tracking), and also improves the electrical breakdown strength of the windings. Since NEMA MW1000 specifications for magnet wire do allow a certain number of "pinholes" in the insulation per length of wire, the varnish or resin ensures that partly conductive films cannot lead to turn-to-turn faults. In addition, the varnish or resin will improve the transmission of heat from the copper to the stator core, since the number of air pockets is reduced. As IFDs permeate the market, the voltage surges they create can lead to destructive PD in any air pockets. Thus filling of the air pockets with varnish or epoxy is becoming more critical, since PD can only occur if air pockets exist.

The materials used for varnish or resins follow the same progression over the years as impregnates for form wound stator coils (Sections 3.1, 3.2 and 3.4). Today, acrylic, polyamide, and polyimide are used as varnishes, and solventless polyesters and epoxies are used as resins. The varnishes are usually applied by dipping the stator in a tank of varnish, and then heat curing the stator. Trickle impregnation is another, more expensive, process that is usually more successful in filling all the air spaces [4.32]. The best process to minimize the possibility of air pockets within the winding (to improve heat transfer and eliminate partial discharges) and to seal the winding against moisture is to VPI the stator using a solventless epoxy or polyester.

4.5 REVOLUTIONARY STATOR WINDING INSULATION SYSTEMS

4.5.1 Superconducting Windings

The development of so-called high-temperature superconductors (HTS or high-TC materials) in 1986, which are superconducting at temperatures above liquid nitrogen at 77°K, has energized the goal of developing motors and generators using these materials. To be efficient, these air core, very high magnetic flux machines, will not have laminated steel cores. The physical stresses that will be placed on the insulation may preclude the use of mica splittings or mica paper as a component, because of the ease with which these materials will delaminate. Although coil-to-coil voltages may not be very high, voltages between phases will be high. Generators that are designed to produce voltages in the transmission voltage range will also have special insulation problems. Some early insulation developments for superconducting machines have used epoxy-impregnated high-strength S glass for the ground insulation. Another alternative may be use of resin-bonded, discharge-resistant polymer film as ground insulation.

Experience with designing and building a number of demonstration or development motors and generators, high-power superconducting magnets for nuclear physics, and thermonuclear fusion reactor coils has provided knowledge of the stresses imposed on large-coil insulation systems and of the materials that work. Development is underway for variable speed 5,000 to 50,000 horsepower superconducting propulsion motors for military ships and industrial uses [4.33, 4.34]. The materials and design challenges are great, but so are the opportunities for reduced machinery size and increased efficiency in converting mechanical energy into electrical power, and the reverse.

4.5.2 PowerFormer™

In the mid 1990s, ABB (now Alstom Power) introduced its version of a combined generator and transformer [4.35]. The PowerFormer has a relatively conventional rotor winding, but the stator winding is designed to have a rated output voltage from 69 kV to potentially 500 kV, thus eliminating the need for a separate power transformer. The British and the Russians had investigated such machines in earlier decades. Such a machine is now apparently more practical due to the tremendous advancements that have been made in power cables, particularly the availability of cross-linked polyethylene (XLPE) cables rated up to 500 kV.

The practical upper limit in rated voltage for conventional form-wound stators seems to be about 30 kV. Even in high-pressure hydrogen-cooled windings, operation above 30 kV causes such excessive partial discharge activity and the need for such a long end-winding, that a normal winding life of 20 to 30 years is difficult to achieve. In spite of all the material and processing advancements made over the past 50 years, it has not been possible to make a mica paper and epoxy groundwall completely void free in order to eliminate the possibility of PD at voltages higher than 30 kV.

The power cable industry has succeeded in making void free groundwall insulation using the extrusion of XLPE or EPR, enabling ever-higher design voltages with little risk of premature failure due to PD. The PowerFormer uses XLPE power cables looped through a special stator core to provide the stator winding. The "coils" near the low-voltage end of each phase use a power cable with a thinner insulation at the phase end of the windings. Modern cable extrusion technology allows a cable of variable thickness to be manufactured. Since the current in the cable is relatively modest, even for a large generator, the cable conductor cross section is much less than needed in a conventional generator of the same output power. The smaller currents result in lower magnetic forces acting on the cable (Equation 1.3), which is necessary since XLPE has nowhere near the mechanical strength of a modern epoxy mica groundwall.

At the time of writing this book, only a few PowerFormers have been made, and service experience is limited. Thus, it is too early to determine what the life of the stator winding insulation system will be.

REFERENCES

4.1   G. H. Miller, "Trends in Insulation Materials and Processes for Rotating Machines," *IEEE Electrical Insulation Magazine, 14,* 5, 7–11, September/October, 1998.

4.2   M. Laffoon, C. F. Hill, G. L. Moses, and L. J. Berberich, "A New High-Voltage Insulation for Turbine-Generator Stator Windings," *AIEE Trans., 70,* Part 1, 721–730, 1951.

4.3   D. A. Rogers, Jr., "Synthetic Resins as Applied to Large Rotating Apparatus," In *Proceedings of the Seventh Electrical Insulation Conference,* IEEE Publication No. 32C79, pp. 100–102, 1967.

4.4   S. J. Stabile, "Extension of the Post Impregnation Concept to High Voltage Windings," in *Proceedings of the 11th Electrical Insulation Conference,* IEEE Publication 73CHO 777-3EI, pp. 117–121, 1973.

4.5   J. Flynn, C. E. Kilbourne, and C. D. Richardson, "An Advanced Concept for Turbine-Generator Stator Winding Insulation," *AIEE Trans., PAS 77,* 358–371, 1958.

4.6   H. N. Galpern and G. H. Vogt, "A New Class F Armature Insulation System for Turbine-Generators," in *Proceedings of the 13th Electrical/Electronics Insulation Conference,* IEEE Publication 77CH1273-2-EI, pp. 215–219, 1977.

4.7   Lister, R. Lefebvre, and L. Kohn, "Epoxy Mica Mat Class F Stator Groundwall Insulation," in *Proceedings of the 16th Electrical/Electronics Insulation Conference,* IEEE Publication 83CH1952-1, pp. 152–156, 1983.

4.8   N. V. Gjaja, C. V. Maughan, and W. R. Schultz, Jr. "A Second Generation Epoxy-Mica Paper Insulation System For Large Turbine-Generator Stator Windings," *IEEE Trans. PAS, 97,* 1, 125–133, January/February, 1978.

4.9   A. L. Lynn, W. A. Gottung, D. R. Johnston, and J. T. LaForte, "Corona Resistant Turn Insulation In AC Rotating Machines," in *Proceedings of the 17th Electrical/Electronics Insulation Conference,* IEEE Publication No. 85CH2133-7, pp. 308–310, 1985.

4.10 R. E. Draper and B. J. Moore, "Development of a Vertical Generator Stator Bar Insulation System for Operation at Increased Stress," in *Proceedings of the Electrical Insulation Conference and Electrical Manufacturing and Coil Winding Conference,* IEEE Publication No. 97CH36075, pp. 597–602, 1997.

4.11 W. McDermid, "A Utility's Evaluation of a Stator Bar Insulation System Operating at Increased Electric Stress," in *Proceedings: Electrical Insulation Conference And Electrical Manufacturing and Coil Winding Conference,* IEEE Publication No. 97CH36075, pp. 603–605, 1997.

4.12 B. J. Moore, R. H. Rehder, and R. E. Draper, "Utilizing Reduced Build Concepts in the Development of Insulation Systems For Large Motors," Proceedings: Electrical Insulation Conference And Electrical Manufacturing & Coil Winding Conference, IEEE Catalog No. 99CH37035, pp. 347–352, 1999.

4.13 M. Markovitz, W. H. Gottung, G. D. McClary, J. R. Richute, W. R. Schultz, Jr., and J. D. Sheaffer, "General Electric's Third Generation VPI Resin," in *Proceedings of the 17th Electrical/Electronics Insulation Conference,* IEEE Publication No. 85CH2133-7, pp. 110–114, 1985.

4.14 P. Bonnet, "Trends in Resin-Rich Type High Voltage Insulation Systems," paper presented at ISOTEC WORKSHOP 93, Section 8, Zurich, November 23/24, 1993.

4.15 J. A. Nurse, "Evaluation of Impregnation Resins for Large High Voltage Rotating Electrical Machines," in *Proceedings of INSUCON/ISOTEC '98—The 8th BEAMA International Electrical Insulation Conference and Exhibition,* pp. 10–19, Harrogate, May 12–14, 1998.

4.16 A. Anton, "New Developments in Resin-rich Insulation Systems for High Voltage Rotating Machines," in *Proceedings of the Electrical Insulation Conference and Electrical Manufacturing and Coil Winding Conference,* IEEE Publication No. 97 CH36075, pp. 607–618, 1997.

4.17 W. Mertens, H. Meyer and A. Wichmann, "Micalastic-Insulation Experience and Progress," in *Proceedings of the Seventh Electrical Insulation Conference,* IEEE Publication No. 32C79, pp. 103–106, 1967.

4.18 N. Didzun and F. Stobbe, "Indirect Cooled Generator Stator Windings with Micalastic Post Impregnation Technique," paper presented at ISOTEC WORKSHOP '93, Section 4, Zurich, November 23/24, 1993.

4.19 B. Doljak, M. Moravec, and O. Wohlfahrt, "Micadur™, a New Insulation for the Stator Windings of Electrical Machines," *Brown Boveri Review, 47,* 5/6, pp. 352–360, 1960.

4.20 R. H. Schuler, "Experience With Micadur™ Synthetic-Resin Insulation for Stator Windings of High Voltage Rotating Machines," in *Proceedings of the 7th Electrical Insulation Conference,* IEEE Publication No. 32C79, pp. 61–65, 1967.

4.21 R. H. Schuler, "Post-Cured Insulation Provides Higher Service Reliability For Large High Voltage Motors," in *Proceedings of the 9th Electrical Insulation Conference,* IEEE Publication No. 69C33-EI, pp. 165–169, 1969.

4.22 R. H. Schuler, "Experience with Post Cured Insulation Systems for Large Form Wound Machines," in *Proceedings of the 13th Electrical/Electronics Insulation Conference,* IEEE Publication No. 77CH1273-EI, pp. 220–224, 1977.

4.23 R. Andersson and T. Orbeck "Evaluation of a New Development in High Voltage Insulation for Large Machines," *IEEE Trans. PAS, 86,* 2, 224–230, 1967.

4.24 R. Andersson and T. Orbeck, "Micapact Insulation," in *Proceedings of the 6th Electrical Insulation Conference,* IEEE, pp. 152–155, 1965.

4.25 M. Tari, N. Iwata, H. Hatano, H. Matsumoto, Y. Inoue and T. Yoshimitsu, "Advanced Technology of Stator Coil Insulation System For Turbo-Generator," in *Proceedings of Insucon/ISOTEC '98—The 8th BEAMA International Electrical Insulation Conference and Exhibition,* pp. 79–86, May, 1998.

4.26 N. Urakawa, S. Takada, M. Tsukiji, W. Bito, A. Yamada, and T. Umemoto, "New High Voltage Insulation System for Air-Cooled Turbine Generators," in *Proceedings of the Electrical Insulation Conference and Electrical Manufacturing & Coil Winding Conference,* IEEE Publication No. 99CH37035, pp. 353–358, 1999.

4.27 S. Hirabayashi, K. Shibayama, S. Matsuda, and S. Itoh, "Development of New Mica Paper Epoxy Insulation Systems for High Voltage Rotating Machine," in *Proceedings of the 11th IEEE Electrical Insulation Conference,* pp. 90–91, 1973.

4.28 F. Aki, K. Matsunobu, M. Taniguchi, S. Isobe, K. Kaneko, and M. Sakai, "Super Hi-Resin Insulation System for Stator Coils of Large Rotating Machines," *Hitachi Review, 23,* 3, 1973.

4.29 S. Fujimoto, K. Tanaka, K. Matsunobu, S. Isobe, K. Kaneko, and M. Sakai, Hi-Mold Insulation System for Stator Coils of Large Rotating Machines," *Hitachi Hyoron, 54,* 611, July, 1971.

4.30 M. Tari et al., "Impacts on Turbine Generator Design by the Application of Increased Thermal Conducting Stator Insulation," *CIGRE,* Paper 11-105, August 2002.

4.31 F. T. Emery, "Preliminary Evaluation of CR Kapton™ -Backed Mica Paper Tape for High Voltage, Coil Groundwall Insulation using Vacuum Pressure Impregnation," *IEEE Electrical Insulation Magazine, 14,* 5, 12–15, September/October, 1998.

4.32 C.E. Thurman, "Trickle Impregnation of Small Motors," *IEEE Electrical Insulation Magazine,* May, 1989.

4.33 M. Rabinowitz, "Superconducting Power Generation," *IEEE Power Engineering Review,* 8–11, May, 2000.

4.34 D. Driscoll, V. Dombrovski, and B. Zhang, "Development Status of Superconducting Motors," *IEEE Power Engineering Review,* 12–15, May, 2000.

4.35 M. Leijon et al., "Breaking Conventions in Electrical Power Plants," *CIGRE,* Sept, 1998.

CHAPTER 5

ROTOR WINDING
INSULATION SYSTEMS

Electrical insulation is required in several types of rotor windings, including round rotors, salient pole rotors, and wound induction motor rotors (Section 1.5). Usually, there is both turn and ground insulation. The voltages present in rotor windings are much lower than those found in stator windings. Thus, the design of rotor winding insulation systems tends to be limited by the mechanical and thermal capabilities of the insulation system.

This chapter outlines where the insulation is needed and the stresses that act on the insulation for the various rotor winding components such as the windings, slip rings, retaining rings, etc. The materials used and how the insulation systems are made are also described.

Rotor winding insulation is exposed to stresses that are different from those in the stator winding. Rotor stresses can include:

- Thermal stress from the I^2R DC current losses in the field winding.
- Centrifugal force from the high rotational speed of the rotor.
- Relatively low electrical stress, since the field and three-phase wound rotor windings rarely operate at more than 1000 VDC. With static excitation systems, voltage surges from thyristor operation may lead to partial discharges in very large generators operating at relatively high DC voltages.
- Oil, moisture and abrasive materials that may be present in the machine, which can either cause electrical tracking between the turns or to ground if the copper is not fully insulated or abrasion of the insulation.
- Expansion and contraction of the copper conductors every time the machine is turned on and off; the copper movement leads to abrasion of the insulation and/or distortion of the copper conductors in the end-winding.

Electrical Insulation for Rotating Machines. By Stone, Boulter, Culbert, and Dhirani
ISBN 0-471-44506-1 © 2004 Institute of Electrical and Electronics Engineers

5.1   ROTOR SLOT AND TURN INSULATION

The most widely used AC motor is the squirrel cage induction motor (Section 1.2.3). The rotor consists of heavy copper, brass, or aluminum alloy bars welded to end rings and embedded in iron laminations. Alternatively, the windings may be cast-in-place aluminum alloy. The winding, without the laminations, resembles a squirrel cage, hence the name squirrel cage motor. The squirrel cage induction motor operates without sliding electrical contacts of any kind. There is generally no applied insulation between the conductors and the laminations, as the difference in conductivity of copper or aluminum versus iron laminations and the low voltage at which the rotor operates obviates the need.

Most other types of rotors in motors and generators operate at higher voltages and have both ground and turn insulation (Section 1.5). Rotors in smaller machines are usually random-wound, using round enameled (magnet) wire (usually with a varnish dip) that serves both as turn and ground insulation. For higher voltage rotors, with either random- or form-wound coils, separate slot and turn insulation is used. This insulation is usually made from formed pieces of materials that are set in place before the windings are installed. Alternatively, some small rotors use an electrostatic coating process to apply polymer powders to the slots for insulation.

Materials selected for slot and turn insulation will vary depending on the temperature class of the winding and the voltage and power rating of the machine. Formed sheet materials may be of aramid paper (such as Nomex™ by Dupont) for medium-size generators or laminates of plastic films and nonwoven layers for smaller machines. A very common slot insulation is called DMD [for a lamination of Dacron, Mylar, Dacron (Dupont trademarks)], with a suitable adhesive. DMD is made by Dupont and other suppliers from polyethylene terephthalate polyester resin in film and nonwoven fiber form. A series of similar laminates are available for use in machines with different temperature classes. Films in use include polyethylene napthanate, nylon, and polyimides. The nonwovens include cotton paper, unbleached wood pulp kraft paper, aramid fiber mat, fiberglass mat, and the combination of polyester and fiberglass in mat form. Similar materials are used for turn insulation in the form of pieces cut from sheets or rolls. For round rotor fields with heavy copper coils, thin laminates of fiberglass cloth and polyester or epoxy resin are used.

The slot insulation for two- and four-pole round rotors (Section 1.5.2) has evolved over time from a mica-splittings-based composite with kraft paper, glass, or asbestos cloth and bonding resins to premolded slot cell or slot armor pieces. Several methods are in use to fabricate these insulation pieces. They include individual moldings made in compression presses, step-press compression molding, and hydraulic or autoclave molding of prepreg lay-ups in sheet metal molds. The long sections from the latter two methods are cut and trimmed into individual pieces. Polyester resins are used to preimpregnate glass fabric for individually compression molded pieces, whereas epoxy resin prepregs with glass cloth are used for the step-press process. This process may also add special layers of aramid fleece and high-temperature-resistant imide films to increase the crack resistance and electrical breakdown strength. The hydraulic and autoclave molding process may also use these materials, although continuous-filament nonwoven fiberglass sheets, preimpregnated with epoxy resins by 3M Company under the Scotchply trade name, have been the most successful.

The resins selected for slot cell insulation must be physically tough, resistant to thermal aging, and have a glass transition temperature (T_G) for the cured material that is above the peak operating temperature. The T_G is the temperature at which the cured resin softens or changes from a crystalline to a rubbery or amorphous state. Operating above this temperature allows the resin to creep out of the reinforcement, destroying the insulation pieces.

Most hydrogenerators and low-speed synchronous motors rated up to about 50 MW have salient pole rotor windings. As indicated in Section 1.5.1, the design used on most motors and generators rated less than a few megawatts is called a multilayer wire-wound type. In this design, the insulated magnet wire, which usually has a rectangular cross section, is used. Each pole is constructed with many hundreds of turns of magnet wire several layers deep, wrapped around a laminated steel pole piece. The turn insulation is the magnet wire insulation. Insulating washers and strips are placed between the magnet wire and the laminations to act as the ground insulation. Often, the entire pole may be dipped in an insulting liquid to glue the various components together.

Salient pole rotors in machines larger than about 50 MW, and those operating at 1200 rpm and above, favor the "strip on edge" design since it can better withstand rotational centrifugal forces (Section 1.5.1). The poles for this type of winding can be made from either laminated or solid steel. In the case of the solid steel pole type, used on larger high-speed machines, the pole tips can be bolted on or integral. In this case, a thin copper strip is formed into a "picture frame" shape and the coils are usually fully processed, including varnish impregnation, before they are installed on the poles. The only exception is the integral pole type construction for which picture frames have to be connected together to form a coil as they are installed on the pole. Separators act as turn insulation to separate each copper frame from one another. On some copper frames, especially near the pole face, an insulating tape may be applied to the copper to increase the creepage distance. The copper picture frames are connected in series to make the coil. As with the multilayer salient pole design, the winding is isolated from the grounded pole by insulating washers and strips.

The bonding varnishes for salient pole designs are selected according to the temperature class of the machine and the hardness and elasticity required for the application. The mica usually chosen for the ground insulation is clear muscovite mica, bonded with shellac, shellac–epoxy, or vinyl–alkyd. Composites of mica splittings, aramid sheets, glass fabrics, and epoxy resins are also used.

5.2 COLLECTOR INSULATION

Unless the rotor winding is of the "brushless" type (in which the DC comes from a rectified AC current induced in an auxiliary winding on the rotor), collector rings are needed to bring the positive and negative DC current to synchronous rotor windings. Collector rings are also needed for wound induction rotor windings. The collector assembly is generally manufactured as a separate item that is heat-shrunk onto the rotor shaft at assembly. A wide choice of materials is available for the insulation needs. A common choice for ring insulation is molding mica. Molding mica is a B-stage material that is applied by heat softening to the collector shell or hub along with additional bonding varnish. After wrapping, the mica is subjected to high compressive force by, for example, wrapping with steel wire under tension. The unit is then oven baked to cure the varnish, stripped of the wire, and machined in a lathe to a closely specified outer diameter. The steel collector rings (copper or copper alloy rings are usually used on wound-rotor machines) are then heat-shrunk onto the mica ground insulation. Modern practice is to apply polyester-resin-impregnated fiberglass rovings, under winding tension, between and beyond each ring. After resin curing, the excess material is machined off and a final coat of sealing varnish is brushed over the fiberglass bands and cured.

The collector brush-rigging insulation is generally made from molding compounds, laminated boards, or tubes made from paper, cotton, or glass fibers suitably bonded and impregnated. The resins chosen for moisture-resistant surfaces of these pieces are very important for good operation.

5.3 END-WINDING INSULATION AND BLOCKING

Round rotors for two- and four-pole designs generally operate in clean environments. Larger generators are usually totally enclosed designs in which the rotor operates in a pressurized hydrogen gas atmosphere to minimize windage losses. Air-cooled machines generally recirculate cooling air through heat exchangers so that only a portion of the air being circulated is outside makeup air. The makeup air is usually cleaned by passing it through filters or centrifugal cleaning units to remove most particulate matter. The clean environment allows the copper edges of the field coils, outside of the slots, to be directly exposed to the cooling gas for maximum cooling efficiency. For larger coils, each turn is separated from the adjoining turns by a strip of sheet insulation. Smaller rotor coils, which are fabricated by edgewise winding, often use taping on every other turn outside the slot section. The turn tape is often B-stage-epoxy- or other resin-impregnated mica paper or splittings, supported by fiberglass fabric.

Air-cooled rotors that are designed for use in unclean environments, such as mining and chemical manufacturing locations, usually have the entire end-winding enclosed in a taped structure made with resin-impregnated fiberglass tape. Some designs have aluminum radiating plates fitted against the end turn insulation and between adjacent coils to help remove heat.

The end-windings are not stable under centrifugal load unless they are supported by retaining rings and are securely blocked together. Missing or badly shifted blocks can lead to coil distortion from conductor thermal expansion under load. Blocking materials are generally machined from compression-molded heavy sheet stock. Although asbestos cloth was used in the past, it has been replaced with fiberglass cloth, impregnated with phenolic, polyester, or epoxy thermoset resins. The blocking between the straight portions of the end-windings protruding out of the slots consists of wedge-shaped blocks, whereas the end blocks are rectangular pieces with curved outer edges. There are usually strips of laminate fastened to the top of the blocks to help prevent them from shifting. All blocks are individually fitted by sanding and may be bonded in place with layers of B-stage, resin-impregnated glass cloth and with brushed- or sprayed-on thermosetting varnish.

After all of the blocks and ties are assembled, the entire endwinding is subjected to radially inward hot pressing. The press completes the cure of resins and varnishes and depresses the outer diameter of the coil package below the diameter of the retaining rings and their insulation, to permit assembly of the rings over them.

5.4 RETAINING RING INSULATION

The end-windings of round rotor coils must be restrained from radial movement by the use of heavy retaining rings that are heat-shrunk onto the ends of the slot section and/or machined sections of the rotor shaft axially outside the coil end-windings. Depending on the design, these rings are machined from either magnetic or nonmagnetic forged steel and are the most highly stressed parts of the rotor, since they must carry their own centrifugal load and that of the field coils under them. The retaining ring insulation is inserted between the radial outer (or top) of the field coils and the retaining rings. A number of materials have been used for retaining ring insulation. Early designs used built-up mica sheets, similar to those described above for brush rigging. Recent practice has used several sections of formed sheet materials made with epoxy and fiberglass cloth. Often, a compressible material, such as aramid paper, is placed between the primary insulation and the top of the coils to provide some cushioning.

The assembly of the retaining rings over the insulation requires the steel rings to be heated by gas torches or by induction. The retaining ring insulation must be clamped to the end-winding, usually with steel bands, as the rings are forced into position. The bands are removed as the rings advance over the insulation.

5.5   DIRECT-COOLED ROTOR INSULATION

The fields and their insulation discussed above are called "indirect-cooled" field windings. The heat generated in the slot part of the windings has to flow through the ground or slot cell insulation, into the steel teeth between coil slots, and into cooling slots also between coil slots. Cooling gas is forced through the cooling slots and their vent wedges to remove the heat.

Field windings can carry significantly more current without overheating by adopting methods for direct cooling of the copper. This greatly increases the efficiency of the heat removal process, allowing increases in the ampere current density of the field windings.

Direct-cooled fields are of several types. They are all characterized by having most of the heat removed by direct contact of the copper turns with the cooling medium. This arrangement adds some complexity to the rotor insulation. Direct cooling does not necessarily eliminate the presence of the rotor cooling slots. Shallow cooling slots may still be used to carry away heat generated in the rotor steel as well as some of the copper I^2R loss.

The least complicated design uses the simple radial flow of cooling gas from subslots under the coil slots, through the slot cell insulation, up through holes in the turn insulation and slot copper and out through the creepage-space blocks and coil-restraining steel or aluminum slot wedges (Figure 1.22). The gas exits into the air gap between the rotor and the inner diameter of the stator core. The holes in the copper turns, subslots, and wedges generally line up perpendicular to the axis of the rotor, thus permitting economical punching of the holes in the copper and turn insulation and machining of holes in the space blocks and wedges.

The insulating components of the simple, radial direct-cooled fields are generally made of the same materials that are used for similar parts in indirect-cooled designs. The most common choices are epoxy fiberglass laminates for the ground insulation, the turn insulation, the space blocks, the end-winding blocking and the retaining ring insulation. The slot section insulating parts are more complicated in cross-sectional shape and thickness, since they contain the gas flow cooling holes. The subslots are narrower than the coil slots and are fitted with insulating laminate covers to support the coils and provide creepage distance to the sides of the steel subslots. The creepage space blocks at the top of the coil stack may be placed within the top of the slot cell insulation. They must be thick enough to provide creepage distance to the steel wedges and the top of the coil slots. The turn insulation and creepage blocks must have very high compressive strength, since they must support the centrifugal load of the coils up through maximum design overspeed conditions.

More complicated designs for direct-cooled fields for the largest generators utilize hollow sections of copper for the slot sections of the coils and partially for the copper under the retaining rings. These sections may be achieved by machining grooves and holes in two sections of an individual turn, which, when joined during construction of the complete coil, provide the openings to connect adjacent turns and the source of the cooling gas. These designs may employ subslots to supply some of the cooling gas or they may rely on scooping gas from the air gap regions where the cooling flow is radially inward through the stator core. The passages through the copper then carry the gas back to the air gap in exit regions where the flow in the stator is in the reverse direction, or radially outward. Although the same insulating materials are used in this design, the detailed machining of the creepage blocks is increased.

The ultimate in direct-cooled fields is reserved for the largest generators, usually over 1200 megawatts. These massive fields use cooling water circulating through hollow field coil turns to remove the heat. This very efficient heat removal, coupled with direct water-cooled stator coils, makes it possible to build, ship, and install these generators. Since the cooling water never reaches the boiling point, the thermal stability of the insulation in these machines does not need to be as great as in typical indirect air-cooled generators. However, the mechanical properties must remain very high, so the choice of insulation materials often does not change.

Mention has been made in earlier chapters of superconducting machines. This technology has advanced rapidly in recent years, and superconductors may now be applied to the DC windings of rotating electrical machinery. Although work on designs is underway, using so called high-temperature superconductors at near liquid nitrogen temperatures, development of cold superconducting rotors with coils made with the niobium–tin intermetallic compound Nb_3Sn and operating at 4°K has been underway since the 1960s.

These machines operate with high air gap flux densities, requiring the use of an air gap stator winding, which is not superconducting. The stators of these generators do not contain the usual teeth that create the coil slots, but retain the back iron to control the path of the magnetic flux. The insulation requirements of these fields are controlled by the very low temperatures in which they operate, as well as the electrical and mechanical stresses of rotating fields. The insulation of these machines has not been standardized since little production has been achieved.

CHAPTER 6

CORE LAMINATIONS AND THEIR INSULATION

6.1 ELECTROMAGNETIC MATERIALS

This chapter covers a class of materials known as ferromagnetics. They are very strongly magnetic and include iron, nickel, cobalt, and some alloys. For motor and generator applications, there are two ways to obtain a magnetic field with ferromagnetic materials. It can be created either by electric currents, as in electromagnets, or by permanent magnets.

For many years, permanent magnets have been primarily used for small motors, in which they provide simplicity of design, reduced size, low cost, and acceptable efficiency. However, their use has been progressively extended to larger machines with the advent of superior permanent magnet alloys having stronger magnetic fields. They are being developed, for example, for use in electric drives in ship propulsion. A detailed discussion of these machines, which are currently in a state of rapid development, is beyond the scope of this chapter. This chapter is restricted to the materials, processes, and insulation of electromagnets.

6.1.1 Magnetic Fields

Magnetic fields are characterized by the magnetic flux (in Webers), the magnetic flux density or B (in Tesla or Webers/m^2), and the magnetic field intensity or H (in Ampere-turns). The latter is also called the magnetomotive force or mmf when the field is generated by an electromagnet. Any standard textbook on electromagnetics presents the relationship between these three quantities [6.1].

6.1.2 Ferromagnetism

Ferromagnetism owes its magnetic properties to the electron spin of the material. This may be thought of as an electron spinning about an axis through its own center, in addition to its

Electrical Insulation for Rotating Machines. By Stone, Boulter, Culbert, and Dhirani
ISBN 0-471-44506-1 © 2004 Institute of Electrical and Electronics Engineers

rotation around the nucleus of the atom. Electron spins and electron charges influence neighboring atoms by trying to keep the spins parallel, while thermal agitation forces try to destroy the lineup. These electron forces result in the formation of small regions, known as domains, which are typically 0.01 mm cubes. Ferromagnetic materials are entirely composed of these domains, each magnetized in a definite direction. Thus, the magnetic material is made up of a large number of elementary magnets or dipoles. These domains, each of which is an elementary magnet, will change to line up with each other when a magnetic field is applied. A ferromagnetic substance is unmagnetized when the domains are oriented at random. In this condition, there is no net mmf across a specimen of the material and no magnetic field is produced outside of itself.

6.1.3 Magnetization Saturation Curve

If one makes a toroidal solenoid, wound on a nonmagnetic core, and energizes the coil with an electric current, the space within the solenoid will become magnetized. As the current is increased, the magnetic field intensity (H) and the magnetic flux density (B) increase in a straight line. Reducing the current reduces the flux density along the same line, reaching zero when no current is flowing. With the same solenoid, if the space within the toroid is filled with an unmagnetized ferromagnetic material, increasing the current will produce a magnetization saturation curve. This curve is also called the B-H curve, the virgin curve, or just the saturation curve (Figure 6.1).

As the magnetic field intensity (H) (the mmf) is increased in ferromagnetic material, the current-induced magnetic field acting on the domains begins to orient them in the direction of the field. Thus, the mmf of each domain is aligned with the external field so that the flux from that field is increased. At first, the domains that are nearly aligned with the external field turn easily in that direction, producing a straight line as the external mmf increases. As the magnetizing force continues to rise, the domains whose individual orientations increasingly diverge from the external field are forced to become aligned, producing a rounded part of the magnetization curve that continues until all domains are parallel. At this point, the ferromagnetic material is said to be saturated. Further increases in external mmf causes the flux density (B) to increase at a much lower rate. The difference in flux between the saturation curve and the line produced by an air-filled solenoid at any magnetizing force is caused by the contribution of the magnetic material. The flux obtained up to the saturation point of the ferromagnetic material is known as the intrinsic flux and is a better indication of magnetic properties than the total flux that can be obtained at very high mmf's.

6.1.4 Ferromagnetic Materials

Ferromagnetic materials can be roughly divided into groups by noting how rapidly their flux density increases with magnetizing force. The materials in the first group give appreciable flux densities at very low magnetizing forces. The group of nickel–iron alloys, such as Nicaloi, Permalloy, and Mumetal, are examples. The next group contains the bulk of industrial magnetic materials and runs from electrolytic iron, the most easily magnetized, to steel and malleable iron castings. The third group contains primarily the permanent magnet materials. Among the traditional materials, cobalt alloys have the highest intrinsic saturation values, followed in decreasing order by electrolytic iron, Armco ingot iron, cold-rolled steel, low-silicon steel, high-silicon steel, and the iron–nickel alloys.

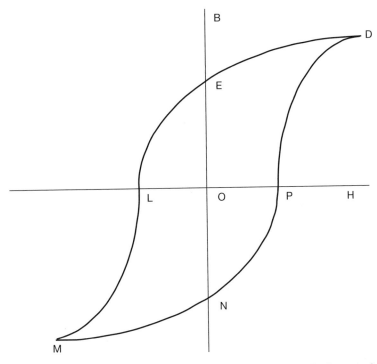

Figure 6.1. The hysteresis loop for a ferromagnetic material. The initial magnetization curve from O to D is not shown. The length O to E is the residual induction, whereas O to L is the coercive force. The width of the loop from L to P is exaggerated compared to materials commonly used in electromagnetic cores.

6.1.5 Permeability

The term permeability refers to the relative ease with which a material can be magnetized. A vacuum or free space has a permeability of unity, a diamagnetic material has a permeability less than unity, a paramagnetic material has a permeability slightly greater than a vacuum and is approximately independent of the magnetizing force, and a ferromagnetic material has a permeability that is considerably greater than unity and varies with the magnetizing force.

6.1.6 Hysteresis Loop

The hysteresis loop is a curve plotted between B (flux density) and H (magnetizing force or magnetic field intensity) for various values of H from a maximum value in the positive direction to a minimum value in the negative direction and back again. When the core of a toroidal solenoid with a ferromagnetic core is first magnetized, the flux density starts at zero and reaches a maximum for the magnetizing force applied, as described in Section 6.1.3. If H is gradually reduced, the flux density does not return to zero, but remains positive even with zero H. The remaining magnetic flux is known as the residual induction in circuits in which there are no air gaps. The more general term remanence refers to the magnetic induction remaining in the circuit, usually containing an air gap, after the mmf has been removed. The residual flux represents work that was done to overcome friction in rotating and aligning the magnetic domains that is not fully recovered when mmf reaches zero, as some alignment remains. A typical hysteresis loop is shown in Figure 6.1.

When the mmf is reversed, the flux passes zero and reaches a negative maximum for the mmf applied. When the mmf is again returned to zero, there is still a remaining negative flux equal in magnitude to the positive flux of the previous half cycle. As the mmf again increases in the positive direction, the flux again passes through zero and increases to the same level obtained in the first application of the mmf. The closed loop, described by the interplay of flux (B) and mmf (H), is a measure of the energy lost during the cycle and shows up as heat in the material. The demagnetizing mmf required to bring the flux back to zero is the coercive force of the magnet.

When a ferromagnetic material is subjected to an alternating mmf, the first hysteresis loops traced out do not necessarily fall on top of each other. When successive loops do retrace preceding loops, the substance is then in a cyclically magnetized condition. For electromagnet core materials, the hysteresis loop is determined when the materials are cyclically magnetized, whereas for permanent magnet materials the value is taken from the first hysteresis loop, since the materials need to be magnetized only once.

6.1.7   Eddy Current Loss

Whenever the flux in a magnetic material changes its value, voltages are induced that cause eddy currents to circulate in the material. The extent of eddy current loss depends on the resistivity and thickness of the material, as well as the frequency and flux density in the application. Eddy current losses increase with increasing grain size up to the point where the grains become as thick as the laminations. Materials such as silicon steel, which have a large grain size, tend to have losses greater than calculated because of this effect.

Eddy current losses increase as the square of the thickness of the material used. To reduce these losses, magnetic structures for use in alternating current applications are usually made with thin sheets or laminations insulated from each other. In practice, it is found that eddy current losses decrease less rapidly than the square law with decreasing thickness. From the losses point of view, there is a limit of lamination thickness below which it is not feasible to go since each sheet must be insulated from the next. As the thickness of the insulation remains essentially the same for varying thicknesses of core plate, the thinner the core plate used, the smaller will be the proportion of magnetically useful space in the core assembly.

6.1.8   Other Factors Affecting Core Loss

Normal core loss is the sum of the hysteresis and eddy current losses. It is determined in practice by the Epstein test, wherein a standardized sample is operated with a standard flux density and frequency, and the power loss in the sample is measured. Other factors relating to design and manufacturing processes occur in apparatus that make total losses considerably higher than indicated by the Epstein test, as discussed below.

Mechanical Clamping. Some method must be used to hold the laminations together. It may take the form of rivets, bolts, or edge welding. All of these techniques tend to short-circuit the lamination insulation and, since they may become part of a closed magnetic loop, during AC operation they can cause a serious loss of electrical energy and undesirable heating. Large machines often use steel end plates that are bolted to keyways that are also used to assemble the laminations in the core. Since these laminations are often insulated after manufacture, they are isolated from electrical contact with the clamping bolts.

Pressure. A core stack must be tightly held together to provide structural rigidity and to minimize any motion between laminations. Loose cores vibrate and are noisy. However, the mechanical strain induced by compression substantially reduces the permeability. Other strains in magnetic materials come about from cold working during the rolling to produce sheet stock, from punching, and from distortions during assembly. These also increase hysteresis and eddy current losses, requiring increased magnetizing current.

Methods to apply pressure to clamp the core laminations together are as follows:

- Welding is used on some machines with single-piece punching designs to allow the core laminations to be held together until they are installed in the stator frame, or to attach the core to frame support bars in designs in which the stator is wound after the core is installed in the frame. The welding is done while the laminations are firmly clamped in a holding fixture and bonds the stack with a series of weld lines that run axially along the outer diameter.

- In segmented cores, the punchings are keyed to bars that form part of the core support structure. These help align the stator slots and restrict core radial movement.

- Long cores are compressed by a special hydraulic press periodically as the stack is built up. A steel end ring is placed below the first punchings and over the ends of the projecting keybar. When the stack is complete, a similar steel ring is placed over the top end of the keybars and the laminations are again pressed while the nuts are tightened on threaded ends or keys are driven into keyways machined into each core support bar. An alternate method is to insert long axial studs that pass through the holes in the core laminations to serve as strength members and to hold the stack in compression. These studs are insulated with tube material and have bolts on either end that are torqued up to apply pressure to the core.

- All medium and large machines have tooth support fingers that are clamped to the ends of the core by the steel end rings to maintain pressure on the tooth section of the core. These support fingers are often spot welded to thicker end punchings to keep them in place until clamping pressure is applied.

- Medium and large cores often have radial vent ducts formed by steel spacers fitted at the center of each tooth clamped between thick laminations on either side of the duct. On large machines, these spacers are often made from "I" beam shaped steel.

Punching and Machining. The punching operation often produces burrs on the cut edges of laminations. They can be minimized by keeping dies sharp and by using carbide dies to prolong intervals between sharpening. For laminations that are insulated after punching is complete, there are several types of deburring operations that can be carried out, including grinding or sanding and chemical or electrochemical operations to dissolve burrs. Those that are not removed can cut through lamination insulation, allowing excessive intersheet eddy currents to flow.

Machining operations that are performed on assembled laminations, such as grinding or broaching, frequently smear the edges together, creating undesirable electrical connections between sheets.

Hysteresis Loss in Silicon Steel Scale. The rolling and annealing processes used to produce magnetic sheet stock produces surface scale as an oxide on iron and steel materials.

Although it is an electrical insulation, the scale is partially magnetic and at high flux levels it carries an appreciable flux. Scale on silicon steel has a very high hysteresis coefficient so that as it begins to carry flux, the total hysteresis loss rapidly increases. Steel mills normally remove most or all of the oxide scale by acid pickling and either replace it with a mill-applied treatment or rely on the machine producer to apply enamel having an inherently low hysteresis loss.

Aging Coefficient. The magnetic properties of ferromagnetic materials change with use and the associated exposure to heat from machine operation. This aging is associated with an increase in core loss and an increase in magnetizing current needed to produce the desired machine performance. A positive value of the aging coefficient indicates a decrease in permeability or remanence. Standardized tests have been developed to determine this coefficient using exposure of a suitable sample to a specified time and temperature. Data is taken before and after the aging period.

6.1.9   Effect of Direction of the Grain

The properties of electrical strip steel vary considerably, depending on the direction of the flux in relation to the direction of rolling. Sheet steel is alternately rolled along the sheet as well as across it, leading to grain development in both directions. Electrical strip steel is rolled in a continuous mill, resulting in the grain being aligned with the direction of rolling. Steels with cross graining show higher hysteresis losses. In making Epstein acceptance tests, it is specified that one-half of the samples be taken with the grain and one-half across the grain.

6.1.10   Effect of Temperature

All ferromagnetic materials will lose their magnetism when heated to a high enough temperature. The temperature at which this occurs is called the Curie point. For iron it is 770°C, for nickel it is 358°C, and for cobalt it is 1120°C. As the temperature approaches the Curie point, the relative magnetization at any specific field intensity decreases. The magnetic properties of iron and iron–silicon alloys are only slightly affected by small temperature changes. As temperature is increased, permeability usually improves, whereas hysteresis and eddy current losses are decreased. As temperature approaches the Curie point, core losses continue to decrease and, when at this temperature, the permeability reaches unity at medium and high inductions.

6.1.11   Effect of Heat Treatment

The type of heat treatment given magnetic materials largely determines their magnetic properties. Annealing is used to relieve internal elastic strains brought on by mechanical operations such as punching, bending, and shearing. Usually, annealing is carried out in an inert or reducing atmosphere and tends to remove carbon. The result is to reduce the area of the hysteresis loop and produce changes in the saturation curve. Annealing will not affect strains caused by impurities or magnetization. Freedom from dissolved impurities is important since they disrupt the orderly arrangement of the atoms and produce lattice strains.

6.1.12   Effect of Impurities and Alloying Elements

Impurities in electromagnetic steel cause an increase in hysteresis loss and a decrease in permeability. Of most concern are the elements carbon, sulfur, phosphorus, oxygen, nitrogen, and manganese. Even a very small percentage of carbon produces a large detrimental effect, resulting in significant permeability loss, an increase in the area of the hysteresis loop, and a lowering of the magnetic saturation point. The next most detrimental element is sulfur, followed by phosphorus and oxygen that, as impurities, have small adverse effects on these magnetic properties. Manganese in very small percentages is only slightly detrimental.

Up to half of one percent of copper is often added to electrical steel to improve its corrosion resistance with little effect on magnetic properties. Silicon and aluminum are alloyed with iron to improve magnetic properties, although with some loss in strength. Both serve as deoxidizers and increase the resistivity of iron, thus substantially reducing eddy current losses. Silicon is more commonly used, since it also reduces hysteresis loss and intrinsic saturation while improving permeability. Recent practice has been to use both elements with the combined alloy content ranging from 0.25% for the least costly grades to about 4% for the high-efficiency grades.

6.1.13 Silicon/Aluminum Steels

The most important magnetic materials for AC applications in motors and generators are the silicon steels. They may be divided into two major types: nonoriented and grain-oriented. There are a number of grades of nonoriented silicon/aluminum steels that have been developed for the cores of electrical machines. The choice of grade depends on the specific application, including the desired magnetic and stamping properties. There is also a choice of lamination thickness or gauge within the AISI or ASTM listed grades or types [6.2, 6.3].

The cost per pound of silicon steel sheet stock increases with the alloy content and decreases with the thickness. The nonoriented grades are used for most applications except for where the highest efficiency is desired. The lowest cost grade for cold rolled motor lamination (CRML) is generally not fully annealed (semiprocessed), with only sufficient postrolling heat treatment to achieve a thermally flattened state. The other grades are available in either the fully annealed (fully processed) state or the semiprocessed state.

Grain-oriented electrical steel is manufactured to have a well-aligned crystal orientation in the rolling direction and is usually obtained in the fully processed state. In past practice, semiprocessed grain-oriented steel was often used, followed by an inert gas atmosphere annealing process after punching to relieve the mechanically induced strains and improve the magnetic properties. The silicon used in flat rolled electrical steels has the effect of aiding grain growth during annealing, thereby lowering hysteresis losses.

6.2 MILL-APPLIED INSULATION

There are a number of electrical steel standard coatings as specified in ASTM A976-1997 [6.4]. A partial listing appears in Table 6.1. In summary, small motors use CRML with a C-0 coating, larger or high-efficiency motors use C-3 or C-3A lamination coatings, and large generators typically use C-4 or C-5 coatings, often with a varnish overcoat.

6.3 LAMINATION PUNCHING AND LASER CUTTING

The traditional method of making core laminations has been to punch or stamp them out of electrical steel sheet stock using tool steel dies. However, electrical steels and some coatings

Table 6.1. Partial List of Standard Electrical Steel Coatings Specified in ASTM A976-1997

C-0	The steel has a natural, ferrous oxide surface.
C-2	An inorganic coating of magnesium oxide and silicates that reacts with the surface of the steel during high-temperature annealing. Principally used in distribution transformer cores; not for stamped laminations due to its abrasiveness.
C-3	Enamel or varnish coating that enhances punchability and is resistant to normal operating temperatures. Will not withstand stress-relief annealing.
C-3A	The same material as C-3, but with a thinner coating thickness to facilitate welding of rotors and stators and minimize welding residue.
C-4	An antistick treatment that provides protection against lamination sticking in annealing of semiprocessed grades.
C-5	The insulation is an inorganic coating consisting of aluminum phosphate. C-5 is used when high levels of interlaminar resistance between laminations are required. The coating will withstand stress-relief annealing conditions.
C-5A	The same coating as C-5, but with a thinner coating thickness to facilitate welding of rotors and stators and minimize welding residue.
C-6	A combined organic/inorganic coating that can withstand stress-relief annealing.

can shorten punching die life as compared to nonelectrical grades. The higher alloy content steels are also more brittle and die design must take into account both the punching characteristics and the need for frequent sharpening. Development of improved tool steels for this application has reduced the need for frequent sharpening and allowed the use of the softer fully processed grades that do not require annealing after punching. Tool steel dies are still widely used for limited-volume applications, such as prototype and repair/replacement production, as well as for cases where few machines of a kind are needed. Much improved dies are now available in carbide for high-volume production. Although more costly to make and requiring increased manufacturing time, carbide dies last longer, need less frequent sharpening, and produce tighter punching tolerances.

Original equipment manufacturers (OEMs) have usually maintained an inventory of old dies for discontinued machine designs so that a rapid response to service needs can be realized. During the last decades of the 20th century, the equipment for computer-controlled laser cutting of laminations became available. The ability to rapidly program the computer for new lamination designs and to quickly start production has led some OEMs to scrap old dies in storage, which are seldom used, and to then rely on laser shops to produce replacement laminations for full or partial cores. Laser cutting is also used for limited new production and prototypes, for modifications of laminations from existing dies or punched stock, and to replace damaged laminations during core repairs.

6.4 ANNEALING AND BURR REMOVAL

Electrical steel is usually formed by the cold rolling process and is annealed to reduce or eliminate the strains in the sheet, as mentioned in Section 6.1.13. The punching operation and any other additional mechanical working of electrical steel will induce strains in laminations that degrade the magnetic properties. When semiprocessed steel is used for punching laminations for higher-efficiency electrical machines, it is common to anneal stampings, usually after burr grinding. The annealing is carried out in an inert atmosphere, such as hydrogen gas, to prevent oxidation of the metal surface.

The edges of laminations acquire a metallic burr during stamping that should be eliminated to prevent the burr from causing electrical shorts between laminations and to improve the stacking factor or density of the stack. Burr removal is most often done by a grinding or sanding operation that also leaves some bare metal where the burr was removed. Laser cut laminations generally have less burr than stamped laminations, and laser burrs contain oxidized metal that is not well attached and is more easily removed.

6.5 ENAMELING OR FILM COATINGS

For large and higher-efficiency machines, OEMs do not rely on the mill-applied insulations described in Section 6.2, although they usually are a part of the final insulation system. On the other hand, small- and medium-size motor manufacturers often use preinsulated core steel to reduce manufacturing costs. The anneal-resistant mill finishes also provide a suitable base for enamel coatings. The most common enamels are formulated with some variety of phenolic resins and may contain finely divided mineral fillers. These enamels are deposited from solvent solution by passing the laminations through rollers that deposit a controlled amount of material. Immediately after coating, the laminations are passed through a heated oven in which the solvents are removed and the enamel is cured. The evaporated solvents have to be removed from the oven exhaust by incineration.

Since the 1970s, the newer polyester and epoxy resins have been in limited use as enamels. The core plate C-3 contains a high-grade, organic-modified, polyester or other resin that provides die lubrication during stamping.

Solventless epoxy enamels have been developed that cure by ultraviolet light at low temperatures and at high line speed [6.5]. The speed of the coating line is primarily limited by the tendency of the laminations to sail off the supporting conveyor belt at high speeds and not on the ultraviolet curing time. The compact coating and curing line is very energy efficient and no solvents have to be removed from the exhaust air stream. The capital and operating costs of such an enameling line are significantly less than the conventional phenolic enameling line.

REFERENCES

6.1   J. F. Young, *Materials and Processes,* Wiley, 1949, pp. 229–255.

6.2   American Iron and Steel Institute, *Electrical Steel Sheet,* Nomenclature Handbook No. 13, Washington, DC, 1979.

6.3   Armco Steel Company, *Selection of Electrical Steels for Magnetic Cores,* Butler, PA, 1985.

6.4   ASTM A976-97, "Standard Classifications of Insulation Coatings by Composition, Relative Insulating Ability and Application."

6.5   E. A. Boulter et al., "Solventless Cured Epoxy Coating for Generator Stator Laminations," paper presented at IEEE International Symposium on Electrical Insulation, June 1980.

CHAPTER 7

GENERAL PRINCIPLES OF WINDING FAILURE, REPAIR, AND REWINDING

Before discussing the specific failure and repair processes for stator and rotor windings, as well as the associated laminated cores, an overview is useful to place this subject in context. As will be apparent in Chapters 8, 9, and 10, there are three dozen separate deterioration mechanisms that can operate on the different kinds of rotors and stators. Furthermore, there are numerous repair procedures in addition to rewinding. This Chapter describes why there are so many failure processes and what causes one process to dominate, eventually leading to failure. Also presented is information on how to select an appropriate repair method from all the possible options. As well, methods to make a machine operational after an in-service failure and how to determine if a rotor or stator rewind is to be undertaken are covered. In keeping with the subject of this book, machine failure due to bearing problems, mounting problems, and cooling system problems (except where they lead to winding overheating) are not discussed.

7.1 FAILURE PROCESSES

Some machine failures, which are identified by a stator ground fault or extremely high vibration, occur as a result of a catastrophic event, regardless of the original condition of the insulation. There is no way to anticipate their occurrence. Such events include:

- Improper electrical winding connections during manufacture or repair, which lead to extremely high circulating currents
- A loose metal object in the machine after manufacture or maintenance, which then cuts through the stator or rotor insulation
- Operating errors such as out-of phase synchronization on a synchronous motor or generator, or inadvertent shutting down of the cooling system (if present)

Electrical Insulation for Rotating Machines. By Stone, Boulter, Culbert, and Dhirani
ISBN 0-471-44506-1 © 2004 Institute of Electrical and Electronics Engineers

Although such failures usually occur soon after the machine has been put into service or returned to service following maintenance, most machine windings do not fail as a result of a catastrophic event. Rather, failure most often is the result of gradual deterioration of the insulation, until it no longer has the electrical or mechanical strength to withstand either normal operating stresses or the electrical/mechanical transients that can occur during normal operation. It is usually the insulation that deteriorates rather than the copper or laminated steel, since the mechanical and thermal capabilities of the predominantly organic insulation systems are far inferior to those of copper, aluminum, or steel.

The failure process, which is a term we will use interchangeably with deterioration process or aging process, is usually slow. Manufacturers have designed most industrial and utility motors and generators to operate for 10 to 40 years before a rewind is needed. Thus, the failure process should take at least this long.

The date on which a failure will occur usually cannot be predicted in the case of age-related failures, because two independent conditions need to occur before failure happens. The first is the reduction in electrical or mechanical capability due to aging. Chapters 8 to 10 will discuss how each deterioration process causes this aging. However, the common insulation systems in use in stator and rotor windings have an amazing ability to continue to operate even though they may be severely deteriorated. Thus, the second condition that is needed before failure is the occurrence of a mechanical or electrical transient. Such transients include:

- A lightning strike to the power system that causes a fast-risetime, high-voltage "spike" that breaks down the stator turn or ground insulation.

- A phase-to-ground fault in the power supply system. Such a fault can give rise to an overvoltage of about twice the normal rated voltage [7.1], which can breakdown aged insulation.

- A hipot test done during maintenance (see Sections 12.2, 12.4, and 12.12), which also applies a larger than normal voltage to the insulation.

- A power supply (or power system) fault that causes a sudden surge in current to go through the stator (and by induction, the rotor), which in turn gives rise to larger than normal mechanical force acting on the windings (Equation 1.3). This sudden burst in mechanical force can crack aged insulation, especially in the end-winding.

- Motor turn-on or out-of-phase synchronization (synchronous machines only), which cause a very high current to pass through the stator winding and rotor. In the stator and in induction motor rotors, this current will cause a large mechanical force that can crack insulation or conductors. In synchronous rotors, there will be high negative sequence current flowing on the rotor surface, leading to rapid overheating and possible melting. Similarly, if only two of the three phase poles close in a circuit breaker, very high current will flow, with similar consequences.

- Operator errors such as forgetting to close field breakers, or opening them under load, will cause very high voltages or currents in the machine. Similarly, turning off or failing to turn on cooling water to the machine, or blocking ventilation channels, can lead to melting of insulation, especially if it has already aged.

The important point to remember is that a brand new winding can often withstand the above events because it has sufficient electrical and mechanical strength. Indeed, as discussed in Sections 12.2, 12.4, and 12.12, manufacturers design new windings to be able to withstand the likely voltage transients and most of the likely current transients. However, as the insula-

Table 7.1. Motor Component Responsible for Failure

Component	Percent of Motor Failures
Bearings	41
Stator	37
Rotor	10
Accessories	12

tion ages, it has less ability to withstand the transient. Eventually, sufficient deterioration will occur so that a transient that could be tolerated in the past now causes failure.

In consequence, the actual date of a machine failure is not only determined by the condition of the windings, but also depends on when a transient will occur. The day when a transient occurs depends on when the operator makes his/her mistake, or when something goes wrong elsewhere in the power system. Thus, it is not possible to predict the time of failure or remaining life in, say, months, because it is not possible to predict when some other external event will occur. We believe that only "risk of failure" can be determined, assuming an event occurs. This theme will be repeated in Chapter 11.

7.1.1 Relative Failure Rates of Components

The main components of a motor or generator are the stator winding, rotor winding, and bearings that support the rotor. Some surveys have been undertaken to find which of these components are responsible for the most failures. By far the most comprehensive survey was a study funded by EPRI on large induction motors in the 1980s [7.2, 7.3]. The study surveyed 7500 motors, and the authors took care to identify the root causes of failures. (In many cases, plant personnel know only that a ground fault relay tripped and assume that the stator winding failed, or plant personnel see a failed bearing and do not realize that the root cause was a rotor problem.) Table 7.1 shows that 37% of the failures were attributed to stator winding problems, and 10% of the failures were attributed to induction motor rotor failures [7.2]. Therefore, windings are a major cause of motor problems.*

A similar but less rigorous failure study was carried out on oil platform motors in the North Sea [7.4]. The stator and induction rotor failure rate as a percentage of all failures was found to be 25% and 6%, respectively. The lower winding failure rates in this study are attributed to the fact that other devices, external to the motor, were included as causes of failures, whereas in the EPRI study, they were excluded.

We are not aware of similar comprehensive studies of generator failure causes. The outage statistics of the North American Electric Reliability Council are not sufficiently detailed to determine the component that is at fault. However, an IEEE working group published data for hydrogenerator failures [7.5]. In that study, stator winding insulation problems caused about 40% of the outages. Rotors caused very few problems.

Published component failure rates for hydrogen-cooled machines seem to be almost nonexistent. However, it is our belief that, contrary to the case for air-cooled machines, rotors are a greater source of failures in hydrogen-cooled machines than stator windings.

---

*A greater number of problems associated with bearings are possible when a motor or generator is first installed and operated. Presumably, this is caused by the difficulty in balancing the motor, or with the driven load. However, once the machine is running smoothly and the number of operating hours increases, the percentage of problems changes as shown in Table 7.1.

7.1.2 Factors Affecting Failure Mechanism Predominance

Since there are many failure mechanisms, it is reasonable to ask what determines which mechanism will actually cause failure in a stator or rotor winding. We believe that there are four main factors that determine the process:

1. Design of the windings
2. Quality of winding manufacture
3. Operating environment of the machine
4. Past maintenance

The first two factors are primarily the domain of the motor and generator manufacturer, whereas the latter two are the domain of the machine user.

Winding Design. The manufacturer is responsible for the machine design. For example, in a stator winding, they set the average electrical stress across the insulation by determining the insulation thickness and the operating voltage. The winding temperatures are set by, among other things, the conductor cross-sectional area and the type of cooling system. The mechanical design is determined by the type of end-winding bracing system and slot wedging/support system. The manufacturer determines what stress levels they are comfortable with based on material properties, aging tests (see Chapter 2), and, of course, past successful experience.

However, the designer must make many compromises when taking into consideration the cost of materials and the labor needed to manufacture the winding. Reducing one type of design stress will often cause another stress to increase. For example, in a form-wound stator, end-winding temperature can be reduced by limiting the amount of blocking material used between adjacent coils. The consequence is that the end-winding will not be as well supported and more likely to suffer from endwinding vibration. Similarly, the groundwall insulation thickness can be decreased in an indirectly cooled winding, which means it will be easier to transmit the heat from the copper conductors to the stator core, consequently lowering the stator winding temperature. However, this same decrease in the insulation thickness will increase the electric stress and, thus, the risk of failure due to electrical aging of the insulation. The machine designer must make these trade-offs. Unfortunately, this results in some stresses being higher than others, and this will favor one failure process over another.

Quality of Manufacture. There are many examples of poor quality leading to specific failure mechanisms being more likely than others. In an induction motor rotor, if the rotor bars are made slightly larger than specified, both the bars and the laminations will be damaged as the bars are driven into the slots. If the insulation on a high-voltage motor stator coil is poorly impregnated, partial discharges will occur that may eventually bore through the insulation. If the bracing and lashings are misapplied or even missing in a stator end-winding, then the magnetic forces during operation or on motor start-up, will lead to vibration and gradual insulation abrasion. There are many more examples. The concept is that one or more failure mechanisms may predominate, depending on what quality deficiency occurs.

Operating Environment. The plant architect/engineer and/or the user determine the operating circumstances for a motor or generator. The operating environment is partly due to the application and partly due to the physical environment. Obviously, if a motor is driving too large a load, the stator windings may overheat, causing thermal failure. If a motor or generator is subject to frequent stops and starts, then some mechanisms such as thermal cycling and surge-induced failure may be more likely. If a synchronous machine is required to supply excessive reactive power, then thermal deterioration due to the increased magnetic flux may occur in the stator winding at the ends of the core. In these situations, the wrong machine is in place for the application. It is interesting to note that the EPRI motor failure survey found that a large percentage of failures resulted from misapplication [7.2].

The physical environment that can affect winding failure mechanisms can include:

- Abrasive particles in the cooling air in open ventilated machines
- Presence of oil, moisture, or salty air
- Presence of chemicals in open ventilated machines
- Altitude in air-cooled, high-voltage machines (see Section 1.4.4, and recall that air pressure changes with altitude)
- Ambient temperature
- Ambient humidity

If any of these environmental factors are extreme, then one or more failure mechanisms will become more likely.

Past Maintenance. If a user has decided not to clean a winding in an oil-rich environment, then the oil may act as a lubricant and increase the chance of winding-vibration-related failures. Similarly, if the oil traps dirt and allows it to accumulate, then cooling passages may become blocked. Consequently, the rotor and stator may suffer thermal-induced failure, unless the windings and cooling air passageways are cleaned. Many other examples are possible. If timely corrective procedures, such as cleaning, rewedging, dip/bake, are not performed, then some failure mechanisms are more likely to occur.

7.2 FACTORS AFFECTING REPAIR DECISIONS

For any particular failure process, there are usually a number of possible repairs. Each repair method will have an associated:

- Cost in parts and labor
- Time to complete (setting the outage/turnaround time)
- Ability to completely restore the winding or just slow down the failure process
- Probability that the repair will accomplish desired result (i.e., reverse, stop, or slow down a failure mechanism)

What repair the user will select often depends complex economic/engineering evaluation that assesses, among other things:

- Criticality of the machine to the plant. If it fails in service, will a process shut down, or will there be no impact in the short term. If there is a large cost impact, then more expensive (and possibly more effective) repairs may be justified.
- Availability of spare machines. If a spare is available, and temporary stopping of the motor or generator is acceptable, then no repair (i.e., replace on failure) or inexpensive repair (probably less effective) may be the best option.

- Cost of the machine. Many small motors (say, less than 50 kW or so) may be so inexpensive in comparison to, say, a stator winding clean/dip and bake, that replacement is cost effective.

For large machines that are critical to plant production, an extensive cost/benefit analysis may be carried out to determine the best repair method. EPRI is funding the development of software tools to help owners of large turbine generators decide the optimum repair options [7.6, 7.7].

7.3   CUTTING OUT STATOR COILS AFTER FAILURE

If an in-service failure occurs, the user has three possible options to return the plant to operation:

1. Replace the machine
2. Rewind the failed component (rotor or stator)
3. Repair the fault

The second option is discussed in the next section. The third option depends on the nature of the failure mechanism and is discussed in Chapters 8, 9, and 10.

For the case of form-wound stator windings, there is a fourth option to return the machine back to service quickly: cutting out a coil (or bar), using a half-coil splice, and/or cutting out entire circuit parallels. In each case, the purpose is to deenergize the coil or bar that has had a ground fault. By isolating the grounded coil or bar from the rest of the stator circuit, the machine can often be put back into service without actually fixing the failed coil. Depending on the operating conditions, and how many coils are cut out, the machine can be returned to service quickly at partial or sometimes full load. If the coil failed because of a local flaw, then the rest of the winding may run its normal life. If the coil that failed was just the first of many because of a widespread deterioration mechanism, then cutting out a coil may allow the machine to be put back into service long enough to order and install a new winding.

In three-phase motors and generators, cutting out a coil or a parallel may give rise to negative sequence currents or other circulating currents, resulting in hot spots in specific areas of the stator. Sometimes, to reduce these currents, it is necessary to cut out more than one coil to balance impedances. The total machine derating that may be necessary to prevent overheating will depend on the total number of coils/parallels removed. The procedures for calculating the derating, determining which coils to cut out, as well as the mechanics involved, are described in detail in an EPRI publication [7.8]. EASA also has published a technical manual describing the process for their member motor and generator repair shops [7.9]. In addition, machine manufacturers and many consultants have the required knowledge.

7.4   REWINDING

If a stator or rotor winding has been destroyed as a result of a failure, or the deterioration is widespread, then a rewind may be the best option. Rewinds are also performed if the machine is to be uprated, because modern insulation systems need less insulation thickness, allowing more copper to be inserted in existing slots. Also, newer materials may allow the machine to be operated at higher temperatures.

If the rewind is needed because the original winding failed prematurely, then it is important to determine the root cause of the premature failure. With an understanding of the failure mechanism, it is often possible to change the design of the winding to reduce the risk of the same failure process happening again. For example, many older motor stator designs tend to run hotter than needed, because the amount of copper in the slot or the arrangement, number, and size of the strands are not optimal. With a suitable redesign, a new winding can be fit in the same slot and run 10–25°C cooler. This will reduce the risk of thermal failure. Similarly, if the subconductors of a round rotor were initially not insulated (see Section 9.1.3), "copper dusting" could lead to many turn faults. This can be avoided in a rewind by making suitable design changes, for example, insulate the subconductors from one another.

A good rewind specification is important if the user is to ensure that a long life is obtained from the rewound rotor or stator [7.10]. The specification should cover, as a minimum, the following:

- Voltage, current, and temperature ratings
- A winding in accordance with NEMA MG1 or IEC 60034 specifications for new windings
- Minimum acceptable increase in core loss if a global VPI stator core is to be reused
- For large stator windings, possible testing such as specified in IEEE 1043 and/or 1310
- Quality control tests during manufacture on dimensional tolerances
- Electrical and mechanical tests during manufacture to ensure there are no shorts
- Chemical testing during manufacture to ensure the bonding resins are within specification
- On complete windings, insulation resistance, surge comparison, power factor tip-up, and other tests described in Chapters 12 and 15
- Require a quality assurance process be in place, such as ISO 9000:2000.

There are several models for stator and rotor specifications in the literature, although these are usually intended for large motors or generators [7.11–7.13]. For those without experience in writing a rewind specification, it is worthwhile to hire an experienced consultant to help prepare the specification if the machine is large enough or critical enough. If one just requires that the rewind meet macroscopic requirements such as voltage, current, load, and speed, then, in a competitive bid evaluation, it is likely that a long rewind life will not result.

REFERENCES

7.1   J. Duncan Glover and M. Sarnia, *Power System Analysis and Design,* PWS Publishing, Boston, 1994.

7.2   E. P. Cornell et al., *Improved Motors for Utility Applications,* Volumes 1 and 2, EPRI Report EL 2678, Oct. 1982.

7.3   P. F. Albrecht et al., "Assessment of the Reliability of Motors in Utility Applications—Updated," *IEEE Trans. EC,* March 1986, 39–46.

7.4   O. V. Thorsen and M. Dalva, "Failure Identification and Analysis for HV Induction Motors in the Petrochemical Industry," *IEEE Trans. IA,* July 1999, 810–818.

7.5   D. L. Evans et al., "IEEE Working Group Report on Problems with Hydrogenerator Thermoset Stator Windings," *IEEE Trans. PAS,* March 1979, 536–542.

7.6   J. Kapler et al., "Software for Cost Assessment of Maintenance Alternatives on Generators with Stator Winding Water Leaks," paper presented at EPRI Conference on Maintaining the Integrity of Water Cooled Generators, November 1996, Tampa.

7.7   J. Kapler and J. Stein, "Probability of Failure Assessment on Generators with Extended Periods Between Maintenance Outages" paper presented at EPRI Conference on Utility Generator Predictive Maintenance, December 1998, Phoenix.

7.8   (a) R. G. Rhudy et al., EPRI EL-4059, Temporary Operation of Motors with Cut-Out Coils, June 1985. (b) R. G. Rhudy et al., EPRI EL-4983, Synchronous Machine Operation with Cutout Coils, January 1987.

7.9   Electrical Apparatus Service Association, *Technical Manual,* 1996.

7.10   J. F. Lyles et al., "Parameters Required to Maximize a Thermoset Hydro-Generator Stator Winding Life, Parts 1 and 2," *IEEE Trans. EC,* September 1994, 620–635.

7.11   B. Lonnecker, "A Guide for Stator Winding Specifications," paper presented at Doble Client Conference, Boston, April 1997.

7.12   IEEE Standard 1068-1990, "IEEE Recommended Practice for the Repair and Rewinding of Motors for the Petroleum and Chemical Industry."

7.13   J. F. Lyles, "Procedure and Experience with Thermoset Stator Rewinds of Hydraulic Generators," in *Proceedings of IEEE Electrical Insulation Conference,* Boston, October 1995, pp. 244–254.

CHAPTER 8

STATOR FAILURE MECHANISMS AND REPAIR

This chapter presents the main aging and failure mechanisms of stator windings, as well as the options associated with each mechanism for repairing the stator or altering its operation to extend winding life. More extensive repairs such as cutting out a coil or rewinding were discussed briefly in Chapter 7. The only failure mechanisms discussed here are those that are due to gradual aging of the winding. Rapidly occurring catastrophic events such as out-of-phase synchronization, improper manufacturing that causes immediate failure, or large objects falling into the machine that destroy the winding are not presented here.

Some of the failure mechanisms will only occur on form-wound stators, and some only in random-wound stators. However, most of the failure processes can occur in either type of stator. Thus, each section will discuss the relevance of the failure process for both random- and form-wound stators.

The symptoms for each failure mechanism are also described. These symptoms are observed with a visual examination of the winding (but no dissection of coils), and with some of the diagnostic tests described in Chapter 12.

8.1 THERMAL DETERIORATION

Thermal deterioration occurs on both form-wound and random-wound stators. It is probably one of the most common reasons stator windings fail, especially if the machine is air-cooled.

8.1.1 General Process

Thermal aging can occur through a variety of processes, depending on the nature of the insulation (thermoset or thermoplastic) and the operating environment (air or hydrogen).

Electrical Insulation for Rotating Machines. By Stone, Boulter, Culbert, and Dhirani
ISBN 0-471-44506-1 © 2004 Institute of Electrical and Electronics Engineers

In air-cooled machines, where the insulation is a thermoset material or a film on modern magnet wire, thermal deterioration is essentially an oxidation chemical reaction; that is, at sufficiently high temperatures, the chemical bonds within the organic parts of the insulation occasionally break due to the thermally induced vibration of the chemical bonds. When bond "scission" occurs, oxygen often attaches to the broken bonds. The result is a shorter and weaker polymer chain. Macroscopically, the insulation is more brittle, has lower mechanical strength and less capability to bond the tape layers together.

For the magnet wire (winding wire) in random-wound stators, brittle insulation resulting from thermal aging is easily cracked as the copper conductors move under magnetic forces during start-up or normal operation. The aged insulation can also easily peel off the conductor. Both mechanisms can lead to magnet wire insulation failure due to abrasion. In both cases, turn shorts may result, which rapidly lead to local overheating at the site of a short, melting the copper and any other nearby insulation, eventually causing a ground fault (Section 1.4.2).

In addition, for form-wound stators, the reduced bonding strength between strands and ground insulation allows the mica tape layers to start separating, resulting in delamination. At this point, two processes can lead to failure:

1. The copper conductors are no longer tightly held together. Eventually the conductors may start vibrating against one another because of magnetically induced forces. Strand shorts or, more seriously, turn faults, if present, will occur as a result of insulation abrasion. This will create local hot spots, which then decompose the ground insulation, leading to a ground fault. Since a turn fault almost always leads very soon to a ground fault, multiturn coils may experience failure sooner than Roebel bars since only 1 or 2 layers of turn insulation need to fail.

2. If the coils are operating at a voltage higher than about 3 kV, partial discharge may occur in the delaminated insulation. Depending on the PD resistance of the mica tape layers, the PD will eventually erode a hole through the ground insulation or the turn insulation, causing a ground fault.

In older form-wound coils and bars made with thermoplastic bonding materials such as asphalt, there may be an additional failure process. As the asphalt's temperature rises above some critical threshold, usually 70 to 100°C, the asphalt softens and may actually flow. The result is less asphalt between the tape layers. This delamination leads to conductor vibration and/or partial discharge, as mentioned above.

Rapidity of deterioration depends on the insulation material and the operating temperature of the insulation. As discussed in Section 2.3, each insulation material is rated for its thermal capability. Most windings made before about 1970 used Class B materials, which implies an average life of 20,000 hours (about 3 years) at an insulation temperature of 130°C. Most modern form-wound insulating materials are rated Class F, i.e., will have an average life of 3 years at 155°C. Modern random-wound machines typically have Class F or H insulation systems that can operate at about 155°C and 180°C, respectively, for 3 years before becoming brittle. Clearly, the better the thermal capability of the insulation, the longer it will last at a given operating temperature.

Since the process is an oxidation chemical reaction, the higher the temperature, the faster will be the chemical reaction, and, thus, the shorter the time to degrade the insulation. As discussed in Sections 2.1 and 2.3, experience shows that for every 10°C rise in operating temperature, the thermal life of the insulation will be reduced by about half. If the insulation is operated at its rated class temperature (say 155°C for Class F windings), then it can be expected that significant deterioration will start occurring after a few years in service. If the insulation is operated at a temperature that is 30°C lower, then about eight times longer life can be expected—about 25 to 30 years. Therefore, the higher the operating temperature, the faster the deterioration process will be. Thermal deterioration can result in failure after just a few months or may take many decades, depending on the insulation materials and the operating temperature.

Generally, thermal deterioration is not too likely in a hydrogen-cooled stator winding, unless there has been a severe operating problem (such as interruption of the stator cooling water). The lack of oxygen slows down the thermal deterioration process. In addition, most manufacturers have designed the machine such that the insulation is usually operating at a much lower temperature than its rated thermal capability.

The maximum insulation temperature is not directly measured in most machines. The standard method used to monitor the temperature in an indirectly cooled form-wound stator is an RTD (resistance-temperature detector) or TC (thermocouple) embedded in the stator slot, between the top and bottom bars/coils. In addition, some large steam turbine generators may be fitted with outlet hose thermocouples to measure the temperature of water flowing out of conductor bars. For an indirectly cooled machine (all motors and most generators under a few hundred megawatts), the RTD or TC temperature will be 5 to 20°C cooler than the insulation at the copper conductors. Therefore, thermal deterioration may happen fastest at the copper, and slowest on the coil surface, unless the thickness of the groundwall insulation reduces the diffusion rate of oxygen to the strands. At the temperatures discussed above, the insulation temperature refers to the insulation temperature at the hottest spot—usually the temperature adjacent to the copper. In random-wound machines, embedded temperature sensors are rare and, thus, the true operating temperature of the stator winding insulation is not accurately known except when a factory load test is followed by a temperature rise measurement (inferred from the copper resistance).

8.1.2 Root Causes

Thermal deterioration is caused by operation at high temperature. There are a number of reasons why high winding temperatures may occur:

- Overload operation, when the load on the motor or generator is greater than what it was designed for. In general, the temperature will increase with the square of the stator current.
- Poor design. For example: the conductors are too small, there are high circulating currents from inadequate transpositions (in form-wound machines); too large a conductor strand copper cross section, giving rise to eddy current losses; unbalanced number of turns in each circuit, leading to negative sequence currents (this also occurs if coils are cut out); inadequate cooling system, etc.
- Poor manufacture. For example, due to strand shorts (in form-wound stators), core lamination shorts, partially blocked ventilation paths because coils are installed too close together in the end-winding.
- For induction motors, insufficient time between motor starts. Each time a motor is started, there is an inrush current about five to six times larger than in normal operation. This inrush current creates an additional I^2R loss. The resulting heat takes time to dissipate. If a motor is restarted shortly after an initial start, the stator (and rotor) winding may still be hot from the first start, and the second start will add to the temperature.

This is particularly true for high-inertia-drive applications such as fan drives, in which acceleration times are long and, as a result, stator and rotor temperature increases during a start are high.

- High harmonic currents from the harmonics generated by inverter-fed drives increase conductor losses and stator core losses.
- Negative sequence currents from voltage imbalance on the phase leads. A 3.5% voltage imbalance can lead to a 25% increase in temperature. The phase-to-phase supply voltage imbalance comes from the power system supply (utilities permit fairly high imbalances), or where one phase may have more load than the other two phases in a plant.
- Dirty windings that block ventilation ducts in the core and fill the space between coils in the end-winding, with the resulting effect of reducing the cooling airflow.
- Dirty, air-locked, or clogged heat exchangers that fail to adequately cool the air (or hydrogen).
- Plugging of a large number of heat exchanger tubes carried out as a temporary solution to leaks that develop in-service or in the presence of significant (>50%) tube wall thinning.
- In direct water-cooled windings, debris or copper oxide in the water channels, obstructing the flow of cooling water.
- Loose coils/bars in the slot, which reduce the conduction of heat from the copper conductors to the core.
- In synchronous machines, operating the machine under-excited. In this case, axial magnetic fluxes are created at the end of the stator core, which induces circulating currents in the stator core ends (and to some extent in the stator conductors), causing local temperature increases at the core ends.
- Too many "dips and bakes," in which the stator is immersed in a liquid resin to tighten up the coils in the slot and repair abraded insulation. This process tends to close up ventilation paths through the core and end-windings of form-wound stators.

8.1.3 Symptoms

Visual signs of thermal deterioration will depend on the type of insulation and winding. Random-wound machines will show evidence of cracked or peeling magnet wire films, as well as discolored or brittle slot liners and bonding varnish. Thermoplastic form-wound stators will have puffy insulation, and the insulation will sound "hollow" when tapped with a hammer. Asphalt may also be oozing out from the groundwall over and above the initial release of excess asphalt that may be present in the form of hardened deposits. Thermoset form-wound stators will also sound hollow when tapped, but this only occurs after severe thermal deterioration. Indirectly cooled form-wound coils and bars will only show surface scorching after severe thermal aging.

In random-wound stators, thermally deteriorated insulation may have a low insulation resistance (Section 12.1) if any of the magnet wire insulation has cracked or peeled away. Also, there may be a low-surge breakdown voltage (Section 12.12). There may also be a small decrease in capacitance and increase in dissipation factor over time (Sections 12.5 and 12.8).

In form-wound stators, thermal deterioration is accompanied by a decrease in capacitance over time, and an increase in power factor and partial discharge over time (Sections 12.10 and 13.4). If thermovision monitoring (Section 13.1.4) has been done over the years, then an increase in machine surface temperature may be evident (under the same operating and ambient conditions). Similarly, if stator winding temperature monitoring is present (Section 13.1.1), as the winding deteriorates, the temperature will increase over time under the same operating and ambient conditions. This occurs because thermal deterioration often increases the thermal impedance between the copper and stator core (due to delamination and/or the reduction in bond between the coils and the core).

If the cooling system is becoming blocked, then the machine temperature will also increase. The location (e.g., core, winding) of the temperature increase will depend on the nature of the blockage, such as an air lock in the hydrogen cooler, debris in the stator cooling water system, or whether the temperature increase is accompanied by an increase in the differential pressure across the affected cooling system.

8.1.4 Remedies

The option selected will depend on the root cause of the overheating. Cleaning a winding to improve airflow will not solve a root problem such as shorted strands leading to circulating currents. Thus, before the repair process is selected, the root cause must be established. Note that thermal deterioration of the insulation itself is not reversible (except if one considers a rewind). All that can be done with a suitable repair or change in operation is to slow the rate of thermal deterioration and, thus, extend life. It is important to note that several utilities have developed epoxy injection methods to restore deteriorated groundwall insulation when faced with the high costs of rewinding [8.1]. Although this novel approach may be suitable in particular circumstances, for example, micafolium windings, the long-term effectiveness of this procedure is not known.

Remedies for overheating include:

- Cleaning the winding and heat exchangers, venting air-locked heat exchangers to improve cooling air (or hydrogen) flow, and more efficiently extracting heat from the core and windings.
- Rewedge/sidepack and/or perform another VPI or dip and bake to improve thermal contact between the stator coils and the core. Note that the latter may increase the winding temperature if the coils are already tight, since the ventilation passages will be slightly more restricted by the addition of the extra insulation film.
- Ensure that the voltages on each phase are within 1% of each other.
- Upgrade heat exchangers (if fitted).
- Install a chiller unit on the cooling air.
- For motors, install protection relaying, which prevents frequent restarts.
- Reduce the maximum permissible load.
- Adjust the power factor in synchronous machines to unity to reduce the stator current.

8.2 THERMAL CYCLING

This mechanism, also called load cycling, is most likely to occur in machines with long stator cores (typically more than 2 m), and in form-wound coils/bars in machines experiencing many rapid starts and stops or rapidly changing loads. Large hydrogenerators subject to peaking duty, air-cooled large gas turbine generators, and pump storage machines are most likely to experience this problem. Random-wound stators are very unlikely to experience thermal cycling related failure.

8.2.1 General Process

There are three variations of this mechanism; which one occurs depends on the type of groundwall insulation (thermoplastic or thermoset) and if the stator has been global VPI'd. The thermal cycling tests described in Section 2.5 model these processes. In all cases below, deterioration will be more rapid with faster load changes, longer stator cores, higher operating temperatures, and/or more frequent load changes. Past experience indicates that the process will normally take more than 10 years to cause failure, but some failures have occurred in a shorter time.

Girth Cracking/Tape Separation. Girth cracking and tape separation are special variations of thermal cycle deterioration in thermoplastic insulation systems such as asphalt mica flake. It has only rarely occurred in generators with core lengths of less than 3 meters. The fresh asphalt and asphalt-modified drying oil varnishes used with this insulation type are either completely thermoplastic or have a low glass transition point, which results in extreme softness and very low physical strength at normal service temperatures when machines first go into service. When asphaltic insulation systems are used in air-cooled generators, service temperature leads to hardening of the asphalt by oxidation to an eventually infusible state, while the drying oil varnishes gradually increase in glass transition temperature and get stronger with time. When this type of generator is used in base load applications, there is very little thermal cycling and service may continue for decades without girth cracks developing.

If a new asphalt–mica-flake winding is put into peaking or other cyclic operation and is over 3 meters in core length, the risk of girth cracks increases. Generators that have run in base load duty for years before being switched to cyclic duty generally do not develop girth cracks.

From the late 1930s through the 1950s, hydrogen-cooled generators insulated with asphaltic mica flake systems were produced. The hydrogen atmosphere inhibited oxidation and hardening of the insulation. The greater length, associated with increased ratings introduced during that period, led to more tensile stress on the insulation from differences in actual thermal expansion and relative motion between the copper strand package and the groundwall. This led to a number of generator failures due to tape separation during the post-World War II period and was the main cause for the development of thermoset stator insulations. Again, conversion of base load generators to cyclic operation could turn a satisfactory generator into one with girth crack problems.

The mechanisms of girth cracking are complex. When a generator with thermoplastic stator insulation is run up to full load in a few minutes, the I^2R loss in the copper rapidly increases its temperature. The high coefficient of thermal expansion of the copper causes it to quickly expand in the axial direction, whereas the thermal lag in the insulation and its contact with the still cool core restrains the insulation from moving with the copper. Initially, the insulation is bonded to the copper and the part of the groundwall outside of the slots easily moves with the copper. The slot section insulation cannot move as readily, since one of the characteristics of asphaltic and some other thermoplastic insulation systems during service is that the insulation expands into the cooling gas ducts and is therefore locked in place. The result is an axial tensile stress within the insulation that may be sufficient to cause tape layers to begin to separate in the regions just outside the slots or within the outer sections of the slots.

The insulation also expands due to thermal softening outside the slots, sometimes causing a bulge to develop that restricts insulation motion with the copper, and moves back into the slots when generator load levels fall and the copper cools.

In a single thermal cycle, as in base load applications, problems are unlikely. However, with many repeating load changes and the related thermal cycling, the insulation may circumferentially separate, following the path of the tape layers as they are wound around the copper package. Asphaltic insulations were always taped by hand. Normally, right-handed persons apply the tape from the left end of the coil or bar as they face it, taping toward the right end. The tape layers are in the form of spirals around the copper and it is these spirals that are stretched out by the girth crack process on the right end of the bar, as taped. The insulation thickness is reduced a small amount each thermal cycle and may eventually go all the way to the copper.

Mica splittings contributed to the development of girth cracks because of the ease of delamination within their crystal structure. Mica paper has much smaller and thinner flakelets and, generally, a thermoset resin binder. An insulation system based on these materials is much stronger and resistant to delamination and tape separation. The tape backers for asphaltic systems were generally a kind of tissue paper, which is quite weak in tension. Modern epoxy mica paper tape uses a much stronger glass cloth backer layer. The result is that girth cracks in modern thermoset mica paper groundwall insulation are unknown, although cracks may sometimes occur in them due to their rigidity when exposed to magnetically induced end-winding motions that are not properly blocked and braced.

Conventional Thermoset Deterioration. With conventional epoxy–mica insulated bars or coils that are fully cured before insertion in the stator, a different thermal cycling process can occur. As above, if the stator current goes from no load to full load in less than a few minutes, the copper temperature rapidly increases, causing an axial expansion of the copper. At the same time, the epoxy–mica groundwall is at a much lower temperature, since it takes some minutes for the heat from the copper to conduct through the groundwall to the stator core. The consequence is that as the temperature increases, the copper expands, but the groundwall expands less. This creates an axial shear stress between the copper and the groundwall.* Although a single thermal cycle is not likely to break the bond between the copper and the groundwall, many thermal cycles will fatigue the bond. Eventually, the copper will break away from the groundwall (Figure 8.1). In some cases, the gap thus created between the copper and the groundwall will enable the copper conductors to vibrate relative to one another, abrading the strand or turn insulation. In addition, if the bar or coil is operating above about 3 kV in air, then partial discharge may occur in the air space. The PD may eventually bore a hole through the groundwall, leading to a ground fault.

Large Global VPI Windings. Since the early 1990s, some generator manufacturers have built air-cooled turbine generators rated up to about 200 MVA, using much the same global VPI process as has been used for decades in motors. In this manufacturing process, the copper is bonded to the groundwall, and the groundwall is bonded to the stator core. At full power, the bars expand beyond the slot. If the load is rapidly reduced, the copper will cool and shrink, as will the groundwall insulation, when the winding cools. However, the stator core retains its dimensions since the difference in stator core temperature between no load and full load is relatively small. Thus, not only is there a shear stress between the copper and the

---

*The greatest shear stress occurs when the windings are cool. When the coils are made, they are cured at a high temperature, and there will be no stress between the groundwall and the copper. As the coil cools, the greater shrinkage of the insulation creates the shear stress.

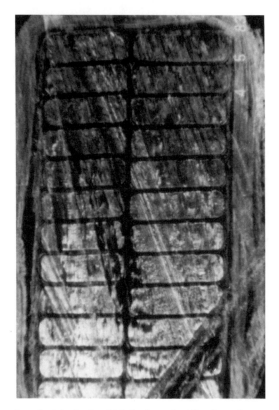

Figure 8.1. Cross section of a stator bar in which the thermal cycling forces have caused the groundwall to shear away from the copper. The black gap around the copper is where the shear occurred.

groundwall, but there is also a shear stress between the surface of the groundwall insulation and the stator core. After many fast load cycles, the coil may break away from the core under the shear stress developed. Once this occurs, partial discharges will occur between the surface of the coil and the core in coils operating at high voltage. In addition, the space that is created between the coil and the core may permit the coil to vibrate, abrading the ground insulation. The latter leads to the slot discharge mechanisms described in Sections 8.4 and 8.5 and a much shorter time to failure. This failure process is known to have caused stator failure in as little as four years.

Some manufacturers have avoided this problem in large global VPI stators by installing a slip plane between the coil and the core. This allows the bar to break away from the core in a controlled manner after thermal cycling. Advanced design is needed to ensure that any air gap created at the slip plane does not have enough electrical stress across it to result in partial discharge by using mica splittings between two layers of semiconductive tape. [8.2].

8.2.2 Root Causes

If any of the thermal cycling mechanisms are occurring, then the root causes of failure are a combination of:

- Too fast a load change for the design.
- Operation at too high a stator winding temperature, since the higher the temperature, the lower the bonding strength; and the lower the bonding strength of the resins/asphalt, the greater is the chance that the interfaces will shear.
- Inadequate design of the insulation system to withstand cyclic axial shear stresses.

8.2.3 Symptoms

In asphaltic windings, the visual symptoms of failure will include puffy insulation at the core ventilation ducts and just outside the slots. If tapped, the insulation will sound hollow. There will also be circumferential cracks in the groundwall just outside the slots.

In modern conventional epoxy–mica windings, there is usually little visual evidence of the problem, although the groundwall may sound hollow when tapped. For global VPI stators, there will be signs of abrasion (powdering/greasing) when looking down the stator ventilation ducts, as well as signs (light-colored powders or other discoloration) of surface PD activity in phase-end coils.

The power factor tip-up and on- or off-line partial discharge tests are the most sensitive tests used to find this problem, since it will be accompanied by intense PD activity. Asphaltic windings may also exhibit low insulation resistance and polarization index if the girth cracks have penetrated to a significant percentage of the groundwall thickness.

8.2.4 Remedies

The effects of thermal cycling are irreversible; thus, the stator winding cannot be restored to new condition if the process has already caused significant deterioration. However, there are several methods available to slow the process:

- Slow the rate of increase and decrease in power. This will let the temperatures of the core, copper, and insulation be in as much thermal equilibrium as possible, reducing the differences in the shrinkage or expansion. The time needed to achieve near-thermal equilibrium can be determined by measuring how long it takes the stator temperature, as measured by the embedded RTDs or TCs, to stabilize after a sudden load increase or decrease. Typically this can range from 10–30 minutes, depending on size and cooling method.
- Reduce the maximum operating winding temperature by reducing maximum permissible load. At lower temperatures, the epoxies, polyesters, and asphalts have greater bonding strength to resist the shear stresses.
- Operate generators near unity power factor to reduce the load current and hence the resistive losses.
- For global VPI stators, inject carbon-loaded (i.e., conductive) epoxy or silicon rubber into the slots between the coils and the core. This will prevent the thermal cycling process from developing into the "loose coil/slot discharge" process. Note that the injection is not likely to stop the PD, since experience shows it is impossible to completely restore the semiconductive layer using postinstallation injection.
- In thermoplastic systems with girth cracks, some users have been able to fill the cracks with silicon rubber or other compatible resins to improve the electrical strength of the coil insulation.

8.3   INADEQUATE IMPREGNATION OR DIPPING

Most random-wound stators are dipped in a resin or varnish after being wound, to seal the winding against dirt and moisture, improve heat transfer to the core, and hold the windings tight in the slot to avoid abrasion from vibration. For similar reasons, in conventional form-wound coils, the groundwall is impregnated by the resin-rich or VPI processes (Chapter 4). In addition to the benefits described for random-wound machines, the impregnation also prevents partial discharge (PD) activity within the groundwall by preventing air pockets from occurring. Similarly, the global VPI process for form-wound stators is intended to eliminate voids within the groundwall to improve heat transfer, prevent conductor movement, and reduce PD.

Inadequate impregnation causing stator failure is more likely to occur in global VPI stators, since it is more difficult for quality control procedures to find poor impregnation at the time of manufacture. Failure can occur in as short a time as 2 years in stators rated 6 kV or more.

8.3.1   General Process

Poorly impregnated random-wound stators are much more likely to fail due to dirt, pollution, oil, and moisture that can be partly conductive. It is not unusual for the magnet wire insulation to have small pinholes or cracks as a result of the rigors of manufacturing. If the impregnation or dipping cycle is poor, then these imperfections in the magnet wire insulation, combined with a partly conductive contamination, leads to shorted turns and, thus, failure. Similarly, if the turns are not bonded together and to the core, the twice-power frequency magnetic forces due to motor start-up or normal operating currents cause the turns to vibrate. This vibration leads to abrasion of the magnet wire insulation and to turn shorts.

Grossly inadequate impregnation of the groundwall in form-wound coils can lead to higher operating temperatures (since the groundwall thermal conductivity is lower), leading to thermal deterioration and/or conductor vibration, resulting in abrasion. However, if the impregnation is poor but not completely missing, the thermal and abrasion processes are less significant. Instead, if the winding is rated at 6 kV or more, partial discharges can occur in coils connected to the phase terminals. PD will occur in any air pockets resulting from the poor impregnation between the copper and the core, as described in Section 1.4.4. The PD will gradually erode a hole through the insulation. If the winding is made from multiturn coils, and the air pockets are located near the turn insulation (Figure 8.2), then the PD only has to erode the relatively thin turn insulation, leading to a turn fault and, consequently, a ground fault due to the high circulating current. For Roebel bars, the PD must erode through the entire groundwall thickness, which takes much longer.

8.3.2   Root Causes

The possible causes of poor impregnation in random-wound stators are:

- A resin/varnish viscosity that is too thin or too thick, or other problems with the resin/varnish chemistry such as contamination and VPI resin gel particles.
- Processing the stator in a manner not in accordance with resin/varnish supplier recommendations, such as not preheating the stator or varnish/resin, inadequate time in the dip tank, and not completely immersing the stator (if appropriate, some impregnation methods such as trickle impregnation do not require this).
- Use of a magnet wire insulation that is not chemically compatible with the resin/varnish.
- A bake temperature that is too high or too low, or a bake time that is too short.

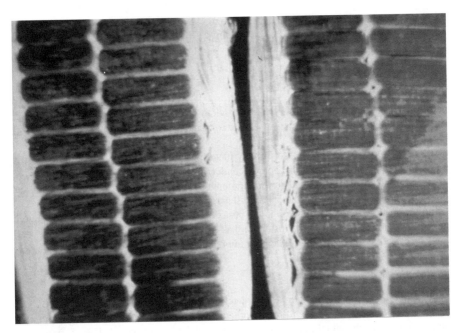

Figure 8.2. Cross section of a 12-turn coil from a 6.6 kV motor stator made using the global VPI process. The dark voids near the copper resulted from poor impregnation with epoxy, just outside of the slot.

In form-wound VPI coils or global VPI stators, the root cause may include:

- Poor groundwall insulation taping, which leaves wrinkles in the tape and, consequently, a void too large to be filled by the resin.
- Improper resin viscosity or chemical contamination.
- Resin that has been through too many VPI cycles and has not been adjusted to restore the proper chemistry specifications and freedom from suspended particles.
- Incompatible resin and insulation tapes.
- Improper preheat and too short a vacuum cycle, or insufficient vacuum, insufficient time in the cure cycle, or using improper curing temperature/pressure.
- Inadequate covering of the stator or coil with resin in impregnation cycle, or not rotating stator so that resin does not drain away from high points.
- Inadequate impregnating pressure and time for the available resin viscosity and insulation thickness.

All of these causes essentially point to poor process control.

In resin-rich coils, poor impregnation may result from:

- Old resin-rich tapes that have already started to cure.
- Improperly stored tapes; i.e., storage at too high a temperature or at too high a humidity.

- Poor taping, resulting in wrinkles that are too large.
- Inadequate pressure or temperature during the cure cycle.
- Poor pressure or pressing.

8.3.3 Symptoms

Poor impregnation is readily observable in random-wound stators, since interstices are not filled with resin, and there is no sheen over the coils and the core. The insulation resistance of the stator may also be too low.

Poor impregnation is difficult to visually identify in form-wound coils/bars because it is most likely to occur adjacent to the copper conductors, far from the surface of the insulation. In severe cases of poor impregnation, the groundwall will sound hollow when tapped. For machines rated 3 kV or above, the partial discharge test (sometimes an over-voltage is needed to test 3.3 and 4.1 kV windings) is the best way to detect poor impregnation. The surge test may also identify if the turn insulation has been poorly impregnated. The PD test will only be effective if the location of the poor impregnation is in or just outside of the slot.

8.3.4 Remedies

A random-wound stator can often be completely restored by a repeat dip and bake if failure has not yet occurred. However, form-wound coils have almost no chance of being properly impregnated once the initial impregnation is done and the stator is cured. This is because the original impregnation will normally block the resin flow to the points near the copper conductors in the second impregnation. Thus, there is no way to extend life by either repairs or operation changes (unless the operating voltage can be reduced or the phase and neutral ends of the windings can be reversed).

8.4 LOOSE COILS IN THE SLOT

This problem is normally associated with form-wound stators using thermoset-type coils or bars, manufactured in a conventional way, i.e., not global VPI. Epoxy bonded insulation is the most susceptible. Loose-coil-induced failure is one of the most likely failure processes in modern large gas and steam turbine generators, as well as hydrogenerators. It is unlikely to occur in global VPI motor stators or thermoplastic windings. Random-wound machines may suffer from this problem if they were poorly impregnated, as described in Section 8.3.

8.4.1 General Process

As described in Section 1.4.3, coils and bars in stator slots are subject to high magnetically induced mechanical forces at twice the power frequency. As the MVA rating of a machine increases, the forces acting on the bars or coils at full power increase proportionally to the square of the current in the bar or coil.

If the bar or coil is held tightly in the slot at full load, these forces have little impact. However, if the coil is not tightly held, and it starts to vibrate, the coil insulation system moves relative to the stator core, primarily in the radial direction (i.e., up and down in the slot). Since the stator core is composed of steel laminations, the serrated edge of the laminations in the slot makes an effective abrasive surface. The coil movement first abrades the semiconductive coating (if present), then the groundwall insulation. Experience shows that once about 30% of the groundwall insulation has been rubbed away, a stator ground fault is likely in both air-cooled and hydrogen-cooled machines, usually in the higher-voltage coils (line end coils).

There are two basic stages for this process, also sometimes referred to as slot discharge. In the first stage, the bar or coil is vibrating but most of the semiconductive coating is intact. In this stage, some contact sparking will occur as the coil or bar semiconductive coating moves away from the grounded stator core and picks up some charge due to the capacitance of the air gap (amounting to a few volts of potential) between the coating and the core. When the semiconductive coating comes back in contact with the core, the stored charge is shorted to ground, causing a spark. This is sometimes referred to as Stage 1 slot discharge.

Stage 2 slot discharge occurs when the semiconductive coating is abraded away and the surface of the coil (at least at the abraded spot) is not grounded, even indirectly. Classic partial discharge will then occur in coils operating at high voltage, since many thousands of volts can build up across the air gap between the core and the exposed groundwall insulation surface. As described in Section 1.4.5, such a high voltage across an air gap leads to electrical breakdown of the air or hydrogen, i.e., PD. The intensity of PD is usually large enough to further accelerate the rate of deterioration of the groundwall insulation.

A good description of the slot discharge process and its variations is found in Reference 8.3. In large machines with high bar forces, failure has occurred in as short a time as 2 years, making loose coils one of the fastest aging mechanisms. The mechanism can occur equally fast in air- or hydrogen-cooled machines. Although the PD activity is usually less in high-pressure hydrogen-cooled machines, the rate of deterioration is similar because the vibration is the principal cause of deterioration, not the PD.

The global VPI process usually keeps the coils and bars tight within the slot, preventing the loose coil failure process. However, if the coils have been poorly impregnated (Section 8.3) and/or are subject to severe load cycling (Section 8.2), the coils and bars have been known to become loose enough to lead to bar vibration and a rapid deterioration rate.

8.4.2 Root Causes

The fundamental cause of this mechanism is the magnetic force. However, coils and bars will not vibrate if they are not at least a little loose in the slot to begin with. With the older thermoplastic insulation systems and, to a lesser degree, the older polyester insulation systems, when the stator current was increased to full load, the heat from the copper losses raised the insulation temperature sufficiently to expand the insulation in the slot. The result was that even if a coil was loose in the slot at low temperatures, it was tightly held in the slot at full load, greatly reducing the chance that the loose coil problem would ensue. Since the coefficient of thermal expansion of epoxy–mica is low compared to asphalt and even polyester, if the coil is loose when it is cold, then it is likely to still be loose when it is hot, when the bar forces are greatest. Consequently, the loose coil problem is most likely to occur in epoxy–mica insulated windings.

Coils and bars may be loose in the slot when they are manufactured because they were installed too loose. Some experts recommend that if only flat side packing and single-part flat wedges are used to constrain the coils or bars in the slot, then there should be no more than about 0.1 mm clearance between the side of the coil and the stator core [8.4]. Initial looseness can be prevented by the use of ripple springs, two-part wedges that can create a positive force down into the slot, incorporation of compressible materials such as silicon rubbers into the slot contents, and/or the global VPI process. Figure 8.3 is a photograph of a coil removed from an air-cooled generator that was loose in the slot due to poor design and installation. About one third of the groundwall was abraded before it failed to ground.

Figure 8.3. Photograph of the side of a coil suffering from insulation abrasion due to loose windings. The abrasion occurred along the length of the coil, since the coil was not installed tightly. The vertical ridges are where the vent ducts occur in the stator core.

Loose coils/bars can also exist in the slot as a result of the following processes that occur during operation of the motor or generator.

- Shrinkage of the insulation. Organic materials tend to shrink as they thermally age. Thus, after a few decades, an initially tight winding may become loose enough to cause abrasion as a result of the gradual shrinkage of the groundwall, sidepacking, and wedges. Note also that in the first year of operation, relatively significant shrinkage can occur as the groundwall insulation completes the curing process during operation. Typically, a modern epoxy–mica insulation system has completed about 95% of its cure during the manufacturing stage. If a new stator winding only has flat side packing and flat wedges for slot support, then it is advisable to check for winding looseness after about one year of operation.

- To combat the problem of gradual shrinkage, many machine manufacturers use ripple springs (warped epoxy–glass composites) under the wedges or as side packing. The springs expand to take up the space created by shrinkage of the other slot components, holding the coils and bars tight. However, the ripple springs themselves gradually lose their ability to hold the bars and coils tight. Some types of ripple springs lose their "spring" in the presence of oil. Oil also acts as a lubricant that facilitates relative movement between the coil and the core. Figure 8.4 shows the side of a stator bar from a large hydrogen-cooled generator, where deteriorated side ripple springs (due to the long-term presence of seal oil in the stator) allowed sufficient coil movement that the glass fibers in the ripple springs themselves abraded the groundwall insulation, resulting in a ground fault.

Figure 8.4. Photograph of a stator bar removed from a large hydrogen-cooled stator. The light-colored arcs are where the side ripple springs abraded the groundwall insulation.

- Wedges can become loose over time, usually due to shrinkage, oil contamination, and/or stator core movement. Unless the coils or bars are forced tightly against the side of the slot, the loose wedges can lead to loose coils or bars in the slot.

8.4.3 Symptoms

Once the rotor of a motor or generator has been removed, it is usually fairly easy to determine if the coils/bars are loose in the slot. If the stator has not been contaminated by oil, when looking down the ventilation ducts in the core, the powder produced by abrasion of the insulation can be seen. Also, in the coils/bars connected to the phase terminal, there will be signs of pitting/burning caused by the PD. If the stator is contaminated by oil, then "greasing," a viscous dark sludge formed from the abrasion products and oil, collects in the ventilation ducts. Often (but not necessarily), the wedges will be loose, yielding a dull thud when tapped with a hammer. If the wedges can be removed, damage to the wedges and sidepacking by the pounding and abrasion is easy to see. A semiconductive resistance test (Section 12.14) will show a high resistance if the semiconductive coating has been destroyed.

Prior to machine disassembly, an on-line partial discharge test (Section 13.4) can be used to determine if the coils are loose in the slot. If the coils are loose, the PD activity will increase dramatically with load. If there is widespread slot discharge, and the machine is air-cooled, then there will be a significant amount of ozone in the cooling air.

8.4.4 Remedies

If loose coils and bars are detected at an early stage in the process, i.e., while it is in the Stage 1 process, then the stator can be restored to new condition by tightening the coils/bars in the slot. One or more of the following can accomplish this:

- Replacing the wedges and sidepacking
- Preferably, replacing flat side packing and flat wedges with ripple springs and/or two-part wedges
- Replacing top ripple springs if they have flattened out
- Injecting carbon-loaded paint, silicon rubber, or epoxy into the slots

In addition, the process can be slowed by operating the machine at reduced load, since this decreases the magnetic forces.

Once Stage 2 slot discharge is occurring, the semiconductive layer has been partially destroyed, at least at some locations. All of the above remedies can be applied and the failure process may be slowed significantly. However, even if the slots are injected with a partly conductive material, experience shows it is impossible to replace all of the destroyed original semiconductive coating. Thus gaps will still remain between the groundwall insulation and the core, resulting in PD. Eventually, this PD may bore a hole through the insulation.

8.5 SEMICONDUCTIVE COATING FAILURE

This failure process involves the deterioration of the semiconductive coating on a stator bar or coil in the slot, in the absence of coil vibration. This mechanism only occurs on form-wound coils and bars that have carbon-loaded paints or tapes as the semiconductive coating; that is, it normally occurs on stators rated 6 kV or more. Since the process involves partial discharges, this problem is far more likely to occur in air-cooled machines. Motors and generators operating at high altitudes, where the air pressure is lower and, consequently, the breakdown strength of air is lower, are more likely to experience this problem. Experience indicates that bars or coils with a paint-based semiconductive coating are more likely to experience this deterioration mechanism as compared to those with a tape-based semiconductive coating.

8.5.1 General Process

As discussed in Section 1.4.5, the semiconducting coating is placed on the surface of high-voltage stator bars and coils to prevent partial discharge. Without the coating, PD would occur in the small gaps that would inevitably be present between the coils and the core. In this failure process, the coating essentially becomes nonconductive by oxidation of the carbon particles, and in some localized areas may be nonexistent. In the coils or bars that are operating at high voltage, PD will occur between the bar/coil and the core. This PD will attack the groundwall insulation. A hole may be bored through the groundwall, resulting in a ground fault. Since all mica-based groundwall insulation is very PD resistant, it can take many decades for this process to progress to failure. Note that this process occurs even if the coils are kept tight in the slot. The mechanism is sometimes called "electrical slot discharge," to differentiate it from the mechanically induced slot discharge described in the previous section.

If the machine is air-cooled, the PD will create ozone. Ozone is a very chemically reactive gas that combines with other gases in air to create nitric acid. This chemical attacks many substances including epoxy, polyester, rubber compounds, and steel. The chemical attack on the materials in the slot often causes a winding that is initially tight in the slot to become loose, leading to insulation abrasion due to coil vibration, as described in the previous section. Unless the coils are retightened, the failure process becomes much more rapid, since abrasion is inherently a faster deterioration process than PD attack.

If the coils/bars are kept tight in the slot, the PD itself may take many decades to cause failure. However, stators with this problem have had to be rewound, not out of fear of imminent failure, but because health organizations have declared the machine to cause an unsafe working environment or the heat exchangers may fail due to corrosion caused by ozone. Most countries regulate the amount of ozone concentration that is allowable in the workplace. An open-ventilated machine that has extensive semiconductive coating deterioration can easily lead to nonacceptable ozone levels (usually around 0.1 parts per million).

8.5.2 Root Causes

Invariably, the root cause of semiconductive coating failure is poor manufacturing procedures, both in making the coating and in testing it regularly to ensure the appropriate coating resistance. It is believed that the process starts when the coating surface resistance is too high in localized areas. This is probably because the carbon black particle density is too low in this area. If a coil with a locally high coating resistance is connected to the phase terminal, then capacitive currents will flow from the copper, through the groundwall insulation, to the stator core via the semiconductive coating. Since the coating is not in direct contact with the core on its top and bottom edges (i.e., the narrow edges in the coil cross section) and at the ventilation ducts, some of the capacitive current must flow parallel to the bar/coil surface. If the resistance is high, small I^2R losses occur where the current is flowing laterally, heating the coating at these localized spots. This local heating, added to the general winding temperature, can lead to oxidation of the semiconductive coating, which further increases the semiconductive coating resistance, exacerbating the problem. In addition, if the resistance is infinite in a spot where currents must flow laterally, then sparking may occur.

A variation of this problem can occur even if the semiconductive coating has a consistent and adequate surface resistance (typically in the range of 500 to 5000 ohms per square). In some manufacturing processes, small air pockets may occur just below the semiconductive coating, between the groundwall and the coating. In coils connected to the phase terminal, small partial discharges can occur in these voids. This PD attacks the semiconductive coating, increasing its resistance. Eventually, the problem progresses to the stages described above.

With both processes, the coating gradually becomes nonconductive in a process that tends to spread aggressively over the surface. As the infinite-resistance area spreads, PD occurs, gradually destroying the underlying groundwall.

8.5.3 Symptoms

During a visual inspection of the stator, this problem is easily seen since the black semiconductive coating turns white, yellow, light gray, or green (depending on the paint base). These light spots can be seen by looking down the core ventilation ducts (if present) in slots containing coils/bars connected to the phase terminals. Removal of the wedges will expose the top surface of the coil or bar, revealing large patches where the coating has disappeared. Figure 8.5 shows two slots in an air-cooled machine from which the wedges have been removed. The lack of semiconductive coating is apparent in the coil in one of the slots, which operates at high voltage. The adjacent slot shows no coating deterioration because it is a neutral-end coil.

Both off-line and on-line partial discharge testing will detect the problem at an early stage. While relatively small in amplitude, the PD activity tends to be widespread. Measurement of the ozone concentration (Section 13.3) is a less sensitive way of detecting the problem.

8.5.4 Remedies

The best way to ensure that semiconductive deterioration does not occur is to ensure that the coating is manufactured correctly. This means that frequent surface-resistivity measurements must be made before coils are installed in the slot. If the problem does occur in an operating stator, there is no repair that can permanently restore the winding to new condition, unless the coils are removed from the slot and a new coating applied. Since removing them from the slot can damage coils and bars, availability of spare bars/coils is critical for this alternative.

Figure 8.5. Photograph of two stator slots taken from the axis of an air-cooled machine. The stator wedges have been removed, exposing the tops of the two coils. The semiconductive coating on the coil connected to the phase terminal has disappeared.

In-situ repairs usually involve injecting a carbon-loaded varnish, silicon rubber, or epoxy compound into the slots containing coils operating at high voltage. This remedy requires the rotor and then the old wedges to be removed. Many OEMs and some repair organizations have developed tooling that injects the compound under pressure down the sides of the slots and via the ventilation ducts. However, it is impossible to inject the compound into all the locations where the coating has deteriorated. Specifically, the injection process usually misses the bottom edge of the coil/bar.

If the failure process remains a purely PD process, then the failure may take many decades to occur. However, if the process develops into a loose coil problem, failure can be much faster. Therefore, it is important to replace the side packing and rewedge to keep the coils tight in the slot. The compound injection process also aids in keeping the coils tight.

There is little that can be done to slow the process by changing the operating regime. However, the process is driven by temperature and voltage. Thus, reducing the load can extend winding life. If the machine can be operated at a lower voltage, then the process will be slowed. Even a voltage drop of a few hundred volts can be helpful. There is evidence that the process is faster under high reactive loads. Reversing the winding, i.e., changing the circuit ring bus connections to place the high voltage end coils at the neutral (and the neutral coils at the phase terminal), may also extend the winding life. However, it is important to install a third harmonic relay to detect any ground fault that may occur near the neutral.

8.6   SEMICONDUCTIVE/GRADING COATING OVERLAP FAILURE

This problem is closely associated with the semiconductive coating deterioration process described above. The problem occurs only in stators in which there is both a semiconductive coating on the coils/bars in the slot and a silicon carbide stress control coating just outside of the slot (Section 1.4.5). The problem is confined to form-wound stators rated at 6 kV and above. Overlap failure is most likely to occur in air-cooled machines, since partial discharge is involved. Where the stress control coatings are tape based (rather than paint based), there seems to be less chance of this problem.

8.6.1   General Process

The silicon carbide coating extends from the semiconductive coating just outside of the slot, along the coil, into the end-winding—typically, 5 to 10 cm. The coating resistance is nonlinear. The resistance is very high in regions of low electrical stress and low in regions of high electrical stress. The silicon carbide coating must be electrically connected to ground by an overlap of the silicon carbide coating and the semiconductive coating. The overlap is typically about 1 cm wide. It is the electrical connection inherent in the overlap that degrades. If the overlap region becomes nonconductive, then the silicon carbide coating is no longer grounded. Instead, the silicon carbide coating "floats," and tends to rise up to the voltage of the copper within the coil, due to capacitive coupling. For example, on a 13.8 kV winding, the floating silicon carbide coating on the phase-end coil will tend to attain a voltage of about 8 kV. A very high voltage then exists between the semiconductive coating (at 0 kV) and the silicon carbide coating at 8 kV, separated by a small air gap (the overlap region). Consequently, the air gap breaks down, resulting in discharges over the surface of the coil between the two different coatings.

A feature of this discharging process, which formally is not a partial discharge process since complete breakdown occurs, is that the discharges are parallel to the insulation surface. That is, the discharges are not perpendicular to the insulation, as is the case for most discharges. Experience indicates that when the discharges are parallel to the insulation surface, the rate of deterioration of the bulk of the insulation is very slow. What groundwall deterioration does occur is usually confined to the surface, where the heat from the discharging vaporizes the organic materials in the groundwall. Thus, it is unlikely for this process to directly lead to groundwall failure, at least in less than 20 years or so.

The overlap deterioration process is important, however, since it can mask more important failure processes when monitoring the winding condition with partial discharge testing. Finally, if overlap problems are occurring, it is often a sign that the more serious semiconductive coating deterioration process might also be occurring within the slot.

8.6.2 Root Causes

Overlap discharging is caused by poor stress control design and/or poor manufacture. The overlap region is an electrical connection. Currents flow through the overlap. The currents originate from the capacitive current between the silicon carbide coating and the high-voltage copper underneath the coating and groundwall insulation. This current flows axially through the silicon carbide coating to the semiconductive coating, and then to the grounded stator core. If the resistance of the connection between the coatings is too high, there is an excessive I^2R loss at the overlap. This heating raises the temperature of the overlap region above the winding temperature, creating a local hot region. This increased temperature further increases the resistance through an oxidation process. Eventually, the overlap has infinite resistance and the discharging begins.

The causes for the initial high resistance of the overlap region are as follows:

- Carbon black particle density in the semiconductive coating is too low.
- Improper silicon carbide particle density or particle size distribution in the silicon carbide coating.
- Inadequate amount of overlap surface area for the capacitive currents that are flowing. Vigilance is especially needed with groundwalls designed at higher electric stress. These "thin wall" designs will have a greater capacitive current flowing through the silicon carbide coating.

Operation with IFDs causes much larger capacitive currents to flow, and thus accelerates the process (Section 8.7).

8.6.3 Symptoms

This problem is easily identified in a visual examination of the end-winding. A white band around the coil or bar circumference occurs a few centimeters outside of the slot (Figure 8.6). The white band will occur on coils/bars operating at high voltage. Coils connected to the neutral will not have the band, since there are no capacitive currents flowing to destroy the overlap. The white band is exactly where the overlap occurs between the semiconductive and silicon carbide coatings. Partial discharge testing, visual observation during blackout, and ozone testing can also identify the problem.

8.6.4 Remedies

There is no permanent repair once this problem occurs. If the end-windings have not been painted with an insulting varnish, then a somewhat effective repair can be made. (Such insu-

Figure 8.6. Photograph of the end-winding area from a hydrogenerator. The white band around some of the coils just outside the slot indicates where the coating overlap area has degraded.

lating varnish coatings are very common.) This involves cleaning the overlap region and then applying a partly conductive paint over the existing semiconductive coating to the coil or bar surface just outside of the slot. The new application of semiconductive paint should extend over the overlap region and cover the silicon carbide layer by at least 1 cm. The purpose is to restore the ground connection to the silicon carbide coating. Great care is needed in applying the partly conductive paint, since it should not encroach too far over the silicon carbide coating (say, less than 2 cm), nor should the paint be spilled over any of the remaining end-winding. In practice it is difficult to apply the partly conductive paint on the bottom coils/bars, or even the bottom edge of the top coil/bar.

If the end-windings have been painted with an insulating varnish, applying the partly conductive paint will not solve the problem. This is because the new partly conductive paint will not be properly grounded. As an alternative, there is a tendency to paint the overlap region (and the complete end-winding) with an insulating varnish. This does not affect the deterioration process in any way, and the white band will reappear within a few months. Thus, an insulating overcoat, while esthetically pleasing for a short time, does not prevent the discharging. Ensure that the varnish used is compatible with the groundwall insulation and will not reduce tracking resistance (Section 8.8).

8.7 REPETITIVE VOLTAGE SURGES

Voltage surges in stator winding insulation systems in motors and generators are transient bursts of relatively high voltage that increase the electrical stress beyond that which occurs in normal service. Voltage surges occur from:

- Lightning
- Ground faults in the power system
- Generator breaker closing under out-of-phase conditions
- Motor circuit breaker closing and opening
- Inverter-fed drives (IFDs)

Until recently, it was believed that a voltage surge either caused the insulation to fail immediately or had no lasting effect [8.5]. That is, repetitive voltage surges did not *gradually* deteriorate and cause the insulation to fail. This seems to be the case with the first four sources of voltage surges.

The introduction of IFDs, however, has shown that gradual aging does occur if the surges are of a sufficiently high magnitude, the risetime of the voltage surge is short, and there are many repetitive surges. Persson was among the first to report that voltage surges from pulse-width-modulated (PWM) type IFDs, using insulated gate bipolar junction transistors (IGBTs), caused random-wound motors to fail due to gradual aging of the stator turn insulation [8.6].

Random-wound motors rated between 400 V and 1000 V and driven by PWM IFDs using modern transistor switching elements such as IGBTs, are most likely to suffer from the surge aging problem. Medium-voltage motors using similar drives may also experience surge-induced deterioration, but turn insulation failure is less likely due to the presence of superior turn insulation. Instead, the semiconductive and grading coatings that are usual for IFD motors rated at 3.3 kV and above can degrade rapidly. Motors driven by drives using SCRs or GTO devices are less likely to induce failure, since the risetime of the surges is relatively slow. Gradual deterioration of the stator winding insulation in large generators due to voltage surges appears to be rare.

8.7.1 General Process

Repetitive voltage surges can induce gradual deterioration of the (a) turn insulation, (b) groundwall and phase insulation, and (c) the semiconductive and grading coatings, if present. Each is described separately.

Turn Insulation Deterioration. Surges produced by IFDs can create voltage transients with risetimes as short as 50 ns [8.7]. By a Fourier transformation, this short risetime will generate frequencies up to about 10 MHz. In Section 1.4.2, we discussed that at power frequency the voltage between adjacent turns in a coil is equal. That is, the power frequency voltage is linearly and equally distributed between each adjacent turn from the phase terminal to the neutral point. However, when a very high frequency is applied to a stator winding, the voltage distribution is nonlinear, with a much greater percentage of the voltage appearing across the turns in the first coil connected to the phase terminal. This nonuniform voltage distribution occurs because the series inductive impedance of the winding is relatively large compared to the low shunt capacitive impedance to ground at this high frequency [8.8]. Consequently, the effect of applying a fast risetime surge to a multiturn stator winding is that a very high voltage appears across the turn insulation, in the first few turns of the winding. As much as 40% of the applied surge voltage can appear across the first turn [8.8].

A high interturn voltage can give rise to a partial discharge if there is an air pocket in the vicinity of the copper turns. In random-wound stators, there are often pockets of air adjacent to the turns of round magnet wire. If a PD occurs in the air, then the discharge will slightly degrade the magnet wire insulation. If a sufficient number of surges occur, then enough cumulative damage occurs from the discharges that turn-to-turn failure results, rapidly progressing to a ground failure (Section 1.4.2). In addition to the gradual aging due to partial discharge, it is possible that deterioration can occur as a result of "space charge" currents [8.5]. Such deterioration does not necessarily require air adjacent to the magnet wire.

Random-wound stators with conventional magnet wire are most susceptible to this process, since the magnet wire insulation is organic and not resistant to attack by partial discharge. Unless special measures such as using PD-resistant magnet wire or increasing the distance between components (Section 8.7.4) are employed, failure can occur within weeks or months of an IFD going into service. Form-wound machines can experience the same surges, but since the turn insulation often contains inorganic glass fibers, or the preferred mica paper, this insulation is much more resistant to PD and consequently less likely to fail.

Ground and Phase Insulation. By definition, in random-wound stators, a phase lead in, say, A-phase may be adjacent to a neutral end turn or a turn connected to the B-phase terminal. IFDs can create relatively high magnitude voltage surges due to voltage reflections between the power cable and the motor surge impedances [8.7]. If there is inadequate space or insulation between phases or to ground, the small-diameter magnet wire in random-wound stators may create sufficient electrical stress in any air spaces surrounding the magnet wire to create PD. As for the turn insulation, the PD may gradually erode the organic insulation, resulting in a phase–phase or ground fault. For this process, the surge risetime is somewhat irrelevant. It is the surge magnitude and the repetition rate that are critical.

Stress Relief Coatings. Since IFDs generate high-frequency voltages, the capacitive currents through the ground insulation are relatively high compared to power frequency. OEMs have found it prudent to employ semiconductive and silicon carbide coatings (Section 1.4.5) on IFD motors rated as low as 3.3 kV to control the stress distribution in the end-windings, and thus to avoid PD at the slot exits. The high-frequency capacitive currents will result in high local I^2R losses in the coatings—far higher than those described in Sections 8.5 and 8.6. The result is that the stress control layers in IFD motors are likely to deteriorate much faster than the layers in 50/60 Hz motors.

8.7.2 Root Cause

This mechanism requires the following conditions to occur:

- A fast surge risetime, usually less than 200 ns, since the faster the risetime, the more voltage appears across the first turn.
- An impedance mismatch between the motor and the cable, which yields surge voltage reflections due to transmission line effects [8.8].
- Air pockets adjacent to the turns connected to the phase terminals, enabling PD to occur.
- Tens of thousands of surges per second. Tests show that a partial discharge occurs very rarely; typically, one PD per several thousand surges [8.9]. Thus, many surges are needed to create enough cumulative damage to cause failure.

A few fast risetime surges (as long as they are withstood) seem to cause no harm.

8.7.3 Symptoms

On random-wound machines, surge-induced PD attack is visible as a whitish powder on the magnet wire insulation, especially between phase-end turns or where magnet wire from different coils or phases comes in contact. If the motor end-windings can be exposed for a short time in no-load operation, then a blackout test (Section 12.26) may indicate the points of light associated with PD. In form-wound stators, there will be no visible signs of surge aging of the insulation since the groundwall insulation covers the turn insulation. However, if the stress control coatings are deteriorating, they generally turn white, and significant ozone will be present.

Methods are available to measure electrically the PD activity either off-line or on-line [8.9]. However, relatively sophisticated methods are needed, since the surge transients are similar to PD pulses, and conventional PD detectors will be destroyed by the high voltage surges. If no PD is detected in the normal surge environment, then failure due to voltage surges is unlikely to occur.

8.7.4 Remedies

Even if PWM/IGBT IFDs are used, it is not likely that a particular motor will see fast rise-time and high magnitude surges. This is because the magnitude and the risetime are governed by many factors, and most practical situations do not combine to yield severe surges [8.7]. Thus, prior to commissioning a new IFD motor, or if several stators have failed and surge aging is a possible cause, it is prudent to measure the actual surge environment. This can be done during normal motor operation with a high-voltage, 10 MHz bandwidth oscilloscope probe and a 100 MHz digital oscilloscope. Specialized instruments are also available [8.7]. The measurement must be done at the motor terminals.

If measurements reveal that the surge environment exceeds specifications,* then one or more of the following may be tried to reduce the severity of the surges and the risk of this failure mechanism:

- Change the cable length and/or grounding, which will primarily change the surge magnitude.
- Replace the cable with a higher surge impedance cable (thicker insulation), or a cable with a more lossy dielectric (e.g., butyl rubber or oil paper are very lossy).
- Install a low-pass filter between the IFD and the motor to lengthen the risetime of the surge. (For conventional motors, the filter is usually a "surge" capacitor of about 0.1 μF. This capacitance, together with the surge impedance of the power cable, usually around 30 ohms, lengthens the risetime of the surge to about 6 μs. More sophisticated filters are needed for IFDs.)

The effectiveness of these corrective measures can be ascertained by repeat measurements of the surge environment. In addition, the capability of a replacement winding to withstand the surge environment can be improved by the following methods:

- Using more PD-resistant magnet wire. Metal-oxide-loaded magnet wire has been demonstrated to withstand the PD during surges for much longer than conventional magnet wire [8.10, 8.11].

*NEMA Standard MG1-1998 Section IV, Part 31 specifies that an "inverter duty" motor rated at less than 600 V should see a risetime no shorter than 100 ns, with magnitude less than 3.1 V_1. V_1 is the rated line-to-line voltage.

- Impregnating the stator to reduce the chance that air pockets will occur at critical locations. Trickle impregnation and the VPI process have been used in this way. In random-wound VPI'd stators, taping the end-windings and fitting "dams" of felt at the ends of the slots, between the slot liner and conductors, has proven effective in significantly reducing the size and number of voids between conductor strands.
- Increasing the distance between adjacent turns in the phase ends and between other coils and phases. In the former case, this can be done by reducing the turns on phase-end coils by one, thus allowing space for more insulation on each conductor, or inserting thicker sheets of ground- and phase-separation material.

8.8 CONTAMINATION (ELECTRICAL TRACKING)

Winding contamination leads to many problems, including increased thermal deterioration (due to blocked ventilation), chemical attack (Section 8.10), and electrical tracking. This section deals only with the latter. Electrical tracking enables currents to flow over the surfaces of the insulation, especially in the end-windings. These currents degrade and eventually cause the groundwall insulation to fail. This problem has a greater probability of occurring in machines with high operating voltage. However, even 120 V motors can fail due to this process, and machines that are dirty or wet are more likely to be affected.

8.8.1 General Process

Cooling air in open-ventilated motors and generators can be contaminated with dirt, insects, plant by-products (fly ash and coal dust in generating stations, various chemicals produced in petrochemical plants, cement dust in cement plants or basements, etc). This contamination mixes with moisture or oil to produce a partly conductive coating on the stator winding. Oil comes from the bearings or the seal oil system, whereas moisture comes from the environment, steam leaks, leaking cooling water systems, and/or from the seal oil system in hydrogen-cooled machines. Even totally enclosed machines can be contaminated by oil or moisture combining with foreign material during manufacture or maintenance. Another source of particulates is carbon dust from brushes during operation of the machine.

There are two processes that can lead to failure, depending on whether the stator is form-wound or random-wound.

Form-Wound Stators. The problem normally only occurs in the end-windings of form-wound stators. If the contamination has some conductivity, say less than a few tens of megohms per square, then currents can flow if a potential difference exists. Figure 8.7 shows the cross section of two adjacent coils in an end-winding, together with an equivalent circuit. Assume the coils are in two different phases. The contaminated surface of the A-phase coil will have a tendency to float up to the voltage of the copper within the coil, by capacitive coupling. Similarly, the B-phase coil surface will tend to float up to the B-phase voltage. Blocks used for stiffening the end-winding structure and for electrically separating the coils bridge these surfaces. The surface of the block represents an "electrical resistance" of the contamination across the block. The equivalent circuit in Figure 8.7 results. If the two coils are phase-end coils in two different phases, then, during normal operation, the full phase-to-phase voltage is applied to the equivalent circuit and current will flow. If the contamination resistance is high compared to the capacitive impedance of the coils in the contaminated area, then the surface of the coils is almost at the

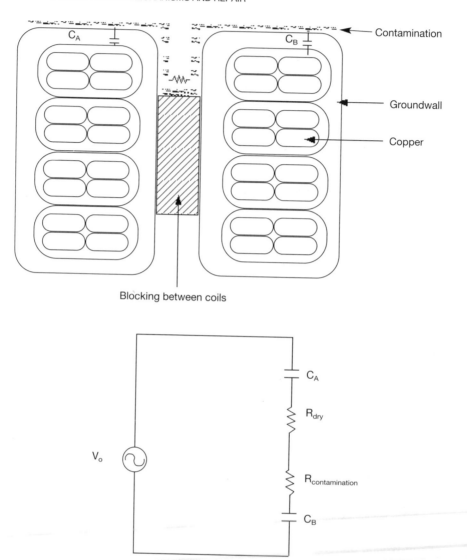

Figure 8.7. Cross section of two adjacent coils in different phases in the end-winding, together with an equivalent circuit, which shows how a small leakage current can flow.

same voltage as the underlying copper conductors, and nearly full phase voltage is applied across the block.

If the contamination had a very uniform resistance across the block, then little deterioration is likely to result, since the current is low (in the nanoamp range) and flowing uniformly across the surface. However, in reality, there are "dry bands" where the resistance is much higher than the general resistance of the contamination. In this case, virtually the entire voltage will then appear across the small dry band, causing electrical breakdown of the adjacent

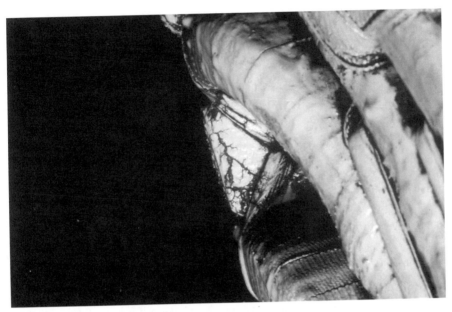

Figure 8.8. A photograph of a carbonized (black) track across a block between two coils in different phases.

air or hydrogen. The discharge degrades and may carbonize the underlying organic resin and tapes. This small area is eventually left very conductive. The electrical stress then transfers to another region of high resistance, where discharges ensue. The result is an electrical track, which slowly grows across the insulation (Figure 8.8). The track often has many branches and appears as a carbonized, black network of valleys across the blocking. Given time, the discharges will also start boring into the groundwall. If the track is between coils in different phases, a phase-to-phase failure results, which allows extremely large fault currents to flow. Alternatively, tracks can appear between coils in the same phase, if one coil is near the phase end and the other is near neutral. Similarly, tracking can occur along a coil between the core and further into the end-winding. This is especially likely in coils without semiconducting and grading coatings.

This mechanism is usually very slow, often taking more than 10 years from the time the winding is contaminated to when it fails.

Random-Wound Stators. The process is somewhat different in random-wound machines. The electrical tracking process is not the main aging mechanism in such windings, although IFD voltage surges could give rise to tracking. Instead, the process requires pinholes or cracks in the magnet wire before deterioration is likely to occur. Pinholes in magnet wire, though rare, can occur. The specification for magnet wire does permit a small number of pinholes after manufacturing [8.12]. In addition, other processes, such as thermal aging, can crack magnet wire insulation, and abrasion of the insulation can expose the copper. When a winding contains such pinholes and cracks, partly conductive contamination allows 50/60 Hz currents to flow between turns at different voltages. Since the distances can be very small, the currents can be relatively large (compared to the currents in the form-wound situation).

The current causes the oil/dirt to carbonize, further lowering the resistance. Eventually, the resistance gets so low that very large circulating currents flow between the turns (Section 1.4.2) and a turn short develops, which is followed quickly by a ground fault.

The time to failure depends on preexisting pinholes or the time it takes for cracks to develop, together with the conductivity of the pollution. In some cases, failure may occur within weeks of being put into operation; in other cases, it may take 30 years.

8.8.2   Root Causes

In form-wound machines, the root cause of failure is presence of contamination. In random-wound machines, failure requires contamination together with either poor manufacture (due to poor magnet wire insulation and poor varnish or resin impregnation) or prior thermal deterioration or vibration.

8.8.3   Symptoms and Remedies

The most obvious symptom in a visual inspection is the presence of contamination. In most situations, this will appear on the winding surface as a dark film that can be wiped away. However, if the winding is clean, occasional moisture can appear on the winding from dripping water or condensation during machine operation at low power. Condensation also forms on the stator winding surfaces after a machine is shut down if it is not fitted with space heaters. Thus, even apparently clean windings can experience contamination failure. In random-wound machines, there may also be evidence of poor dipping, or deteriorated magnet wire insulation from other causes.

The insulation resistance and polarization index tests are a very reliable way of detecting this problem, in any type of machine. In form-wound machines, the capacitance, power factor and partial discharge tests will all detect the presence of contamination, providing that these values were obtained from the noncontaminated windings.

8.8.4   Remedies

This is perhaps one of the simplest problems to repair. Even if the problem is well advanced, the stator can often be restored to as-new condition by a thorough cleaning, together with a dip and bake, paint, or VPI treatment. Note that it is essential for the dipping varnish or paint to be resistant to tracking. Some varnishes have been found to accelerate the tracking process rather than retard it [8.13]. If electrical tracking has occurred, it is prudent to remove the tracks with grit cloth.

There are a wide range of cleaning methods. Smaller stators can be put in a special steam-cleaning unit. Larger stators with modern insulation systems can be cleaned with high-pressure water or steam. Sometimes, dry ice or broken walnut shells or corncobs driven through a nozzle by high-pressure air can effectively clean encrusted dirt. Note that water and abrasives themselves can degrade the insulation, especially if the insulation is of an older type (Sections 8.9 and 8.10). The dry ice cleaning method is particularly effective for high-speed motors and generators, in which there is concern that the debris from the particulate cleaning (i.e., walnut shells, etc.) may block small cooling passages, especially in the rotor. Since the dry ice evaporates, blockage is not a problem. However, some users have reported that dry ice cleaning does not appear to remove oil films on the windings.

Where the dirt layer is relatively thin, solvent or other cleaning materials can be poured on lint-free rags and wiped over the coils. This may also be necessary for the removal of heavy buildup of contaminants in large generators where other methods may not be suitable. Solvents such as trichlorethylene are very effective, especially in removing oil. However, the most effective solvents produce vapors that are a health hazard. They can only be safely used with breathing apparatus. A trade-off is often obtained by using less-effective citrus-based cleaning fluids without special equipment but with increased labor cost.

The following are some of the measures that can be taken to eliminate the problem by preventing the source of the contamination:

- Install air-inlet filters and ensure that they are maintained regularly (for open-ventilated machines).
- Specify or upgrade to a totally enclosed cooling circuit, especially if the machine must run in a dirty environment.
- Ensure bearings and seal oil systems are properly maintained, so that they do not leak oil on the winding.
- Repair nearby steam leaks or other sources of contamination as soon as they occur.
- Ensure that brakes are applied after a machine has already slowed down considerably, or change airflow patterns so that brake residue cannot find its way into the winding.
- In machines with brush gear mounted inside the enclosure, periodically vacuum away any carbon dust that accumulates.
- Pay attention to the dew point.

8.9   ABRASIVE PARTICLES

Abrasive particles in the cooling gas stream can grind away stator winding insulation. This problem is most likely to occur on open-ventilated machines operating in environments where sand or other abrasives exist. Both form-wound and random-wound stators are vulnerable.

8.9.1   General Process

Sand, fly ash, glass fibers, or any other small, hard particles, if they are contained in the machine enclosure, will be blown through the stator winding at high velocity as part of the normal cooling air or hydrogen flow. These particles will abrade some of the insulation due to the energy of their motion. With a sufficient number of particles, and a high enough velocity, most types of groundwall insulation (in form-wound stators) and magnet wire (in random-wound stators) are relatively easily abraded because of their lower mechanical strength in directions that are not normal to the surface. Eventually, enough insulation may be abraded so that copper is exposed and a ground fault occurs. Any partly conductive contamination of the winding will accelerate the process, since it will promote the flow of current.

Abrasion is most likely in the end-windings but, in severe cases, the insulation in the stator core ventilation ducts may also abrade.

8.9.2   Root Causes

The fundamental cause of this problem is operation of an open-ventilated machine in an environment that contains abrasives. The problem, therefore, is a poor motor or generator specification. The affect of abrasive particles can be accelerated if the user does not maintain the filters that are intended to block ingress of particulates into the machine.

Abrasive material attack can also occur as part of the stator cleaning process. Materials such as dry ice (frozen carbon dioxide), walnut shells, corncob bits, etc. blown under pressure through a nozzle are sometimes used to clean encrusted dirt from the stator windings. Insulation can be abraded very quickly if the nozzle is held over any one spot too long.

8.9.3 Symptoms and Remedies

In a visual inspection, signs of abrasion in the end-winding are usually easy to see. When the problem is in an advanced state, i.e., copper has been exposed, then the insulation resistance, polarization index, or hipot tests may detect the problem. (If the end-winding is clean, then only the hipot test may be effective.) In the early stage, no diagnostic test is effective for identifying the problem.

The obvious remedy is to specify or upgrade the machine so that it is not open-ventilated. If a random-wound stator has encountered abrasion, then a dip and bake or VPI treatment will restore the insulation. If only minor surface abrasion has occurred on a form-wound machine, a VPI treatment or varnishing of the end-winding may be sufficient. If more serious abrasion is present, then the groundwall thickness, at least in the end-winding, may be partially restored using a spatula to apply a thixotropic (very high viscosity) epoxy prior to a VPI or varnishing treatment. This repair will probably be ineffective if abrasion has occurred in the slot.

If provisions cannot be made to prevent the introduction of abrasives into the machine, then it may be wise to overcoat the winding with a thin coating of an abrasion-resistant material. Silicon rubber and certain polyurethanes have good abrasion resistance. Note, however, that it is important that the covering does not promote electrical tracking [8.13].

8.10 CHEMICAL ATTACK

Chemical attack describes the deterioration of the insulation that can occur if the insulation is exposed to an environment in which chemicals such as acids, paints, and solvents, as well as oil and water, are present. This problem can also occur if inappropriate cleaning methods are used, or if the winding is given a VPI or dip and bake treatment with a noncompatible resin. All types of stator windings can suffer from this problem.

8.10.1 General Process

Most types of older insulation systems are prone to chemically induced degradation due to the presence of solvents, oil, water, or other chemicals. For example, magnet wire insulation such as polyester can soften or swell from exposure to certain chemicals [8.14]. Groundwall insulation using asphalt, varnish, and some early polyesters as bonding agents are prone to softening, swelling, and loss of mechanical and electrical strength. For example, when polyester is in contact with water for long periods of time, hydrolysis and depolymerization results in weak and soft insulation, greatly reducing the mechanical and electrical strength of polyester [8.15].

Softening of insulation makes it susceptible to cold flow, i.e., the insulation gradually becomes thinner in places where mechanical pressure is applied. Eventually, enough insulation material may migrate so that the insulation can no longer support normal operating or surge voltages. Swelling leads to delamination of the insulation, and even peeling of the insulation in magnet wires. If the mechanical or electrical strength of the insulation decreases significantly due to chemical deterioration, a high-voltage surge or the transient mechanical forces during motor turn-on or synchronous machine malsynchronization can puncture or crack the insulation. The result is a turn or ground short.

Modern stator winding insulation systems are more resistant to most kinds of chemical attack. Epoxy is relatively resistant to solvents, moisture, or oil. However, if the epoxy is exposed to oil and water for many years, for example, from a water coolant leak problem (Section 8.13), then it will eventually degrade. Modern magnet wire films such as polyamide–imide or polyester with a polyamide–imide overcoat have excellent resistance to most chemicals.

8.10.2 Root Causes

The root cause of chemical attack is the presence of oil, water, solvents or other reactive chemicals. Oil is used in all large motors and generators. If the bearing lubrication system is poorly maintained, then excess oil will drip or be blown onto the windings. In hydrogen-cooled machines, a leaky seal oil system is a likely source of oil. Water can come from cooling water leaks and openings in the machine enclosure that expose outdoor machines to rainstorms. In addition, if an open-ventilated machine operates at low load or is shut down in high-humidity areas, moisture can condense on the winding surfaces if space heaters are not fitted or not working. In hydrogen-cooled machines, excessive moisture can enter the machine via the seal oil system, where the oil is in contact with atmospheric air. The problem is made worse if the hydrogen dryer is not functioning.

Materials used to clean a winding are a source of chemicals. It is common to clean stators with steam or high-pressure water (Section 8.8.4). This should not be done with older types of insulation, since the water may be absorbed into the insulation. If water or steam is used for cleaning modern stators, then the stator should be dried in an oven or by circulating current through the stator as soon as possible. Cleaning with solvents or caustic compounds can also degrade older insulation systems, although this is rarely a problem with modern stators.

Windings in motors and small generators with grease-lubricated bearings can be contaminated with this lubricant if grease relief plugs are not removed during regreasing. If this is not done, then grease is forced between the bearing cap and machine shaft and enters the stator end-winding region.

8.10.3 Symptoms

In a visual inspection, the insulation may be discolored by chemical attack. The insulation may have swollen and, if tapped, may sound hollow. If it is easy to scrape away the insulation with a small knife, then chemical attack may be occurring.

There are few diagnostic tests that can be done from the machine terminals that will indicate that the problem is occurring. If the problem is in an advanced stage, the insulation resistance and polarization index test results may be low. Chemical analysis from small samples of the insulation and any local debris or dirt will indicate if chemical deterioration is occurring.

8.10.4 Remedies

The most important means of preventing winding failure by this mechanism is to specify a totally enclosed machine, ensure that oil leaks or grease ingress are not occurring, and that improper materials are not used for cleaning.

For existing machines, the following actions may delay failure:

- Clean the stator regularly if the machine is operating in an environment where chemicals are likely to be found.
- Ensure that the filters are effective.
- Use as benign a cleaning material as possible.
- Repair oil or water leaks as soon as possible.
- Ensure effective heater operation (where fitted, or consider retrofitting heaters), to prevent moisture condensation on the windings when the unit is cold.
- Ensure that old grease is expelled through grease relief plugholes when new lubricant is being added.
- Repair or replace an ineffective hydrogen dryer.

8.11   INADEQUATE END-WINDING SPACING

In large stators, space is left between adjacent coils in the end-windings to ensure that sufficient cooling air flows over them, to aid in limiting the winding temperature. In form-wound stators, particularly windings rated at 6 kV and above, adequate spacing is also needed to prevent partial discharge activity. If the spacing is too small, PD can occur, which may lead to ground and/or phase-to-phase faults. The problem does not happen on random-wound stators, and is unlikely to occur in high-pressure, hydrogen-cooled machines.

The partial discharge problem is most likely to occur in air-cooled machines rated at 6 kV or higher. It may also occur on stators rated at 3.3 kV and above that are using a thin groundwall design. Although in a different location, the PD can also occur between circuit ring busses, coil connection insulation, or in the terminal leads if they are too closely spaced, especially if they are in different phases.

8.11.1   General Process

As was discussed in Section 1.4.4, partial discharges occur when a gas is subject to an excessive electric stress. Figure 8.9 shows a cross section of two adjacent coils or bars in the end-winding of a form-wound stator. The cross section can be thought of as three capacitors in series between the copper conductors: the capacitance of the A-phase groundwall, the capacitance of the air gap between the coil surfaces, and the capacitance of the B-phase groundwall. As a simplification, each of these capacitances can be calculated from a parallel plate model of a capacitor, in which the capacitance depends on the insulation thickness (or the distance between the coil surfaces in the case of the air gap) and the dielectric constant of the insulating material (see Equation 1.5). Using normal capacitive divider relationships from basic circuit theory, the percentage of voltage across the air gap can be calculated, knowing the thickness and the dielectric constants. From this, the electric stress is calculated (Equation 1.4). If the stress exceeds 3 kV/mm in an air-cooled machine at sea level (or a lower stress at high altitudes), the air breaks down, creating a partial discharge. Partial discharge most likely occurs when adjacent coils are connected to the phase terminals and in different phases, since the full rated voltage will drive the process.

Given sufficient time, the PD will erode a hole through the groundwall insulation. Since the discharges are usually occurring in air, ozone is created, which further accelerates the insulation deterioration process because the ozone creates an acid that also etches the insulation. The

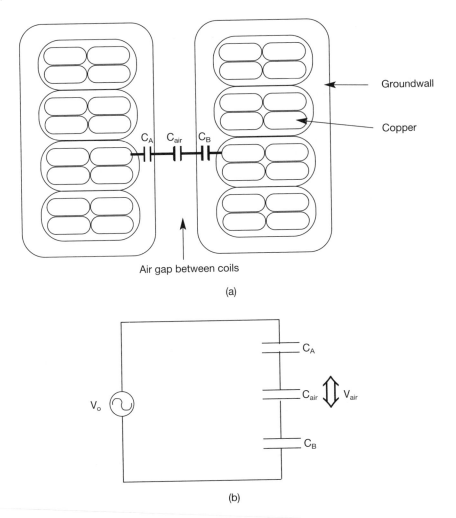

Figure 8.9. Cross section of two adjacent coils in the end-winding where PD will occur if the distance between the coils is too small.

ozone may also corrode heat exchangers and damage rubber gaskets. The time to failure is usually 5 years or more, since mica-based groundwall insulation is resistant to PD attack.

If the space between the coils is too small, typically less than 5 to 7 mm in a 13.8 kV winding with an average design stress of 2.5 kV/mm, PD will result if the full phase-to-phase voltage is across adjacent coils. PD is more likely at high altitudes, since the breakdown stress of air is lower in such locations. In addition, thin groundwall insulation designs increase the capacitance of the groundwall, reducing its capacitive impedance and applying more voltage across the gap. Consequently, such designs are more prone to this problem. High-pressure, hydrogen-cooled machines are less likely to have PD, since the breakdown strength of a high-pressure gas is many times higher than the strength at one atmosphere.

The problem will occur wherever adjacent coils (side-to-side or between the top and bottom layer of coils) are too close. This problem can also occur if high-voltage circuit ring busses are too close. An interesting variation has occurred in some hydrogenerators with Roebel bar windings and "end caps" used to insulate the connection between bars. The end cap insulation is often made by placing a polycarbonate (or similar material) mold over the connection and filling it with a thixotropic epoxy. The epoxy often contains fillers, but these are not especially PD resistant. If two adjacent end caps are insulating high-voltage connections in different phases, and the end caps are too close, PD will occur between them and damage the end caps. Since the mold and the epoxy are not resistant to PD, failure occurs in a period ranging from months to years. If the dielectric constant of the epoxy is too high, the capacitive reactance of the end cap will be low, increasing the stress across the air gap. Sometimes such PD only occurs when the generator is operating at high temperature, since the dielectric constant of the end caps is usually higher at high temperature.

8.11.2   Root Causes

Poor design and/or poor manufacture are the most likely cause of these problems:

- Insufficient space between coils for the altitude at which the machine is expected to operate. The purchaser may also be responsible if the OEM was not made aware of the intended location for the machine.
- Use of a thin groundwall design, without increasing the normal space between coils or making other provisions in the end-winding.
- Inconsistent coil shape after manufacturing. If one coil is longer than its adjacent coils, the longer coil must be closer to the adjacent coil than designed for, after the first bend in the end-winding.
- Poor installation of the coils in the slot. Even if the coils have identical shapes, if the first bend in a coil occurs farther out of the slot than the other coils, then it will close the space between the coils, after the first bend in the end-winding. An inadequate number of space blocks between adjacent coils may also make it harder for winders to keep consistent spacing between coils.
- Insufficient blocking or poor use of conformable packing may enable coils to bend in the end-winding, after high fault current or high motor starting current flow through the stator. This may reduce the distance between coils operating at greatly different voltages.
- Poor chemical control over the epoxy (in the coils or the end caps), which allows a high dielectric constant at operating temperature.

8.11.3   Symptoms

Usually, this mechanism is easy to spot in a visual inspection of the end-winding. Intense white powder (result of ozone attack) will occur between the coils (or end caps or circuit rings) connected to, or close to, the phase terminals (Figure 8.10). The most likely region for the white powder is where there are two adjacent phase-end coils in different phases. However, the powder can occur whenever any two adjacent coils have a sufficiently high potential difference. The white powder will not occur between neutral end coils, or where there is a

Figure 8.10. PD between adjacent coils in different phases results in white powder on the coil surfaces.

larger than average space between adjacent coils. The powder may assume shades of gray or brown in the presence of dust, debris, and oil.

On-line partial discharge tests are the most effective diagnostic means of finding the problem. (Off-line tests usually do not have sufficient voltage in the end-winding to initiate the PD.)

8.11.4 Remedies

To avoid this problem, testing and ensuring conformity to the specified distance between coils are needed at the time of stator manufacture. If the problem is discovered at an early stage of operation, very effective repairs are possible as discussed below.

- A good repair is to inject a conformable insulating material into the affected areas between the coils. Since a solid insulation has 100 times more electric strength than air, PD is prevented. Silicon rubber* or resin-filled felt pads are the most common materials used. For larger machines, resin-bonded glass fiber blocks wrapped with resin-filled felt are often used. Rigid epoxy-reinforced blocks or even mica flake are usually not as effective since air spaces often remain, which will give rise to very intense PD. Also, rigid materials may break away from the coil after load cycling, again creating

---

*There is some concern about the use of silicon rubber in a machine that has slip rings. The acetic acid released during cure by some types of silicon rubber can impair the operation of some types of slip rings. However, since a number of OEMs have been using silicon rubber for decades to secure coils in the slot, carbon brushes have been developed that are not affected by the acetic acid.

an air gap that may discharge. Only gaps showing PD activity should be filled, since such filling impairs the end-winding air cooling.

- If the end-winding coils are flexible, insert blocks to increase the spacing between adjacent coils with a large potential difference. This procedure needs special care due to the risk of damage to the windings. Similarly, changing the bus supports can increase the spacing between adjacent circuit ring buses.

- For end cap PD, replace the caps and use a low dielectric constant epoxy to fill the new caps. Alternatively, increase the space between the caps. Sometimes it is better to have less insulation in the end cap. Another repair is to hand tape the connection with a resin-rich epoxy–mica paper tape. Although discharges may still occur, the mica paper has a much greater PD resistance.

- Reverse line and neutral ends.

8.12   END-WINDING VIBRATION

Normal 50 or 60 Hz current flowing through the stator coils and bars creates large (100 or 120 Hz) magnetic forces. If the end-winding is not adequately supported, the coils vibrate, gradually abrading the insulation. The problem is most likely to occur on form-wound two- and four-pole machines, since such machines have long end-windings, which may have resonant frequencies close to the frequency of the magnetic forces. End-winding vibration is one of the most common failure mechanisms of large steam turbine generators rated at several hundred megawatts and above. Global VPI stators or random-wound machines are less likely to suffer from high end-winding vibration. However, any random-wound or form-wound stator can fail due to this problem if the end-windings are not adequately supported.

8.12.1   General Process

Section 1.4.7 discusses extensively the need for adequate stator end-winding support. If the support is inadequate, the bars or coils will begin to vibrate. In form-wound machines with long end-windings, this vibration in the radial and circumferential directions will pivot at the stator slot exit, since the coils/bars in the slot are presumably tightly held by the slot support system. Thus, the coil/bar insulation may eventually fatigue crack just outside of the slot. The fatigue cracking can occur even if the coils themselves are tightly held together. The cracked insulation will enable a phase-to-ground fault.

If the blocking and bracing is loose, then coils and bars can also vibrate relative to one another in the end-winding. The bars and coils will then rub against each other, the blocks, surge rings, support cone, and/or other end-winding support structures. The rubbing will abrade through the insulation. Fiberglass roving is especially hard and very effective in cutting through the groundwall insulation. If not corrected, sufficient groundwall insulation can be abraded so that a phase-to-ground failure occurs.

There are additional consequences of end-winding vibration in directly cooled stator windings. In water-cooled stators, the 100 or 120 Hz end-winding vibration can fatigue crack the brazed connections of one bar to another and/or the water nozzles. This can allow water to leak into the insulation (Section 8.13). Also, some failures have occurred because hydrogen becomes entrained in the stator cooling water. The hydrogen bubbles combined with excessive vibration lead to cavitation of the copper conductors. The thinning copper tubes eventually crack, allowing large amounts of hydrogen into the cooling water.

In direct hydrogen-cooled windings, end-winding vibration has led to broken copper strands due to the copper fatigue. The strand arcs at the break, causing localized overheating of the insulation. The insulation eventually melts, precipitating a ground fault. Arcing has also occurred at broken resistor connections (due to vibration). The resistors are installed to give a fixed potential to the hollow hydrogen tubes within the bar. Direct hydrogen inner cooled stator windings require a longer end-winding than other designs (everything else being equal), since the exposed metallic tubes require longer creepage distances. This makes such windings more susceptible to end-winding vibration problems.

8.12.2 Root Causes

The following are possible reasons for end-winding vibration to occur in form-wound windings:

- Inadequate initial design. Many large steam turbine generators had end-winding vibration problems in the 1970s and 1980s, before OEMs developed more effective support systems. Such systems had to allow for load cycling in which the winding tended to expand and shrink axially. Sometimes, the end-winding design was such that it had a resonance point near twice the power frequency. This was more likely on a machine originally designed for 50 Hz operation but modified for 60 Hz. In other types of form-wound stators, an insufficient number of rows of blocking were installed. Such windings are particularly susceptible to large fault currents, but even the 100 and 120 Hz operating forces allow enough movement so that fatigue cracking or abrasion occurs.
- Poor manufacturing procedures. These include blocking and bracing that is not installed uniformly around the end-winding circumference or an inconsistent distance from the stator core. Lashing and roving can be installed nonuniformly or not be properly saturated with polyester or epoxy resin. The coils or bars may have different shapes or be installed with different lengths of the end-winding extending from the slot. This may create very small spaces between the coils in some locations, making it impossible to insert blocks that can brace the coils. There are many other manufacturing-related problems that lead to loose end-windings.
- A prior out-of phase synchronization or excessive motor starting current, which can create sufficient magnetic force to break the lashing and allow blocks to come loose.
- Long-term operation at high temperatures. This results in thermal aging of the insulation and organic end-winding support materials (Section 8.1). These materials will shrink, enabling looseness to develop.
- Excessive oil in the end-winding. The oil can reduce the effectiveness of some lashing materials, allowing them to slacken. In addition, oil acts as a lubricant that seeps between blocking points, enabling small relative movements that would not normally occur without a lubricant. The low-level vibration very slowly abrades material, increasing the vibration amplitude.

Random-wound machines suffer end-winding vibration deterioration usually only if the stator is poorly impregnated or the coils were poorly lashed in manufacture. In addition, if the stator has seen strong thermal deterioration, then the organic components will probably shrink enough to allow relative movement to occur.

8.12.3 Symptoms

In a visual inspection of a form-wound stator, insulation vibration is usually manifested as "dusting" or "greasing." If there is little oil present in the machine, a fine, light-colored powder is created by the relative movement between a coil and the blocking or lashing. If oil is present, the oil mixes with the powder to create a thick black paste that accumulates at blocking and lashing points. Dusting or greasing can occur at any location in the end-winding where two surfaces meet. It may occur at the high-voltage or neutral end of the winding. If the coils are covered with a varnish, the varnish may be cracked.

Conventional nondestructive evaluation (NDE) methods can be applied in an off-line test to determine if the end-winding is loose. This is made easier if a fingerprint result was taken when the stator was new (and tight). If the winding is loose, when hit with a hammer and monitored with an accelerometer attached to the winding, there will be little damping of the response. Although reasonably expensive, large turbine generators have been equipped with end-winding vibration monitors (Section 13.5) to detect movement. In direct water-cooled machines there will be an increase in the hydrogen content in the water if the problem is so advanced that strands have cracked. Also, in direct-cooled windings, bar outlet temperatures increase a few degrees from normal due to broken strands.

Other than overvoltage (hipot) testing, there are no electrical tests that can detect this problem, unless it is in a very advanced stage. In severe cases where the copper strands may have fatigue cracked, on-line PD detection may be effective.

In random-wound machines, the problem can only be deduced from a visual examination. There may be signs of dusting or greasing at lashing points, and the magnet wire insulation may have cracks.

8.12.4 Remedies

For form-wound stators, operating the machine at a reduced load will extend the life, since vibration increases with the square of the load current. Since the forces in the end-winding are affected by the operating power factor, there may be some reactive power loads that will minimize end-winding vibration. Some of the repairs that can be done include the following:

- Complete end-winding support system replacement with a superior system. This has been needed on some very large two-pole direct-hydrogen inner-cooled windings [8.16].
- Removal and reinstalling of blocking and lashing materials.
- Installation of additional blocking and bracing, although higher end-winding temperatures may result in air-cooled machines due to reduced coolant flow.
- If originally the stator was global VPI'd, or the form-wound stator did not have a semiconductive coating in the slot, then a global VPI treatment may slow the process. Note that a global VPI treatment should not be applied to a conventional stator that has a semiconductive coating in the slot. The VPI resin may insulate some coils from the core, giving rise to massive slot discharge.

Random-wound machines should be cleaned and given a dip and bake treatment to solidify the end-winding.

8.13   STATOR COOLANT WATER LEAKS

This problem is only associated with large hydro and steam turbine generators with directly water-cooled stator windings. Such machines are usually larger than 200 MVA. The water used for cooling leaks onto or into the insulation, degrading it over time.

8.13.1   General Process

Since the first direct water cooled stators were built, water leaks have occurred. Most leaks are due to poor workmanship—maintenance personnel did not correctly attach the Teflon™ hoses to the nozzles at each end of the stator bars. The stator winding pressure and vacuum tests prior to generator startup would help eliminate a majority of such defects in assembly (Section 12.27). Any remaining joints that are marginal can cause water to spray onto the end-winding when the machine is in service. This contamination leads to electrical tracking (Section 8.8) or even immediate flashover if there is a large leak. Such gross water leaks are usually easily detected by the presence of gas in the coolant alarm, liquid in a generator alarm, an increase in the water make-up rate, and an increase in hydrogen dewpoint

A more serious problem occurs if a very small leak develops at the end of the bar, but before the nozzle, in the region termed the "water box" or the "clip." Such small leaks are not easily noticed because they may not trigger the alarms mentioned above. Crevice corrosion of the brazed connection between the stator bar copper strands to the waterbox is the primary cause of these small leaks. Porosity in the brazed connections and cracks in strands can also result in the leaks.

In spite of the hydrogen pressure being higher than the water pressure, osmotic forces allow a trickle of moisture to be drawn into the groundwall insulation. Although very resistant to moisture, the ground insulation epoxy–mica tape layers tend to delaminate and swell due to the constant ingress of moisture. The water can then migrate along the bar through this delamination or in interstices between copper strands. Dissection of bars removed from affected generators has shown the presence of moisture all the way down to the end of the bar opposite to the leaking end.

The small water leak will have three effects on the groundwall insulation. First, the electric strength of the epoxy–mica insulation is reduced if the insulation has been saturated with water. Thus, the stator will be more prone to fail if an overvoltage event occurs (such as a ground fault in the switchyard) or during an AC or DC hipot test (Sections 12.2 and 12.4). Second, the delamination may enable partial discharge to occur in bars connected to the phase terminal if the delamination has progressed from the brazed connection to the slot. Third, water will reduce the mechanical strength of the insulation. If a current surge occurs due to a fault on the system or if the unit is synchronized out-of-phase, then the groundwall is more likely to rupture.

Since this failure mechanism was first recognized around 1990, more than 250 stators have been tested and about one-half of them found to be leaking. Until 2001, 45 of the stators with leaks failed hipot tests, although there were very few, if any, in-service failures. Consequently, dozens of stators have either been repaired or rewound.

The deterioration process is very slow, often taking decades before there is a high risk of failure, since each step of the process requires time. The sequential steps are:

1. Corrosion in the brazed connection begins. This may take from years to decades, depending on the water chemistry, the design of the waterbox, the brazing materials used, and the skill with which the brazing is applied.

2. The water must leak from the brazed connection into the slot area. This probably takes years, depending on the stator temperature, the leak rate and the rate of corrosion.
3. Either a hipot test or a system disturbance must occur.

8.13.2 Root Causes

Poor brazing at the connection of the copper strands to the water box during manufacturing or rewinding of the stator has caused the majority of such leaks. Deficiencies in the brazing materials and the brazing procedures gives rise to a process called "crevice corrosion" [8.17]. Water fills small voids or fissures at the face of the brazed connection to the strands and between the strand and the waterbox. Crevice corrosion then begins when the following conditions are satisfied:

1. Uneven application of the braze results in small pockets at the face of the strand packet and between the strands and the waterbox.
2. The pockets or gaps in the braze allow water to stagnate locally.
3. Phosphorus in the braze combines with water and oxygen to produce phosphoric acid, which corrodes the braze and enlarges the pockets.

Eventually, tiny cracks develop in the braze, allowing water to leak between the adjacent strands or between the strand packet and the waterbox until the water comes in contact with the groundwall insulation.

Generators designed by one major manufacturer (and its licensees) have experienced most of the problems. A new brazing design by this manufacturer on units shipped since 1984 may eliminate this problem. Other designs use stainless steel tubes down the stator bar to carry the water, rather than using hollow copper tubes. Such designs have not developed water leak problems of this type to date. Even stator designs having hollow copper strands have been free of this problem when a different brazing process has been used and the brazed connections have a smooth surface so the water has no place to stagnate.

In an attempt to retard the rate of corrosion, conversion of high-oxygen (>2 ppm) stator cooling water systems to low-oxygen (<50 ppb) operation has been considered [8.18]. However, whereas aeration may play a role in the initiation of corrosion in the brazed joints, test data does not indicate that deaeration will halt the attack once it is underway [8.19]. A decision for such a conversion also has to address the risk of copper oxide formation leading to plugging of the strands if the oxygen regime cycles through or drifts into the high corrosive range of 100 ppb to 400 ppb of dissolved oxygen.

8.13.3 Symptoms

It can be difficult to determine from a visual inspection of the windings if the problem exists, since there is usually no evidence of the problem at the surface of a coil. However, in an advanced stage, the groundwall insulation may sound hollow when it is tapped with a hammer. The closer to the stator core the hollow sound occurs, the greater the risk of failure.

On-line and off-line tests are best at detecting the problem at an early stage. If the machine is hydrogen cooled, during operation there will be a high concentration of hydrogen in the stator cooling water. Also, the hydrogen makeup rate will be increased from normal. Proprietary on-line diagnostic systems are available for monitoring hydrogen leakage into the stator cooling water system. Off-line tests and repairs can then be planned to avoid a forced outage and to minimize outage time. In off-line testing, vacuum and pressure decay tests are convincing means to determine if the problem exists (Section 12.27). Capacitance mapping (Section 12.5) and helium or SF_6 tracer gas leak tests can help to locate the bars that may have experienced leaks.

8.13.4 Remedies

To prevent the process from occurring or to slow the process once it has started, it is very important for users to follow the manufacturer's instructions for maintaining the cooling system chemistry. In particular, the oxygen content must be monitored and adjusted to keep it within the specified limits. In addition, if the machine is a hydrogen-cooled turbine generator, the hydrogen pressure should be maintained above the water pressure to reduce the water leak rate even if cracks have already developed. If a winding is known to have severely delaminated insulation due to water leaks, it is very important to avoid hipot testing, out-of-phase synchronizations, or (if possible) current or voltage surges from the power system, unless one is prepared to handle the consequences of a ground fault. Even very degraded epoxy–mica insulation can withstand normal operating stresses for years.

Replacing individual bars and coils can help restore normal operating life of the winding. However, before bar replacement or rewinding is needed, implementing the following repairs may extend the life:

- Adding certain chemicals to the cooling water. The chemicals may collect at the cracks and be able to stop the leaks at the braze, without blocking cooling water flow elsewhere [8.20].
- Using localized epoxy to plug leaks in individual bars. Multistage epoxy compounds are injected in the pores of the braze connection to eliminate pockets where water can stagnate. While it is necessary to cut back insulation to expose the repair site, this method does not require the application of heat to the brazed connection.
- Performing a global epoxy repair. In addition to the known leaky brazed connections, all the brazed connections are repaired with the epoxy injection method as a preventive measure. This major undertaking may not be justified when compared to the additional assurance and possible uprating capability resulting from the rewind option.
- Replacing leaking water boxes and brazing with modern materials using modern methods. In addition to the need to cut back insulation, this method requires the removal of the existing braze and the application of a significant amount of heat, hence increasing the risk to insulation integrity.

The long-term effectiveness of some of these repairs has not yet been established. A computer program has been developed to evaluate the costs and benefits of each of these repair alternatives [8.21].

8.14 POOR ELECTRICAL CONNECTIONS

In a typical stator winding, there are many electrical connections. If the resistance of the connections is too high, overheating joints thermally degrades the insulation, eventually causing failure. Form-wound stators are more likely to encounter such problems, due to the large number of joints that are necessary between coils and bars. However, any poorly made stator can suffer from the problem.

8.14.1 General Process

In form-wound stators, the connections between copper leads in coils and bars are usually brazed or soldered. Motor and generator bus bar connections and leads in the terminal box are usually bolted to the power system cables or bus. If the connections have too much resistance, the leads will become hotter than necessary. The increased copper temperature further increases the resistance and, if the connections are hot for a long period of time, oxidation is accelerated, again further increasing resistance. Eventually, the copper may become so hot it melts, resulting in failure. In addition, the hot connections will thermally deteriorate the insulation over the connections (if present) and the adjacent coil insulation. If a current surge occurs, for example, from motor start-up, the degraded insulation may crack, exposing the copper conductors. The stator then has a high risk of failure from pollution, water, or voltage surges. Since the connections are generally far removed from the grounded stator core, the insulation over the leads can be extremely degraded, yet still not result in immediate failure.

Random-wound motors generally have far fewer connections, but there are still some connections between coils and in the motor terminal box. Since the insulated connections are usually in close proximity to the core or other coils, if the connection overheating has progressed to the state that the insulation is melted or so brittle it has fallen off, a ground fault can follow soon after.

Depending on how high the resistance is, and the proximity of the connection to ground or other conductors, failure may occur from minutes to decades after the machine is put into service.

8.14.2 Root Causes

Usually, poor workmanship is the main cause of connection overheating. Overheating can be caused by poor brazing, poor soldering, or inadequately tightened or designed bolted connections. In some cases, the connection between coils or bars may fatigue crack over time if there is excessive end-winding vibration. Some manufacturers have been known to use clips instead of brazed intercoil group connections. These clips can come loose, resulting in a high-resistance connection.

8.14.3 Symptoms

In a visual inspection of the winding, connections that pose a risk of failure are usually apparent because the insulation looks discolored (scorched). When touched or rubbed with a knife, the insulation may crack or be easily peeled off.

If one has access to an infrared imaging camera, then it may be observed that parts of the machine enclosure may be hotter than in the past, under the same load and atmospheric conditions. One can sometimes open the terminal box while the machine is in operation, and "shoot" the terminals to see if they are overheated. Unfortunately, except for hydrogenerators, it is impossible to locate bad connections between coils on-line using an infrared camera. The trend in winding conductivity (Section 12.3) over time is the best way to detect poor connections off-line.

Manufacturers sometimes use infrared camera scans to detect high-resistance winding connections. Sufficient current is passed through the windings to cause resistive heating, which can be detected by a thermovision camera.

8.14.4 Remedies

If overheating connections are suspected, the life of the stator can be extended by reducing load and ensuring maximum cooling air or hydrogen circulation over the affected areas. Otherwise, overheating connections should be separated and rejoined, and new insulation installed, during a suitable outage.

REFERENCES

8.1   S. Cherukupalli, G. Holdorson, and J. Lasko, "Rejuvenation of a 45 Year Old Generator Winding by Resin Injection," paper presented at EPRI, IEEE and CIGRE Rotating Electric Machinery Colloquium, Sept. 8–10, 1999.

8.2   W. Schier et al., "Impregnable Configuration of a Carrier Body and Winding Elements," International Patent Application No. PCT/EP92/01874, 1992.

8.3   R. J. Jackson and A. Wilson, "Slot Discharge Activity in Air-Cooled Motors and Generators," *Proc. IEE, 129,* Part B, 159–167, 1982.

8.4   J. F. Lyles, "Procedure and Experience with Thermoset Stator Rewinds of Hydraulic Generators," in *Proceedings of the IEEE Electrical Insulation Conference,* Chicago, pp. 244–254. October 1985.

8.5   G. C. Stone, R. G. VanHeeswijk, and R. Bartnikas, "Investigation of the Effect of Repetitive Voltage Surges on Epoxy Insulation," *IEEE Trans. EC,* 754–759, December 1992.

8.6   E. Persson, "Transient Effects in Applications of PWM Inverters to Induction Motors," *IEEE Trans. IAS,* 1095, Sept. 1992.

8.7   G. C. Stone, S. Campbell, and S. Tetreault, "Inverter-Fed Drives: Which Motors are at Risk?", *IEEE Industry Applications Magazine,* 17–22, Sept. 2000.

8.8   B. K. Gupta et al., "Turn Insulation Capability of Large AC Motors: Part 1—Surge Monitoring; Part 2—Impulse Strength; Part 3—Insultion Coordination," *IEEE Trans. EC,* 658–679, December 1987.

8.9   S. R. Campbell and G. C. Stone, "Examples of Stator Winding Partial Discharge Due To Inverter Drives," in *IEEE International Symposium on Electrical Insulation,* Anaheim, April 2000, pp. 231–234.

8.10   A. L. Lynn, W. A. Gottung, and D. R. Johnston, "Corona Resistant Turn Insulation in AC Rotating Machines," in *Proceedings of IEEE Electrical Insulation Conference,* Chicago, October 1985, p. 308.

8.11   W. Yin et al., "Improved Magnet Wire for Inverter-Fed Motors," in *Proceedings of IEEE Electrical Insulation Conference, Chicago,* September 1997, p. 379.

8.12   National Electrical Manufacturers Association Standard MW1000-1997, "Magnet Wire."

8.13   J. Dymond, K. Younsi, and N. Stranges, "Stator Winding Failures: Contamination, Surface Discharge and Tracking," in *Proceedings of IEEE Petroleum and Chemical Industry Conference,* September 1999, pp. 337–344

8.14   T. Shugg, *Handbook of Electrical and Electronic Insulation Materials,* 2nd ed., IEEE Press, 1995.

8.15   E. M Fort and J. C. Botts, "Development of a Thermalastic Epoxy for Large HV Generators," in *Proceedings of IEEE International Symposium on Electrical Insulation,* June 1982, pp. 56–59.

8.16 J. M. Butler and H. J. Lehamnn, "Performance of New Stator Winding Support System for Large Generators," in *Proceedings of American Power Conference*, 1984, p. 461.

8.17 M. P. Manning and R. T. Lambke, "Analysis of Stator Winding Clip-to-Strand Leaks," paper presented at EPRI Conference on Maintaining the Integrity of Water-Cooled Stator Windings, Atlanta, June 1995.

8.18 "Conversion to Deaerated Stator Cooling Water in Generators Previously Cooled with Aerated Water: Interim Guidelines," EPRI, Palo Alto, CA, Report No. 1000069, 2000.

8.19 A. K. Aggarwal, J. A. Beavers, and B. C. Styrett, "Mechanism for Corrosion of Brazed Joints in Stator Bar Clips of Water Cooled Generators," paper presented at EPRI Conference on Maintaining the Integrity of Water-Cooled Generator Stator Windings, Tampa, November 1996.

8.20 T. Sallade, R. Boberg, and B. Schaefer, "Development Statues of an Anaerobic Sealant Process for Global Repair of Leaking Stator Clips," paper presented at EPRI Conference on Maintaining the Integrity of Water-Cooled Generator Stator Windings, Tampa, November 1996.

8.21 J. Kapler, "Generator Stator Winding Leak Decision Advisor," paper presented at EPRI Conference on Maintaining the Integrity of Water-Cooled Generator Stator Windings, Tampa, November 1996.

CHAPTER 9

ROTOR WINDING FAILURE MECHANISMS AND REPAIR

The aging processes and failure mechanisms affecting rotor winding insulation and, where appropriate, the winding conductors are described in this chapter. Root causes of the failures and symptoms characteristic of the failure mechanisms are also discussed, as are the repairs and refurbishment options available for dealing with these conditions. The repairs covered in this chapter tend to be common for all types of deterioration processes. Thus, repairs are described for each type of rotor. Extensive repairs for large, round-rotor synchronous machines, such as rewinding, are discussed in Chapter 7.

The failure mechanisms cover the four major types of machine rotor designs: round rotors, salient pole synchronous rotors, wound rotors, and squirrel-cage induction motor rotors. Separate sections on each of these rotor types will include both the common and the specific failure modes involved.

Experience has shown that visual inspection is an invaluable tool for not only identifying the failure mechanism present but also for determining the optimum repair solution. However, since it is a major task to remove the rotor and disassemble the components in large motors and generators, other symptoms, measurements, and tests that can help with the assessment are also presented.

9.1 ROUND ROTOR WINDINGS

9.1.1 Thermal Deterioration

Thermal aging of insulating materials due to high temperatures has been the most studied and is perhaps the best understood because it was the most common underlying reason for winding failure of older types of insulation. Before the 1970s, the slot cell insulation often included composites made with kraft paper, mica splittings, and glass, glass/asbestos, or all asbestos cloth, all bonded with natural resins like shellac or solvent-based phenolic materials.

Electrical Insulation for Rotating Machines. By Stone, Boulter, Culbert, and Dhirani
ISBN 0-471-44506-1 © 2004 Institute of Electrical and Electronics Engineers

The organic components thermally aged in service and allowed the inorganic components to be displaced by cyclic mechanical forces experienced in operation, leading to cracks and gaps in the ground insulation and electrical failure. Similar failures often took place in the turn insulation tapes or strips. Materials such as rubber with mineral fillers, thin cotton phenolic laminates, and tapes with mica splitting and glass backing with early synthetic resin binders were often used. These tapes were applied by hand before the coils were wound into slots and not cured until all coils were assembled. Insulation abrasion and displacement were common during winding, often resulting in turn insulation failure. Although a few turn shorts can often be tolerated, they cause unbalanced heating of rotors in service and excess vibration due to thermal bowing of the rotor (Section 1.5).

Starting in the 1950s, modern thermosetting resins and glass fabrics began to replace the older systems. End-winding blocking was changed from phenolic-bonded asbestos cloth laminates to glass cloth with either polyester or epoxy laminating resins. Similar changes were made in the slot cell insulation. These changes raised the temperature class of the insulation into the Class F (155°C) range and significantly reduced the incidence of thermal aging. Smaller fields are insulated with nonwoven aramid sheets and tapes, whereas larger fields often use nonwoven glass laminates for insulation and blocking.

General Process. Glass laminates bonded with epoxy or polyester resins are commonly used for both the turn and the ground insulation in direct-gas-cooled rotors. Chapter 5 describes these and other types of materials in more detail, including materials used for end-winding insulation and blocking. Thermal degradation of these materials may be treated as a chemical rate phenomenon (described by the Arrhenius relationship, Equation 2.1) and includes loss of volatiles, oxidation, depolymerization, shrinkage, surface cracking, and embrittlement (Section 2.3).

Modern glass laminate insulation systems are made from Class F (155°C) materials for operation as Class B (130°C) systems. Since the average rotor winding operating temperature is in the range of 60°C to 90°C, there would appear to be an adequate temperature margin. However, the margin is reduced at hot spots in the winding, which cannot be measured directly (even the average rotor temperature is an indirect measurement derived from rotor amps and volts; see Section 13.1.2). Depending on the type of the cooling-gas flow system, the estimated hot spot temperature could exceed 130°C. Thus, thermal aging becomes a factor, particularly where Class B materials have been used. The thermal degradation is less likely on hydrogen-cooled rotors because of the lack of oxygen, which accelerates chemical aging, and because of the operating temperature margin usually available.

As described in Chapters 2 and 8, the higher the temperature, the faster will be the chemical reaction, resulting in shortened life of the insulation under thermal degradation.

Root Cause. All organic insulating and bracing materials deteriorate with time due to the heat from the windings, which promotes chemical changes leading to degradation of the material properties. The rate at which component materials deteriorate is a function of their thermal properties and the temperatures to which they are subjected. If the thermal ratings of component materials have been properly selected (see Chapter 5 for the application of various materials in insulation systems), the thermal aging and associated deterioration will occur gradually over an acceptable service life. On the other hand, the following may lead to an unacceptable rate of aging:

- Overload operation or high cooling-medium temperatures, leading to operating temperatures well above design values

- Inadequate cooling, which can be general; e.g., insufficient cooling air, hydrogen, or water, local dead spots in the cooling circuit due to poor design or manufacturing procedures, or blockages in cooling systems due to debris or movement of slot packing strips, insulation strips, etc.
- The use of materials that have inadequate thermal properties and consequently deteriorate at an unacceptable rate when operated within design temperature limits (unlikely with present day systems since most of these are thermally qualified to some IEEE/ANSI or IEC Standard)
- Overexcitation of rotor windings for long periods of time
- Stator winding negative sequence currents due to system voltage unbalance, faulty breaker operation, inadequate protection setting, etc. This leads to circulating currents in the rotor pole face and rotor wedges.

Symptoms. In operation, the first indication of thermal aging may occur only at an advanced stage of deterioration as a shorted turn alarm if a shorted turn detector (Section 13.6) is fitted, higher bearing vibration due to a shorted turn (Section 13.10.2), or a ground fault alarm or a higher shaft current in case of deteriorated ground insulation.

When shut down, a low insulation-resistance reading, low impedance reading, or a spike in the surge test curve (Section 12.21) can signal thermal degradation that has progressed to the stage where the insulation has failed or is about to fail.

Visually, thermal degradation is easy to spot if it has advanced to the point of rupture. The damage will exhibit a characteristic irregular-shaped hole burned through the material and the accompanying discoloration in various shades of brown and gray. In practice, this symptom can be detected quickly if the damage has occurred in the end-winding region. However, a significant amount of disassembly may be required if the damage is in the slot region toward the bottom turns.

When the damage has not yet caused rupture, the symptoms will show up as flaky surface, fine cracks, blistering, embrittlement, and discoloration. Gentle tapping with a blunt instrument is likely to result in pieces of insulation falling away from the parent material in the affected area.

9.1.2 Thermal Cycling

Operating temperatures have a direct aging effect on the insulation materials, as seen in the previous section. However, when the operating temperature undergoes variation due to load changes, start–stops, etc., additional stresses are set up that exacerbate the thermal aging process. The temperature changes cause the expansion and contraction of the winding copper relative to the insulation, giving rise to aging due to abrasion. Unless the rotor winding design can accommodate this movement of the copper, additional built-up stresses can damage not only the insulation but other rotor components.

General Process. During operation, copper losses from the winding, and to a lesser extent windage and iron losses from the rotor forging, cause an increase in temperature of the rotor components. The physical location of the components in the rotor is altered due to the axial thermal expansion caused by the increase in temperature. When the unit is shut down, the rotor cools down and the components contract to their original position, provided that the copper has not been stretched beyond its elastic limit and there is no restriction to their movement. Depending on the duty, this expansion–contraction cycle may be repeated hundreds of times during the life of the unit. The axial movement of the copper tends to abrade the ground insulation, especially toward the end of the rotor slots (Figure 9.1). Wear imposed on the winding insulation and other components, due to this repeated back and forth movement, results in mechanical aging due to thermal cycling.

Peaking units are more susceptible to thermal cycling damage compared to base loaded units because of the higher number of start–stop cycles. Longer rotors are likely to experience a higher level of damage due to the larger amount of expansion and relative movement. Similarly, air-cooled units, which generally operate at a higher temperature, will be affected more than hydrogen-cooled units.

Root Cause. A generator rotor has copper, aluminum, steel, insulation, blocking, and sealing components that are required to maintain their relative positions in the rotor to allow reliable operation within operating limits. Over the years, the spatial relationship between these components is altered due to thermal expansion and contraction, and as a result of start–stops and load changes.

The metallic components expand mainly due to the resistive heat from the winding. Expansion causes the components to undergo the following displacement mechanisms:

- The rate and the amount of growth is different for each component due to differences in coefficients of expansion. This results in the first level of displacement between the components.

- Although most of the components such as the slot copper and the wedges are axially oriented, critical components such as end-winding interturn connections and radial leads run in different directions. Components moving in two or more planes produce

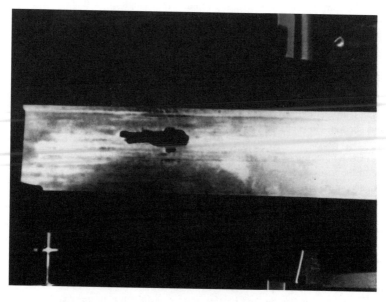

Figure 9.1. Hole abraded through the ground insulation from a slot liner near the end of a rotor slot.

stress risers at their connections, which represents the second level of displacement between the components.

- Last, when the rotor cools on shutdown, the components may be prevented from returning to their original locations because of distortion, copper exceeding its elastic limit, other small dimensional variations, and wear. This causes the third level of displacement between the components.

Consequently, instead of the "neutral" as-assembled status of the components, thermal cycling results in progressive displacement, increased mechanical stresses, insulation damage due to wear and mechanical loading, and interference of cooling passages. The faster the load changes, or the faster the excitation current changes, the greater will be the difference in thermal expansion between conductors and the insulation, and, therefore, the more likely that relative movement will occur.

Symptoms

Abrasion. Rubbing of copper against turn insulation, slot liner, and slot packing will show up as dusting, particularly in the slot exit area. Such damage can lead to looseness of the turns in the slot if not corrected. Once the coils become loose, the insulation may be further damaged by other aging mechanisms.

Turn Shorts. Repeated expansion–contraction of the copper causes pushing and pulling of the turn insulation with which it is in close contact. The insulation strips have a tendency toward ratchetting or "walking away" under the displacement mechanisms set up by thermal cycling, as described in the previous section. In a new rotor, the bonding glue between the insulation and copper would normally prevent this from happening. However, the bond becomes weak under the action of operating temperatures. Moreover, relative movement due to thermal cycling can not only shear the bond but can also crack the insulation in areas where the material may be inherently weak or where the movement is restricted. This is most noticeable in the end-winding area, where the turn insulation loses alignment with the edge of the coil and begins to creep out from between the turns. Cracking and creeping of the turn insulation eventually leads to turn shorts.

Radial Lead Shorts. Insulation of the radial lead is designed to be sturdy because of the requirements of handling the dual stresses of electrical and mechanical duty. However, the considerable forces of displacement due to thermal cycling can be large enough to wear out the insulation, resulting in chipping and cracking, leading to a short to ground.

Blocking of Ventilation Holes. Cooling gas travels through axial and radial ventilation ducts in the rotor to control the operating temperature. In the slot section, the ducts are formed by holes in the copper turns, turn insulation, packing, and wedges to form continuous passageways. The holes are normally elongated to prevent any restrictions due to misalignment from one section to the next. Thermal cycling causes relative movement between the slot contents and pushes the turn insulation out of position. Eventually, the ventilation holes in the copper and insulation become misaligned, restricting the passage of the cooling gas. Although some misalignment is normal and will not affect the cooling, it can cause uneven heating of the rotor in extreme cases. The extent of any such blockage can be determined by shining a flashlight down the ventilation holes to observe if blockage has occurred and recording the results for each winding slot.

9.1.3 Abrasion Due to Imbalance or Turning Gear Operation

Two- and four-pole rotors in large generators can weigh as much as 100 tons. Moreover, they have components such as the windings that can move independently in the axial, radial, and transverse directions. Ensuring that the rotor runs smoothly within acceptable vibration limits under all operating conditions is important to ensure reliable performance over the long term. High vibration can lead to relative movement of the rotor winding components, which, in turn, can lead to insulation and copper abrasion.

Considerable effort is required to ensure a mechanically balanced rotor system, starting with the design stage, to factory assembly and testing, site assembly, and setup, to operating practices and monitoring. Nevertheless, this balance is upset at times due to a number of factors, resulting in an increase in rotor vibration, which can cause further damage and even lead to shutdown of the generator.

General Process. Although rotor vibration is a mechanical phenomenon, its origin can be electrical or thermal in nature. The weight of the rotor is usually supported by a journal bearing at each end. Apart from the requirements for supporting this dead weight, the uniformity of weight distribution is an important mechanical consideration in the design and assembly of rotor components. Because of electrical and thermal stresses that build up during operation, additional forces are superimposed on the distributed weight of the rotor. Initially, these forces may be small; however, the underlying mechanisms being progressive in nature, the forces can become large enough over time to affect the weight distribution of the rotor.

Rotor dynamic performance is sensitive to changes in the weight distribution, particularly for two-pole designs in which the longer and thinner rotor shaft is more susceptible to bending forces. The increased vibration due to rotor unbalance can cause damage to the rotor components, leading to further imbalance. Ultimately, if relative movement occurs between the copper and the insulation, or the insulation and the rotor body, the insulation may abrade, leading to shorts.

Large turbine generator rotors have to be operated on turning gear at very low speeds (a few rpm) to prevent catenary "sets" in the shaft during unit shutdown. Operation at turning gear speed with low radial centrifugal forces on the windings can cause copper dusting abrasion, particularly when there are two or more conductor substrands that are not insulated from one another and the slot side packing is not tight enough to prevent sideways strand movement. The copper particles can lead to turn or ground shorts.

Root Causes

Rotor Vibration. Operating machines may exhibit high rotor vibration due to mechanical, thermal, or magnetic imbalances, which are caused by component movement or turn shorts (Section 13.10.2). This vibration, in turn, may lead to insulation abrasion.

The movement and distortion of winding conductors or bracing materials can result in mechanical forces on the rotor. Although some relative displacement of these components is normal during the cycle of operation and shutdown, excessive motion can result in permanent set and distortion. When free movement of components is restricted because of distortion, the weight distribution of the rotor is compromised due to the additional mechanical loads imposed. The balance of the rotor is upset, causing vibrations to increase.

Turning Gear Operation. At normal operating speed, centrifugal forces prevent relative movement between conductors or strands. However, during very low speed operation, these centrifugal forces on turning gear are negligible. Consequently, relative lateral movement between uninsulated subconductors will take place during rotor rotation if the side packing in the slot is not initially tight enough or is compacted by the side pressure induced by the conductors.

This relative movement between two adjacent copper strand surfaces results in fretting (abrasion), which generates small copper particles that migrate to the top of the slot under centrifugal action at synchronous speed. Much of this copper dust passes out of the slot via ventilation ducts, but some gets trapped in areas bounded by the top conductor, the space between the top creepage block, and the sides and top of the slot. If sufficient copper dust accumulates in one of these critical areas, a ground fault can occur as a result of the top turn shorting to the rotor body via the copper dust particles. It is also possible that turn-to-turn shorts could occur as a result of this mechanism.

Failure from this aging mechanism can be prevented by improved designs that eliminate significant sideways movement of subconductors, e.g., tapering conductors to fit slot contours, inserting insulation between each subconductor, or operating at turning gear speeds that produce sufficient centrifugal force to prevent relative sideways movement.

Common Symptoms. Increasing bearing vibration (Section 13.10.2) over time indicates that relative motion may be occurring and, ultimately, shorted turns have occurred. If the abrasion has progressed to shorted turns, then a rotor flux monitor (Section 13.6) will detect the problem.

The operation of the field ground protection indicates that the ground insulation of the rotor has failed. This can be due to a weak or damaged slot liner, failure of the winding lead insulation, bridging of the conductor bars and the rotor forging by foreign material, or by a broken or dislodged rotor component. It can also show that a severe condition of a turn short has deteriorated to the extent that it has established a path to ground.

When the rotor is dismantled, small copper particles may indicate relative motion due to turning-gear operation. Relative movement due to rotor vibration may produce signs of insulation abrasion without overheating.

9.1.4   Pollution (Tracking)

Insulation is used to electrically separate the copper winding at up to about 500 volts from the grounded rotor body. The insulation takes many forms and shapes due to the highly distributed nature of the rotor winding, the need to allow for movement due to expansion, and the geometry of the rotor forging, end rings, balance rings, support rings, slip rings, and connection hardware.

During assembly, care is taken to ensure that the isolation is maintained at these numerous interfaces of live copper and ground that can be potentially bridged to create a short. Normally, the creepage path to ground is more than adequate. However, during service, contamination at these critical interfaces can reduce the creepage path to such an extent that turn-to-turn shorts could develop or the winding could short to ground.

General Process. Hundreds of interfaces exist in a rotor where insulation separates the live parts from the grounded components such as the forging, wedges, retaining rings, and balance rings. Intermittent surface discharge between turns or from the winding components to grounded parts occurs when these insulation interfaces are compromised due to surface contamination. The discharge results in a chemical reaction of the components involved, producing carbon and other chemicals. The products of the reaction lodge themselves across the interface, creating a path of reduced resistance along which subsequent discharges occur.

Eventually, this path is worn deeper and wider after repeated discharges until their frequency increases to the point of a continuous discharge. This damages the insulation surface further and produces the characteristic "burn-in" impression commonly referred to as tracking.

Root Causes. Ingress of carbon dust from the brushes, dust in the cooling air, coal dust, fly ash, copper dust from abrasion (Section 9.1.3), iron dust, etc. is the primary source of the pollution that results in tracking in the rotor insulation. This can combine with moisture from cooler leaks and oil mist from seals and bearings to produce a partly conductive coating on the surface over time. The rotor speed helps to orient this foreign material along the many discontinuities of the rotor surface, particularly in the end-winding region. When sufficient quantities are deposited, currents may flow and contact surface discharges may be initiated between live and grounded parts or between turns, which eventually worsen into surface tracking.

Open-type machines are particularly vulnerable to tracking because of their exposure to the elements and pollutants. Hydrogen-cooled generator rotors are inherently protected from many pollutants that affect the open-type generators. Other precautions such as epoxy painting of the rotor body to obtain a smooth surface further help to keep any pollutants from lodging themselves on the surface. However, it is very difficult to maintain a pristine environment within the generator at all times. Failure of hydrogen sealing and coolers and variation in the dew point of hydrogen do occur from time to time, and abrasion products are generated within the machine due to the relative movement between parts. Hence, hydrogen-cooled generator rotors can also be affected by tracking.

Common Symptoms. Symptoms of tracking are not apparent during operation until the damage worsens to turn-to-turn or ground faults in the winding. Such faults can be detected by vibration and flux monitoring (Sections 13.6 and 13.10.2). Upon disassembly and inspection, the presence of surface contaminants would indicate failure due to tracking. Other sites may display the characteristic "burn-in" prints on the surface insulation, which are still at an early stage of tracking damage. Severe faults to ground will result in a low insulation resistance (Section 12.1) and severe turn-to-turn tracking may be detected by the rotor surge test (Section 12.21).

9.1.5 Repetitive Voltage Surges

In generator rotors, the applied voltage of around 500 V DC is rather modest compared to the rating of insulating materials used, particularly in modern designs. The applied voltage further divides between the turns to result in as little as 10 V between turns, which is a small fraction of the insulation voltage rating. The insulation would last indefinitely under these minor electrical stresses. However, transient overvoltages from the excitation supply can be orders of magnitude higher, and may lead to insulation degradation.

General Process. Events internal or external to the excitation system can induce large transient voltages in the rotor windings. While the occasional spike may not be harmful, repetitive spikes can reduce the life of the insulation due to gradual deterioration caused by partial discharges. Thus, this is similar to the stator winding aging process caused by inverter-fed drives (Section 8.7.1). Insulation that has been already weakened by other aging mechanisms is particularly vulnerable to repetitive voltage surges. As the insulation between turns is the thinnest and is subject to high levels of mechanical stresses, the voltage surges are most likely to cause turn-to-turn faults.

Root Causes. Static excitation systems inherently produce voltage surges due to their wave-chopping circuitry. Typically, six voltage surges are created per AC cycle. The magnitude of the surges can be higher than expected due to voltage reflections and oscillations caused by inductance and capacitance. While a correlation between the surges and insulation failure has not been firmly established, the design of any excitation system should be reviewed with respect to the levels of voltage spikes produced.

For faster response to disturbances on the power system or to sudden load changes, excitation systems are often designed with capability of "field forcing." Higher levels of output are thus available over short periods of time, which can increase the voltage to several times the rated level. As such operation often occurs during periods of disturbances on the power system, it can worsen the impact of the voltage surges striking the rotor winding.

The rotor is also exposed to events affecting the stator because of induced effects across the air gap. Hence, disturbances on the system such as breaker opening, equipment maloperation, and lightening are transferred to the rotor winding during these transient events, elevating the voltage across the insulation to orders of magnitude beyond the rated voltage.

Connecting the generator to the power system is also a potential cause of transient overvoltages on the rotor winding. Missynchronization, motoring, and load shedding can give rise to overvoltage surges that can spike to ten times the applied excitation voltage, depending on the angle between the voltage waves involved.

Common Symptoms. Repetitive voltage surges can result in turn-to-turn shorts. The first such short may not be evident during operation unless the generator is fitted with a flux monitor (Section 13.6). Otherwise, several shorted turns may develop before the effect can be felt through higher vibration levels (Section 13.10.2).

An excitation system prone to producing large voltage spikes, transient events on the power system, or synchronizing difficulties experienced at the time of the higher rotor vibration would indicate damage to the insulation due to voltage surges.

When the rotor is examined, the punctured insulation may show no signs of overheating or abrasion.

9.1.6 Centrifugal Force

At speed, the rotor generates large mechanical stresses in the form of centrifugal forces that can exceed 1,500 tons at the wedges and 15,000 tons at each retaining ring. Significant tangential forces are also present, particularly during startup and shutdown of the generator. Rotor insulation systems may crack or yield under these forces, leading to turn or ground faults.

General Process. Because of the rotor speed, centrifugal forces are developed that can exceed 8,000 times the weight of the component. The self-weight of the insulation materials used will result in significant stresses under these forces, which are generally compressive in nature and well within their compressive strength. However, the contribution of the copper bars to the total stress on the insulation materials is orders of magnitude larger.

Insulation materials made from quality stock and applied with adequate margins can endure these crushing forces over long-term operation. However, where the materials are weakened due to inadequate quality control or other aging mechanisms such as thermal aging, the insulation can bend, buckle, and crack under the influence of the large centrifugal forces. This can lead to turn-to-turn faults or ground faults. The insulation materials involved include slot liners, turn insulation, slot packing and pads, bracing materials, and connection insulation. The effects of centrifugal forces are a function of the design of the winding bracing system, the properties of the materials used, and the frequency of start and stop cycles.

Root Causes. High continuous stresses can cause yielding, distortion, movement, and cracking of winding, insulating, and bracing materials if their design, constraint provisions, or mechanical properties are inadequate. When winding bracing or conductor bonding materials crack, the winding becomes loose and abrasion of insulation due to relative movement occurs. This eventually leads to turn-to-turn or ground faults. The winding conductors or connections may also become distorted and fracture, causing open circuits in the winding, and, in extreme cases, pieces of the conductor may break loose and fly outward into the stator winding, causing faults. Also, slot armors may migrate radially outward, leading to rupture of the insulation or exposure of the conductors.

Cyclic stresses may cause loosening or fatigue failure of winding insulation system components. This will result in the same types of failures as those induced by overstressing. In case of rotor retaining rings, fatigue failure induced by cyclic stressing can lead to fracture, fire, explosion, and severe consequential damage to the generator.

Slot wedges or wedge retaining grooves in large generator rotors may also fracture as a result of cyclic stressing. In rotors with short slot wedges, the damage to the wedge grooves (T-slots) becomes more pronounced at locations that correspond to the wedge ends.

Common Symptoms. Direct evidence of damage due to centrifugal forces may be determined during rotor overhaul. Early stages of such damage can be detected during inspection of various rotor components. Depending on the extent of the disassembly, the following signs may indicate the development of this failure mechanism in various stages:

1. If the retaining ring has been removed, the interface of the ring insulation and the end-winding bars underneath can provide an important clue. An imprint of the shape of the end-winding on the insulation surface is normally seen as an indentation. The depth of the indentation should be uniform around the circumference and not exceed 25% of the thickness of the insulation at any point. Any variation should be investigated for high centrifugal loads. The same action applies in the case of cracks emanating from the edges of the indentations or for pieces missing from the insulation.

2. The top of the slot liners (slot armor) may show fretting, cracking, or chipping due to high centrifugal force pushing the liner against the retaining rings. Cycling stresses may combine to hasten the damage to the liner. This is also symptomatic of inadequate blocking of the slot liner.

3. End-winding turn insulation is prone to damage under centrifugal forces, particularly where the insulation from the straight slot section is joined to the curved insulation of the end-winding. Lap joints are used in this area to ensure adequate creepage distance to prevent shorts. This weak spot can break down under cyclic centrifugal forces if not properly established. Only the top turn insulation can be inspected for this condition unless the winding bars are to be removed for further inspection or for other reasons.

4. If the wedges are removed from the slots, the top pad and any packing strips used should be inspected for damage. Where short wedges are used, the surface of the top pad may show abrasion or cracking in the circumferential direction, corresponding to the ends of the wedges.

5. Chipping or cracking of the outer edges of the end-winding blocks may indicate damage due to centrifugal forces.

During operation, the above failure mechanisms may not be evident in their early stages. When the damage has progressed to cause a change in operational behavior, the rotor is likely to exhibit increasing vibration and turn shorts, which can be detected by flux monitoring (Section 13.6).

9.1.7 Remedies

Restoring the section of the insulation that is damaged is the basic repair for thermally failed insulation. This can be accomplished in several ways, depending on the resources and materials available. The effort required can range from minimal disassembly to the removal of several turns, depending on the location and severity of the damage. Also, further modification, refurbishment, or changes to maintenance and operating practices may be necessary to address the root cause of the damage and to ensure reliable operation in the future without reoccurrence of the problem. It is generally advisable to rebalance a rotor that has been subjected to extensive repair. Rebalancing can be used to overcome thermal bowing caused by a few turn shorts without having to repair the shorts.

Actions that may be required to restore the rotor insulation are summarized below.

Section of turn insulation. Replacement of the damaged section of the insulation. This should only be considered for turn insulation and when the damage is in the end-winding region. Continuity of the insulation without gaps must be assured. The same considerations apply to winding lead connections. It is also suitable for insulation under the retaining ring that is normally replaced with new material before the ring is reinstalled.

Full-length turn insulation. Replacement of the entire length of the turn insulation is necessary when the damage is in the slot section of the winding. This is to ensure integrity of the insulation under other stresses in operation and to avoid the inadvertent reduction in creepage distances. If turn insulation damage is widespread, all the rotor turns should be removed and the interturn insulation replaced. In such a situation, the opportunity should be taken to replace the slot liner at the same time.

Repair of the slot liner. Repair of the slot liner may be possible if the damage is confined to the area outside the slot. In the case of cracks, the end of the crack should first be drilled out to prevent the crack from propagating towards the rotor forging. Pieces of epoxy glass laminates (G10 or G11 glass) can then be bonded to the damaged area with epoxy glue to strengthen the liner. Where limited access does not allow drilling a small hole at the apex of the crack, the repair can proceed as above, but the next rotor inspection would have to be scheduled earlier than planned to determine the effectiveness of the repair. The same approach can be taken when the cracks are close to the rotor forging but a decision is made for a local repair rather than a replacement of the liner due to other constraints.

Replacement of the slot liner. Replacement of the slot liner is the most extreme form of insulation repair as it amounts to a full rotor rewind. If the liner cannot be restored by local repair or in the case of ground faults, all the insulation should be replaced, including turn insulation, retaining ring insulation and winding lead insulation. The end-winding blocks may be reused provided they still fit and are not damaged. Any copper smearing on the blocks or windings due to rubbing against the winding should be cleaned up.

Reuse or replace the copper. One of the most difficult decisions facing machine owners in the case of a rotor rewind is whether to reuse or replace the copper. Although some companies would start with new copper, others would tend to reuse the existing copper since the difference in cost can be significant. The following factors may be considered when evaluating options:

- The condition of the copper
- Remaining life of the generator
- Operating regime, e.g., base-load, peaking, or reserve
- Effort involved in removing existing insulation (if bonded)
- Effort involved in restoring existing copper (extent of deformity)
- Number of brazing joints involved
- Opportunity and benefits of uprating
- The cost of the new copper and winding manufacture
- Delivery schedule for the new copper winding.

In addition to the repairs mentioned above, the aging process can be slowed by taking the following action:

- Reduce load
- Slow down the rate of load changes
- Reduce excitation
- Clean rotor to eliminate blockage
- Ensure that negative sequence currents are minimal

9.2 SALIENT POLE ROTOR WINDINGS

As discussed in Section 1.5.1, there are two types of windings in salient pole rotors. These are the "strip-on-edge" type used in large machines, especially if they are high speed, and the "multilayer wire wound" type used in large slower-speed and small high-speed motors and generators. Some of the coil insulating materials used are dependent on the type of winding construction. On the other hand, the insulation between the coils and poles can be the same for both types while the turn insulation is quite different.

A weakness in some designs is the susceptibility of the series connections between poles to fail from fatigue. Field windings tend to move relative to one another during machine operation. This results in flexing of the connections between them, which can lead to failure from fatigue, even if they have a laminated construction. This will also cause failure of the insulation on these connections. The top series connections are most likely to experience to this problem.

Failure mechanisms for slip ring insulation found on some salient pole rotors is covered in Section 9.3.5.

9.2.1 Thermal Aging

All insulating and nonmetallic bracing materials deteriorate with time due to the heat from the windings. The rate at which component materials deteriorate is a function of their thermal properties and the temperatures to which they are subjected. If the thermal ratings of component materials have been properly selected, the thermal aging and associated deterioration will occur gradually over an acceptable service life.

General Process. Modern salient pole winding designs using Nomex™ ground and turn insulation in strip-on-edge windings, glass laminate pole washers, Dacron-glass-covered high-temperature enamel turn insulation in wire wound poles, and thermosetting bonding resins provide insulation systems that have a thermal rating of at least Class F (155°C). If these materials are operated at Class B (130°C) temperatures, they should have a more than adequate thermal life. The materials most susceptible to thermal degradation are organic bonding and backing materials, whereas inorganic components such as mica, glass, and asbestos are unaffected at the normal operating temperatures of electrical machines.

The thermal life of insulation at hot spots in windings is significantly reduced since the margin between operating temperature and thermal rating is much less. This effect is more critical in older Class B insulation systems and the presence of such hot spots is very difficult to detect.

Root Causes. The following are the most common causes of thermal aging in salient pole windings:

- Overloading or high air temperatures leading to operating temperatures well above design values
- Inadequate cooling, which can be general, e.g., insufficient cooling air or cooling water, or local dead spots in the cooling circuit due to poor design, manufacturing, or maintenance procedures
- The use of materials that have inadequate thermal properties and consequently deteriorate at an unacceptable rate when operated within design temperature limits
- Overexcitation of rotor windings for long periods of time
- Negative sequence currents in stator windings due to system voltage imbalance, etc., which leads to circulating currents on the rotor

Common Symptoms. Severe thermal deterioration can give rise to shorted turns and/or ground faults, which can be detected by bearing vibration monitoring (Section 13.10.2). Visual symptoms of thermal aging in salient pole rotor windings are dependent on whether it is general or localized.

General Thermal Aging
- Loss of bonding between conductors and brittleness in the bonding varnish or resin
- Shrinkage of pole washers and intercoil bracing insulating materials
- Brittleness and darkening of the insulation system materials
- Looseness of the windings on the poles due to ground insulation shrinkage (indicated by dusting at the interfaces between components)

Local Thermal Aging
- Signs of local overheating of insulating materials with the remainder of the materials being in good condition

9.2.2   Thermal Cycling

Insulation aging from thermal cycling occurs mainly in synchronous motors that are started and stopped frequently. Thermal cycling is rarely a problem in salient pole generator rotors.

General Process. There are two heat sources within a rotor when a synchronous motor is started. One mainly applies to motors that are started directly-on-line, causing heating due to currents flowing in the pole tips of solid pole rotors or the damper winding in those with laminated poles. The other is the I^2R heat generated in the windings once excitation is applied.

Frequent starts and stops cause winding expansion and contraction as a result of the presence or loss of these winding heat sources. Relative movement due to the different coefficients of thermal expansion in the various components leads to insulation abrasion.

Root Causes. The thermal cycling resulting from frequent starts and stops leads to the cracking of the resin or varnish bonding the insulation system components together. This causes loosening and relative movement between these components, which leads to increased looseness and abrasion. Also if the windings are restrained from returning to their cold position, they may become distorted. Poor design or too-rapid or too-frequent load cycles for the design are the root causes.

Common Symptoms. If the following symptoms are found with no evidence of significant winding insulation overheating in a synchronous motor that is frequently started, then thermal cycling is the most likely cause:

- Cracking of the insulation bonding the conductors together and the winding to the pole ground insulation
- Distortion of the windings
- Windings loose on poles

9.2.3 Pollution (Tracking and Moisture Absorption)

Salient pole rotor windings, especially strip-on-edge types, are generally susceptible to failure from contamination by conducting materials because they rely on adequate creepage distances between bare copper conductors to prevent shorts. Such problems are not confined to machines with open-type enclosures since oil leaking from bearings and moisture from condensation or leaking air coolers can contaminate windings. Such problems can be avoided in wire-wound types by encapsulating the windings to keep contaminants out.

General Process. When contaminants such as moisture, coal dust, and oil–dust mixtures cover the surfaces of salient pole windings they can produce conducting paths between turns and to ground. This can lead to turn-to-turn failures (especially in strip-on-edge types) and ground faults. Certain chemicals can also attack insulating materials to cause them to degrade.

Earlier insulation systems containing materials such as asbestos, cotton fibers, paper, etc. bonded by organic varnishes are much more susceptible to failure from moisture absorption.

Root Causes

- Ingress of contaminants such as coil dust, fly ash, iron dust, etc. into open-type machines, especially in the presence of oil leaking from bearings
- Ingress of moisture from the atmosphere (open enclosures), from condensation (all enclosures), and leaking coolers (TEWAC enclosures)
- Ingress of chemicals that attack the pole winding insulating materials

Common Symptoms. Pollution may cause a low insulation resistance (Section 12.1) or cause a failure in the voltage drop test (Section 12.20). The rotor may also appear to be greasy, wet, or have a liquid film. If severe, dark carbonized tracks may be present between turns or to ground.

9.2.4 Abrasive Particles

As with stator windings (Section 8.9), rotor windings operated in environments containing abrasive dusts will also experience insulation failures from dust impingement.

General Process. Abrasive dust from the surrounding atmosphere carried into the interior of a motor or generator by cooling air will abrade the rotor winding insulation surfaces. This may eventually expose the conductors in multilayer wire-wound poles, resulting in turn shorts. Also, the ground insulation in both types of salient pole windings and their interconnections may be eroded to cause ground faults.

Root Causes. The use of open-enclosure machines without inlet air filters in locations where there are abrasive materials in the environment.

Common Symptoms. Winding faults in conjunction with erosion of the coil and connection insulation where there is evidence of abrasive materials such as coal dust, iron ore dust, and sand inside the machine enclosure.

9.2.5 Centrifugal Force

Among the most common causes of failure in salient pole rotor windings are the continuous centrifugal forces imposed on them by rotation and the cyclic centrifugal forces induced by starting and stopping.

General Process. The radial and tangential centrifugal forces imposed on rotor winding insulation system components tend to distort the coil conductors and crack the coil insulation if they are not adequately braced. If the pole winding bracing is inadequate or becomes loose, the resulting coil vibration and movement of the coils on the poles will cause abrasion of the conductor and ground insulation. Inadequate interpole bracing in large, high-speed machines will lead to coil distortion, whereas erosion from loose windings will occur mainly during starts and stops. Winding looseness can also lead to pole washer and intercoil connection cracking from fatigue. Mechanical winding stresses will become excessive and cause serious winding damage if the rotor is made to run over speed.

Root Causes

- Inadequate intercoil bracing due to poor design, or shrinkage of materials from thermal aging
- Inadequate radial coil bracing due to poor design, failure of pole washers, or shrinkage of coil insulation/bracing from thermal aging
- Fatigue failures of intercoil connections from relative coil movement
- Frequent starts and stops causing low-cycle fatigue failure of winding components
- Inadvertent overspeed of rotor causing overstressing of winding conductors and insulating materials

Common Symptoms. Operating symptoms of mechanical aging due to centrifugal forces may be confirmed by a visual examination of the rotor windings and bracing components. If there are signs of abrasion, fracture, etc., without indication of thermal or electrical aging, then mechanical aging is the likely cause. A detailed examination of the type of fracture in the winding or bracing materials may also indicate whether it resulted from overstressing (yielding) or cyclic stressing (fatigue). If a turn short has occurred, the bearing vibration (Section 13.10.2) level may increase.

9.2.6   Repetitive Voltage Surges

The normal DC voltage applied to rotor windings does not cause rotor insulation aging. Also, normal voltage levels in a rotor winding are usually so low that they will not induce insulation aging even in weakened materials. Hence, electrical stress is not an important cause of aging. However transient overvoltages induced by fault conditions on the stator side or faulty synchronization can cause rotor winding insulation puncture.

General Process. High transient overvoltages may be induced into rotor windings by phase-to-phase stator winding short circuits, faulty synchronization, asynchronous operation, or static excitation systems. Such transient voltages, in conjunction with weak insulation or insulation that has been degraded by thermal or mechanical aging, can cause failures, which are predominantly turn-to-turn. These overvoltages are most severe in salient pole windings due to their design configuration.

Root Causes. All of the following causes of voltage transients in rotor windings lead to failures of weakened winding insulation, especially that between turns on the winding poles:

- Faulty synchronization due to automatic synchronizer defects and manual synchronization errors.
- Asynchronous operation due to loss of excitation while the stator is energized or inadvertent energizing of the stator, with no excitation, while the machine is shut down.
- The introduction of static exciters about 30 years ago brought with it some concerns about the possible long-term aging effects of their pulse voltage transients on field winding insulation. These high-frequency voltage spikes, generated by thyristor commutation at a repetition rate of 6 pulses per cycle, can reach magnitudes of 3 to 4 kV during field forcing (Section 9.1.5). Operating experience at levels in use to date in salient pole rotors has not revealed these voltage spikes to be a significant electrical degrading mechanism on healthy turn or ground insulation in salient pole machines. They may accelerate the failure of weak or degraded insulation, however.

Common Symptoms. (a) Turn faults in synchronous machine field windings induced by voltage transients and static exciter voltage spikes are sometimes indicated by high rotor vibration due to magnetic or thermal imbalance. (b) Confirmation of rotor winding shorted turns by tests and/or visual inspections with the knowledge that they have had transient voltages imposed on them.

9.2.7   Remedies

This section deals with repair methods for salient pole rotor winding coils, ground insulation, bracing materials, and connections. Damper (amortiseur) winding repairs are covered in Section 9.4, which deals with squirrel-cage induction rotors.

Strip-on-Edge Coil Repairs. If the coil turn insulation has degraded and the conductors are in good condition, the coils can be dismounted from their poles and the insulation on them incinerated in an burnout oven. The copper conductors are then cleaned before new turn insulation is applied. The new turn insulation, which is usually a Nomex™ type of material (Section 4.4.2), can then be installed and bonded to the conductors with a thermosetting resin. The complete coil is then hot pressed to consolidate the conductors and the turn insulation.

If the coil conductors have been distorted due to excessive centrifugal forces, then it is best to replace them. The new coils would be insulated in the same manner as described above.

The ground insulation between the coils and pole body and the top and bottom pole washers should also be replaced. New pole washers should be made from a single piece of epoxy or polyester-resin-bonded glass fiber sheet material. It is important that, once the new coils are installed, they be tight on the poles and, for high-speed machines, the intercoil bracing must also prevent any coil bulging due to centrifugal forces.

If the coil top washers crack but the coils themselves are in good condition, and if the poles have bolt-on-tips, then new washers can be installed without disturbing the coils. This is done by removing the pole tips and old top washers, replacing the washers, reinstalling the pole tips, and torquing down their mounting bolts.

Multilayer Wire-Wound Coil Repairs. It is not practical to reinsulate the conductors on this type of coil since the wire is purchased with the insulation already applied. For replacement of this type of winding, the poles, with coils installed, must be removed and placed in a burnout oven in which both the turn and ground insulation is incinerated. Once this is done, it is easy to remove the coils from their laminated poles, which can be reused if they were not damaged by the burnout (Section 10.1). The condition of the pole lamination and damper winding must be checked (Section 14.3) and any repairs to these completed before installing the new winding.

Once the poles have been cleaned, new ground insulation and top washers are installed. The new coil can then be wound directly on to the pole using a preinsulated magnet (winding) wire, which normally has an enamel covering overlayed with Dacron™–glass blended-type fibers. A thermosetting resin is applied between each coil layer as it is wound and, once complete, the coils are dipped in resin and baked in an oven to cure the bonding resin. If a sealed coil construction is required, a layer of insulation is applied to the outside of the coil and bonded to it with resin. Once this is done, the bottom washers can be installed and the poles remounted on the rotor.

Coil Connection Repairs. Damaged coil connections and/or their insulation can be replaced without disturbing the pole windings. Before doing this, it is important to determine why the failure occurred, since an improved connection design may be required to prevent recurrence of such failures, e.g., better support against centrifugal forces.

Winding Contamination and Damage from Abrasive Particles. Providing the coil conductors and insulation are in good condition, contaminated windings can be refurbished without major expense.

If the contamination is moisture, the winding should be dried out in an oven or by passing current through it. Insulation resistance and polarization index tests, as described in Section 12.1, can then be performed to confirm adequate dryout. Once the windings are dry, they can be coated with a heat-curing or air-drying varnish to seal their surfaces from moisture.

Oil and dust contamination require more aggressive treatment to remove them. If the rotor can be sent to a service shop, it can be steam cleaned, dried out, and varnish treated. Cleaning with dry ice or other abrasives (corn cob, walnut shells) is often effective. Since the equipment used to perform this process is portable, it can be done on-site. However, experience has shown that dry ice cleaning may be less effective if oil is present.

The extent of insulation damage from abrasive particles can vary significantly. Therefore, required repairs can be minor or major in nature. If only minor insulation damage has been inflicted, repairs can be implemented by applying a surface treatment of insulating varnish or resin using a dip and bake process. On the other hand, abrasion that has completely removed the conductor and/or ground insulation will require rewinding or winding insulation replacement.

9.3 WOUND INDUCTION ROTOR WINDINGS

The advent of the solid-state variable-speed drives for SCI motors, which can provide high starting torques and low inrush currents, has made the wound-rotor induction motor less popular since the 1990s. There are, however, significant numbers of these machines still in service for applications such as crane drives, wood chippers, etc. As indicated in Section 1.5.3, there are two basic types of construction: random-wound and bar-lap/wave-wound. The ground insulation is similar for both types but the turn insulation is different. In the random-wound type, enamel or enamel overlaid with glass fibers is used, whereas the bars in the lap- and wave-wound types are usually insulated with a varnish or resin-bonded tape.

These three-phase windings operate at similar voltages to the random-wound stators described in Section 1.3.1. The thermal aging and failure mechanisms are, therefore, similar. As a result of the low operating voltage, continuous electrical aging is not a factor. However, electrical transient voltages induced in these windings can cause them to fail if their insulation has been weakened by thermal degradation and mechanical forces, or they have become contaminated. The high continuous and transient centrifugal forces imposed during operation of this type of rotor can also induce winding failures. Also, the slip rings used to allow external connections of starting resistors or slip energy recovery solid-state converters to the rotor windings have ground insulation on them that can be shorted.

Since thermal aging failure mechanisms are similar to those found in random-wound stator windings (Section 8.1), these are not covered in this section. Also, since failures due to contamination are similar to those found in salient pole rotors (Section 9.2.3), these mechanisms are not covered. The following are electrical and mechanical failure mechanisms that are specific to wound-rotor windings.

9.3.1 Transient Overvoltages

In a wound-rotor induction motor, there is a transformer effect between the stator and rotor windings. Consequently, power-system surge voltages imposed on the stator winding will induce overvoltages in the rotor winding. This overvoltage may puncture the turn or ground insulation.

General Process. Providing there is adequate turn and ground insulation on the rotor winding, such voltages should not cause electrical aging; that is, partial discharge is unlikely. Transients will, however, accelerate the failure of insulation that is initially weak, or that has been degraded by thermal or mechanical aging.

Root Causes. Turn-to-turn, phase-to-phase, or ground faults can be induced by transient rotor winding overvoltages if the insulation dielectric strength has been significantly reduced by aging. The contribution of these voltages to insulation failures is difficult to verify unless their magnitudes and frequency of occurrence can be determined.

Common Symptoms. These are the occurrence of rotor winding turn or ground shorts in motors operating on power systems in which surge voltages are known to be present.

9.3.2 Unbalanced Stator Voltages

Unbalanced stator winding power supply voltages will induce negative sequence voltages and currents in the rotor winding. These can lead to insulation overheating.

General Process. Negative sequence currents induced by unbalanced stator winding voltages increase rotor winding heating in all phases and, therefore, induce accelerated thermal aging of both the turn and ground insulation.

Root Causes. Power distribution system design or unbalanced phase impedances (e.g., due to high resistance connections) cause unbalanced voltages to be applied to the motor stator. Due to the transformer effect between the stator and rotor, the negative sequence fluxes generate negative sequence currents in the rotor winding. These significant additional currents cause the rotor winding to operate at a substantially higher temperature, which accelerates insulation thermal aging.

Common Symptoms. Signs of general rotor winding overheating in conjunction with evidence of significant, i.e., more than 3%, stator winding supply voltage imbalance while operating within the nameplate rating of the motor are observed.

9.3.3 High-Resistance Connections—Bar Lap and Wave Windings

If a joint between two conductors has been poorly soldered or brazed, it will present a high resistance to the current flowing through it under load and this will produce overheating of the joint insulation.

General Process. The excessive amount of heat produced by high-resistance bar-to-bar connections induces rapid thermal aging of the insulation around the connection and on adjacent connections until a turn-to-turn, phase-to-phase, or ground fault develops. In many cases, the heat generated is sufficient to melt the solder or brazing material in the joint.

Root Causes

- Poorly brazed or soldered connections between rotor winding bars that are not found during the manufacturing process
- Failure of the joint between bars due to low-cycle fatigue from frequent starting

Good quality control checks of a completed winding such as surge testing and the use of thermal imaging devices to detect "hot" joints during winding manufacture should minimize the possibility of failures of this type.

Common Symptoms. There will be signs of excessive heat and burning in the area of the connections, together with pieces of molten solder or brazing material sprayed around the inside of the motor by centrifugal forces.

9.3.4 End-Winding Banding Failures

Application of banding over the rotor end-windings is required to brace them against the high centrifugal forces imposed on them during operation. Up until the early 1950s, end-winding banding consisted of a number of turns of round steel wire applied tightly over an insulating layer, which was required to give mechanical and electrical separation from the conductor insulation. The round wires were bonded together with a low-melting-point solder. The development of prestressed resin-coated fiber material then prompted motor manufacturers to start using this material because of its superior mechanical and thermal capabilities, as well as its elasticity.

General Process. If the steel wire or resin-coated fiber materials fail from overheating, overstressing, or poor manufacture, the end-windings fly outward under the influence of centrifugal forces. This results in a rotor winding ground fault and, often, a consequential stator winding failure.

Root Causes

- Excessive winding temperatures, causing wire-banding solder to soften or melt
- Excessive winding temperatures, causing resin-coated fiber banding to thermally age and the resin bonds between layers to become brittle, crack, and fail
- Poor manufacturing processes that result in inadequate mechanical strength to withstand high imposed centrifugal forces.

Common Symptoms

- If the cause is winding overheating, there will be evidence of thermal degradation in the remaining coil insulation in the slot region.
- If the cause is poor manufacture, there will be no evidence of winding insulation overheating.

9.3.5 Slip Ring Insulation Shorting and Grounding

The three slip rings in a wound-rotor motor must be separated from the shaft by a layer of insulation applied between the two. The spacing between the rings must be sufficient to provide an adequate electrical creepage distance. Also, the two outer rings are usually connected to the winding leads via studs that pass through the other rings. These studs must be electrically isolated from the rings and this is normally done by fitting insulating tubes over them. The failure mechanisms in this section also apply to salient pole machines with slip ring connections to their field windings.

General Process. If the slip ring enclosure is contaminated with oil, dust from brushes, moisture, or a combination of these, then shorting between rings and the shaft and/or between the rings can occur. If this happens, serious damage can occur to the shaft, the rotor windings, and the slip rings. Also, if the shaft or stud insulation fails due to thermal aging or mechanical stresses, these types of failures will also occur.

Root Causes

- Failure to periodically clean up carbon dust resulting from brush wear
- Ingress of dust, oil from bearings, and moisture into the slip ring enclosure
- Failure of slip ring shaft insulation due to thermal or mechanical stresses
- Mechanical failure of the slip ring connection stud insulation, causing phase-to-phase failures

9.3.6   Remedies

This section deals with the rewind and repair of the windings, bracing, and slip ring insulation found in this type of winding.

Failed Windings.   If the failure is due to the end-winding banding coming loose or a serious winding insulation fault, then the winding damage is so severe that there is no alternative but replacement. A complete rewind is required if random-wound rotors experience an insulation failure since it is impractical and uneconomic to repair these.

Individual top coil legs in bar-type windings with failed insulation can be rewound if there is a slot opening large enough to remove the old coil. This will necessitate removing the end-winding banding and replacing it after the repair. If the insulation on a bottom coil fails, a rewind is necessary since other coils will have to be removed, damaging the insulation in the process.

Contaminated Windings and Slip Ring Insulation.   If the windings are in otherwise good condition, they can be cleaned and refurbished in a similar manner as described in 9.2.7 for salient pole types. If there is any doubt about the condition of the end-winding banding, this should be replaced. A similar approach can be taken with slip ring assembly insulation.

Failed Clip Joints in Bar-Type Windings.   Providing the failure is caught early enough and it has not caused damage to the insulation on adjacent coils, the clip can be removed, the joint cleaned up, and a new clip brazed or soldered in place. Once this is done, the joint can be reinsulated. It is important to check for other high-resistance joints during such repairs since they should also be repaired. This can be done by surge comparison testing (Section 12.12) or the passage of current through the winding in conjunction with an infrared thermography scan.

Damaged End-winding Banding.   As indicated above, failure of end-winding banding is a catastrophic event since it results in almost complete destruction of the windings. If deterioration in this banding is detected, it should be replaced by prestressed resin-coated fiber material, which has very good properties for this application.

Failed or Contaminated Slip Ring Insulation.   Failure of the insulation between the slip rings and the rotor shaft should be replaced. Since the slip rings are normally shrunk onto this insulation, they have to be heated to expand them to allow their removal. Nomex™-type sheet material is a good replacement for this insulation due to its good mechanical, electrical, and thermal properties.

If the insulation on slip ring connecting studs fails, it has to be replaced. Resin-bonded glass fiber tube material is best for this application since it has excellent mechanical strength and adequate thermal and electrical properties.

Contamination on slip ring insulation can be easily removed by a suitable cleaning agent such as isopropyl alcohol.

9.4 SQUIRREL CAGE INDUCTION ROTOR WINDINGS

Although most squirrel-cage rotor windings are not insulated (Section 1.2.3), they can fail due to various aging mechanisms, which can be thermal, electrical, or mechanical in nature. Their susceptibility to failure is dependent on the type of winding construction, the motor application, operating duty, winding geometry, and the material of construction. Although motors used in centrifugal pump applications very seldom experience winding failures, those used in high-inertia drive systems such as power station induced-draft fans are most susceptible, especially if they are frequently started directly-on-line. Bonnett gives a general review of SCI rotor failure processes in reference 9.1.

9.4.1 Thermal

Every time a motor is started directly on-line, a significant amount of heat is generated in its rotor windings due to the high stator winding starting current (5 to 6 times full-load current) that flows through them. The amount of heat generated and, therefore, the maximum rotor winding temperature is a function of the torque margin between the motor and driven-equipment torque/speed curves and the combined inertia of the driven equipment and motor rotors. The longer the rotor takes to come up to speed, the greater will be the temperature the rotor winding experiences. Little rotor winding heating occurs once the rotor is at operating speed.

Another factor that determines the maximum rotor bar temperature is "skin effect" that results from leakage fluxes across the rotor slots. This creates nonuniform current distribution in the bars during start-up, when the frequency of the current in the rotor bars is close to the 50 or 60 Hz frequency. As a result, the current at the top of the bar is much higher than that at the bottom. Hence, the tops of the bars reach much higher temperatures than the bottoms during motor starting and the bars bow outwardly due to nonuniform thermal expansion. This temperature increase is more significant in deep, thin bars and "inverted T" bars, which are much narrower at the top than the bottom.

Repetitive starts (three or more) over a short period of time can produce excessive rotor winding temperatures in motors used in high-inertia drive systems, since the heat created during the first start(s) may not have radiated or been conducted off the rotor. Thus, the rotor temperature may still be high from the first start when the second start causes a further increase in temperature.

Motors with speed control devices such as fluid drives, magnetic clutches, and solid-state variable-speed drives do not experience high rotor temperatures. This is because the motor is up to speed before the load is applied, or the starting current is controlled to a much lower value than for a directly-on-line, loaded start.

General Process. The nonuniform heating and high temperatures in the bars of motors with long run-up times or those that are subjected to repetitive starts cause distortion and high mechanical stresses in the bars and end-windings due to differential expansion. That is, the tops of the bars expand more than the bottom of the bars, distorting the bars and leading to internal mechanical stresses that may crack the bars, especially after many starts. These effects are much more pronounced in machines with fabricated aluminum or copper alloy bar windings, which have bars that extend beyond the end of the core to connect to the shorting rings (also called short-circuit rings). The type of winding construction and the bar and shorting ring materials are also a factor in determining the effects of this distortion from heating during starts. The stresses in the end-winding region of fabricated rotor windings are further increased by the centrifugal forces (Section 9.4.2).

Diecast aluminum windings are least susceptible to distortion since they have no bar extensions. However, such windings can melt due to the heat produced by repetitive starts. Fabricated windings made from aluminum alloy are most susceptible to failure from distortion since the mechanical properties of this material deteriorate significantly at temperatures above 100°C. Having said this, copper/copper alloy windings can also fail if the end-winding and bar design are poor and allow too much distortion during starting.

As already indicated, the tops of "inverted T" shaped bars become very hot if used in motors for high-inertia drives. This excessive heat can cause the bar tops to crack and break off.

Root Causes

- Poor rotor winding design and/or winding material selection in conjunction with a high-inertia drive application
- Frequent starting, especially if the motor thermal overload protections are set too high or jumpered out to prevent tripping
- The use of inverted "T" rotor bar designs in high-inertia applications

Common Symptoms

- Cracked or broken bar-to-short-circuit-ring connections at their interfaces or beyond the end of the core. Core burning may also be present due to the passage of current through it (Section 10.2.2).
- Cracked or broken short-circuit rings
- Noise and increased starting times due to cage winding breaks
- Increasing vibration levels (Section 13.10.1) and power frequency current sidebands (Section 13.7)
- Melting of diecast aluminum winding bars where frequent starting has been confirmed

9.4.2 Cyclic Mechanical Stressing

There are a number of contributors to cyclically induced stresses in squirrel cage rotor windings and these can cause failures due to "low-cycle fatigue." This failure mechanism occurs mainly in motors driving relatively high-inertia loads that are frequently started, such as an induced-draft fan in a fossil generating station that is being operated in a "two shifting" mode.

The locations of these forces are illustrated in Figure 9.2. In relation to this figure, the forces are:

F_c = Electromagnetic forces on the bars

F_b = Thermal expansion/centrifugal forces on the bar extensions outside the core

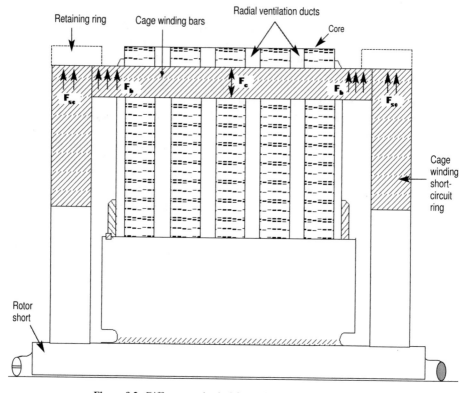

Figure 9.2. Different mechanical forces that occur in an SCI rotor.

F_{sc} = Thermal expansion/centrifugal forces on the shorting rings

F_r = Force on the retaining ring (if fitted)

General Process. Each time a motor is started, the mechanical forces described above act on its windings to increase the mechanical stresses. Some of these forces are transient and die away once the rotor is up to speed, whereas others remain while the motor is operating. The cause of cyclic fatigue is the change in the rotor winding stresses between stand-still and running conditions. These forces, which mainly affect fabricated rotor windings, are described in more detail as follows.

Electromagnetic Forces on Bars (F_c). Every time a motor is started, electromagnetic forces are imposed on the bars in the slot region. These forces occur at two times rotor current frequency and are proportional to the square of the rotor current. Consequently, they are only significant during starting, but they can cause fatigue failures if the bars are loose in their slots. This is most likely to occur in fabricated rather than cast rotor winding designs.

Forces on Bar Extensions Outside Core (F_b). These forces are highest during starting and result from a combination of thermal distortion and centrifugal forces on both the bars and the short-circuit rings to which they are connected. The net result is deformation of the end-winding structure, as illustrated by Figure 9.3. The cyclic deformation that occurs due to starts and stops can cause failures in bar extensions, short-circuit rings, and the joints between them due to fatigue. As indicated in Section 9.3.1 above, fabricated aluminum alloy windings lose their mechanical properties at lower temperatures and are therefore most susceptible to failures from these cyclic stresses.

These stresses become more significant in fabricated rotors windings that are loose in their slots. In this situation, the whole winding moves axially in one direction to change the end-winding geometry. Usually, bar or joint failures occur by this process at the end where the bar extensions have been shorted.

Forces on Short Circuit Rings (F_{sc}). Each time a motor is started, the short-circuit rings are stressed by thermal expansion from current heating and centrifugal forces, causing them to expand radially. As well as stressing the rings themselves, this expansion imposes stresses on the bar extensions. Again, changes in stress between standstill and running can cause failure of the shorting rings from cyclic fatigue.

Forces on Retaining Rings (F_r). In large two- and four-pole motors, F_b and F_{sc} can be controlled by fitting nonmagnetic retaining rings over the bar ends and shorting rings, as illustrated in Figure 9.4. This induces high mechanical hoop stresses in the rings.

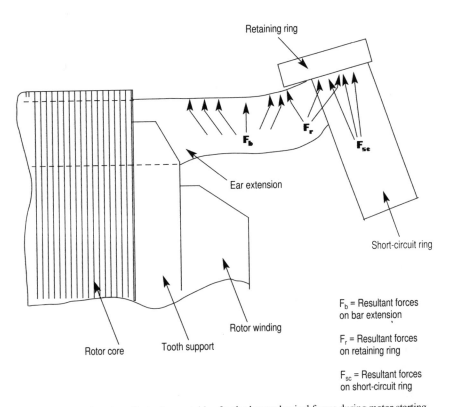

Figure 9.3. Distortion of rotor bars outside of a slot by mechanical forces during motor starting.

Figure 9.4. Photograph of a retaining ring over a short-circuit ring.

Root Causes

- Poor winding geometry associated with frequent starts and stops and relatively high load inertia
- Numerous sucssessive directly on-line starts
- Loose rotor bars in the slot region

Common Symptoms

- Cracked or broken bars or bar-to-shorting ring joints in the end-windings. This is usually at the end where the bar extensions are shortest if there has been axial migration of the whole cage winding
- Cracked or broken shorting rings
- Cracked or broken bars in the slot region, associated with loose bars or thin bar top sections
- Burning of the core laminations
- Increased starting time and noise during starts
- Increasing vibration levels and stator winding current harmonics

9.4.3 Poor Design/Manufacture

There are a number of design and manufacturing deficiencies that can cause failure on their own, accelerate the failure mechanisms described in Sections 9.4.1 and 9.4.2, and/or give unacceptable motor performance. These are:

- Poor end-winding geometry and support
- Loose bars in slots due to excessive clearances from poor design or manufacturing practices

- Poor quality bar-to-shorting ring brazed or welded joints in fabricated windings
- Air pockets in diecast aluminum rotor windings

General Process and Root Causes

Poor End-Winding Geometry and Support. The stresses described in Sections 9.3.1 and 9.3.2 can be minimized by good design that takes account of bar extension lengths, shorting ring dimensions, bar-to-shorting ring interface configuration, selection of the brazing or welding technique, and materials to be used. Also, as already indicated, retaining rings are used on large two- and four-pole rotors to help control end-winding cyclic stresses.

Any weakness in the rotor winding design can lead to premature failure that, in some cases, can occur after only a few months of operation.

Loose Bars in Slots Due to Poor Design or Manufacturing Practices. Any design deficiencies that lead to the loss of the tight bar fit in the slot or the use of undersize bars will lead to bar looseness and failures from this cause. If only some bars are loose, then nonuniform bar expansion can occur during motor operation and this can create sufficient force to bend the rotor and put it out of balance. Two-pole motors are most susceptible to this problem. Examples of this are the use of bars with rounded corners instead of sharp ones, loose slot liners that migrate out of the slots, and omitting the bar swaging operation during manufacture. Small cracks in the tops of the bars can be introduced if the swaging tool used is too sharp. These cracks may grow during motor operation.

There are various methods used to ensure tight bars, including:

- Having an interference fit between bars and slots at the slot corners
- Driving interference-fit steel wedges into keyways in the core laminations below the bars
- Deforming the bars by swaging the tops once they are in the slots
- Using custom fitted steel liners between the bars and slots.

Poor Quality Bar-to-Shorting Ring Brazing or Welding. The quality of the brazed and welded joints between bars and the short-circuit rings are very dependent on the skill and experience of the person who performs this operation during rotor winding manufacture or repair. In addition, quality assurance practices used to verify the integrity of such joints play a big part in detecting poorly made ones and taking corrective actions. Failure to manufacture high-quality joints will result in rapid joint cracking and fracture resulting from the high stresses imposed on them during motor starts and stops, as well as centrifugal forces during operation.

Poor Shorting Ring Design and Manufacture. Some manufacturers roll rectangular copper, copper alloy, or aluminum alloy to form a shorting ring. The two ends must then be brazed or welded together. Since these joints sometimes fracture, such a design is not recommended for high-inertia loads.

Air Pockets in Diecast Aluminum Rotor Windings. This defect can be hard to confirm since, in most cases, the air pockets are not on the surface of the winding where they can be seen. This defect causes the following problems:

- If the void creates an open circuit in a bar, then the motor will exhibit the symptoms associated with broken bars, leading to high vibration and stator current sidebands (Sections 13.7 and 13.10.1).

- The increased rotor winding heating from nonuniform distribution of current in the rotor winding due to the presence of voids can cause excessive shaft temperatures. These high temperatures can cause the inner rings of rolling-element bearings to expand to such an extent that the bearing clearances are reduced and the bearing fails.
- Voids in bars create nonuniform rotor cage winding expansion that produces sufficient force to bend the rotor and put it out of balance.

Common Symptoms

Poor End-Winding Geometry and Support

- Premature breaks in bar extensions between the end of the core and shorting ring
- Lack of retaining rings in large high-speed rotors
- Shorting ring cracking

Bars Loose in Slots Due to Poor Design or Manufacturing Practices

- Noise during starting due to effects of high electromagnetic forces
- Bar breaks inside core
- Signs of arcing between bars and core
- Axial cage winding migration
- Radial movement of bars when pressure is applied

Poor Quality Bar-to-Shorting Ring Brazing or Welding

- Premature failure of bar-to-shorting-ring joints
- Increased starting time and noise during starting
- Metallurgical analysis indicating poor quality brazing

Poor Shorting Ring Design or Manufacture

- Broken brazed or welded ring joints
- Cracks in one-piece rings

Air Pockets in Diecast Aluminum Rotor Windings

- High rotor vibration levels that vary with load
- Premature bearing failures due to loss of clearance from high shaft temperatures

In all cases, bearing vibration (Section 13.10) will increase as well as specific frequencies in the stator winding current (Section 13.7). Incipient cracks may be detected by a dye penetrant test (Section 12.23).

9.4.4 Repairs

It is appropriate to remember that sometimes a few cracked rotor bars can be tolerated, and perhaps do not need to be repaired. These few cracked bars may have occurred as a result of an unusual operating mode. However, if the vibration, starting noise, stator current signature, or other factors are worsening over time, repairs should be made.

Squirrel-cage winding repair techniques are very much dependent on the type of winding. Fabricated windings can sometimes be repaired by replacing only some of the components. On the other hand, diecast rotors can only be repaired by replacing them with a complete new winding of an equivalent fabricated type. In many instances, it is more economical to replace the complete rotor if a diecast winding fails or has a significant air pocket content.

Given that in most cases diecast rotor windings are not repairable, the following repair methods apply to fabricated windings only.

Shorting Ring Cracking or Failure. Providing there is no indication of cracked bars, the damaged short-circuit ring can be machined off and replaced by a new one. Weld or braze repairs to shorting ring cracks are not recommended since the high mechanical stresses imposed on these are likely to cause them to fail. If the shorting ring has been formed by rolling a strip of copper or copper alloy into shape and welding or brazing the ends together, consideration should be given to installing new cast or forged single-piece rings, which are much more robust. Before a new shorting ring is installed, the areas of the bars to which it is to be brazed or welded should be machined to provide a clean surface, which helps ensure good bonding.

Cracked or Broken Bars. If cracked or broken bars are found, it is advisable to replace all the bars in the rotor since the others are likely to develop similar defects. If possible, the reason for the bar cracking problem should be investigated. If it is design-related, the design or rotor bar material should be changed to reduce the possibility of further failures. If it relates to manufacturing deficiencies, then the manufacturing techniques used for the new winding should be reviewed to ensure that they will not cause similar failures. If broken bars have caused core damage that can only be repaired by lamination replacement, then all bars will have to be replaced. Except in an emergency, cracked bars should not be repaired by welding or brazing since this is only a short-term fix.

Loose Bars. If the rotor slots are open at the top, this problem can usually be repaired by swaging the tops of the bars with a round-tip chisel. If the slots are totally enclosed, this technique cannot be used. Dipping the rotor in a polyester or epoxy thermosetting resin is not a recommended repair technique since the resin bonds between the bars and core will eventually crack, especially at the high temperatures occurring during starting. This leads to the presence of both loose and tight bars, which expand differently during motor operation to cause shaft bending and high vibration levels. Two-pole rotors are most sensitive to this problem. Thus, for completely enclosed rotor bars that are loose, a rewind may be needed.

REFERENCE

9.1 A. H. Bonnett and G. C. Soukup, "Cause and Analysis of Stator and Rotor Failures in Three-Phase Squirrel Cage Induction Motors," *IEEE Trans. IA,* 921–937, July 1992.

CHAPTER 10

CORE LAMINATION INSULATION FAILURE AND REPAIR

This chapter discusses the most common causes of stator and rotor core failures in both induction and synchronous machines, together with proven repair methods. Core design and manufacture are discussed in Chapter 6. Core lamination insulation shorting and mechanical damage can occur from a variety of aging and failure mechanisms that can be thermal, electrical, mechanical, design, or manufacturing related. Stator cores used in large turbine generators, hydrogenerators, and motors, as well as squirrel-cage motor rotors with fabricated windings are most susceptible to failures from these causes. Some of these failure mechanisms will only occur in specific types of machines, whereas others are applicable to all types.

Note that some lamination insulation failure can be tolerated in machines, particularly in hydrogenerator and motor stators as well as SCI rotors. Such insulation failure will allow currents to flow between laminations to increase the local temperatures. However, if the insulation failures are very local, they may not expand in scope. Stator and rotor core insulation testing is discussed in Chapter 14.

Section 10.5 presents repair options that will arrest the aging process to prevent failures, or at least extend the life of the affected core. However, if aging goes undetected for a long period of time, the resulting degradation may be so advanced that a partial or complete core replacement will be required to give reliable operation for the remaining life of the machine.

The symptoms of each failure mechanism are listed below. Some of these can be detected from a visual inspection, whereas others require tests (Chapter 14) to provide early detection and to assess the severity of degradation in core condition.

10.1 THERMAL DETERIORATION

Degradation of the core condition due to the effects of thermal aging can occur in all rotating machine laminated cores. However, it is most likely to occur in large turbine generator, hy-

Electrical Insulation for Rotating Machines. By Stone, Boulter, Culbert, and Dhirani
ISBN 0-471-44506-1 © 2004 Institute of Electrical and Electronics Engineers

drogenerator, and motor stator cores. The rotor cores in squirrel-cage induction motors with fabricated windings used in high-inertia-driven equipment applications are also susceptible to failure from overheating as a result of broken rotor bars (Section 9.4).

10.1.1   General Process

Core overheating will cause accelerated aging of the core insulation if its thermal rating is exceeded for an extended period of time. This is most likely to occur if an organic varnish is used since this will dry out due to loss of solvents by evaporation of low-molecular-weight components. Once this occurs, the varnish becomes brittle, cracks, and eventually breaks down. As a consequence, interlamination shorts will develop, eddy currents will increase, and this will eventually lead to core melting due to even higher temperature operation. In large hydrogen-cooled turbogenerators, condition monitors may be installed (Section 13.2). These monitors detect the presence of materials driven off the core by excessive heating.

10.1.2   Root Causes

The root causes of core insulation thermal degradation can be subdivided into the categories general overheating, local overheating, and inappropriate stator winding burnout procedures. These are discussed separately below.

General Stator Core Overheating. The most common causes of this are:

- Loss of cooling water for hydrogen or air coolers in totally enclosed machines
- High ambient air temperatures for open-ventilated air-cooled machines
- Blockage of air inlets in open-ventilated air-cooled machines due to pollution or debris
- Complete or partial blockage of cooling-air passageways due to the accumulation of oil, dirt, etc.
- Turbine generator operation at reduced hydrogen pressure

General overheating of cores in large machines, especially turbine generators, may also cause core slackness that results from thermal expansion of the core steel. When core temperatures exceed design limits, the radial and axial forces exerted on the core support structure may be high enough to cause permanent deformation of these components. As a consequence, the core will loosen when the condition causing the overheating is removed. This core looseness will lead to insulation failure from abrasion. Core looseness has sometimes occurred as a result of extended full-flux core testing (Section 14.2). General thermal aging of core insulation also causes loosening of the core due to shrinkage and weight loss of the organic components.

Local Core Overheating. The most common causes of this are:

- Inadequate cooling of certain areas of the core due to poor design or blockage by debris (e.g., cooling air or hydrogen flow is too low or nonexistent in these areas)
- Manufacturing errors (e.g., some cooling-medium passages have been blocked due to missing holes or cut-outs in the core support structure)
- Broken rotor bars within the core in SCI rotors (Section 9.4). The interrupted currents in the bars flow through the adjacent core, overheating the core and its insulation.

It is unlikely that local overheating of the core will cause looseness. However, as with general overheating, it will cause accelerated thermal aging of the lamination insulation and eventual shorting of the core insulation. It should be possible to verify this cause by an analysis of the cooling gas circuit and temperature data from tests or the generator temperature monitoring system if inadequate design is the cause, or by inspection if cooling-gas passages have been blocked.

Winding Burnout at High Temperatures. Stripping an old global VPI stator winding from the core during the rewind of a motor or small-to-medium-size synchronous generator poses a major thermal threat to the core insulation. This stripping is usually accomplished by placing the stator in an oven at high temperatures to lessen the force needed to pull the coils from the stator. Weaknesses in the interlaminar insulation could be inadvertently introduced during this burnout. During subsequent service, these weak spots may overheat due to local or general high interlaminar currents that cause premature core failure. To avoid core insulation damage in modern stators, the winding burnout should be performed in a precision temperature-controlled oven with a water spray quenching system. In no case should the maximum oven or component temperature exceed 360°C (680°F) for C-4 organic [10.1] and 400°C (750°F) for C-5 inorganic core insulation. If the core insulation type is unknown, the burnout temperature should be limited to 360°C (680°F). The core insulation in machines built before about 1970 should be carefully evaluated for the type of core insulation used as it may deteriorate even at 360°C. For example, some machines built in the early 1900s had paper-insulated cores.

The pullout force required to remove an old stator winding decreases significantly with the temperature. For the same "burnout" time in the oven, the pullout force can decrease by an order of magnitude for an increase of about 100°C in burnout temperature. In addition, the time needed in the oven could be reduced to less than 25% by increasing the oven temperature by 100°C. The repair shop and motor owner, therefore, have a large incentive to use high burnout temperatures to remove a stator winding from a core, to shorten turnaround times.

Following a rewind involving a burnout of the original stator winding, the operating losses in the motor or generator are likely to change. Provided that the number of stator turns and the copper cross section per turn are not altered, the change in losses will be mainly in the core and in stray load losses. During repair, the core interlaminar insulation may have been degraded by high temperature, mishandling, burrs, or assembly pressure. This would increase the eddy current losses in the iron due to higher circulating currents induced in the laminations. Overheating of the laminations could lead to local melting of iron and eventual core failure. This increase in losses can be minimized if the tests described in Chapter 14 are performed before and after removal of the original winding, and any core insulation damage is repaired before the new winding is installed.

10.1.3 Common Symptoms

General Stator Core Overheating. Signs of general deterioration of the core insulation are associated with lamination shorts distributed generally throughout the core. Shorts in visible areas of the core will be indicated by discoloration, broken teeth, and fused metal. Those in the core body may be detected by special tests (Chapter 14), which should be performed to provide early detection of core insulation deterioration that is not visible. Core insulation dusting may indicate a loose core. Core shorts that develop at the edges of slots may increase the temperature enough to cause thermal aging of the winding insulation and resulting ground faults in the stator winding.

Local Core Overheating. Small burn marks or lamination discoloration, common symptoms of localized core shorts that can be related to inadequate cooling in their vicinity, should be detected early. Overheating can, of course, result from core insulation degradation initiated by electrical or mechanical aging mechanisms, as described in Sections 10.2 and 10.3.

Winding Burnout at High Temperatures. Higher full-load and no-load currents associated with significantly higher core and stray losses after repair work could be due to excessive burnout temperatures. It is important to obtain baseline core loss and temperature data before repair so that the effectiveness of a repair can be assessed. The "EL-CID" test or rated core flux test (Chapter 14) can be used to identify shorted laminations.

10.2 ELECTRICAL DEGRADATION

Electrical aging occurs when the voltage across the lamination insulation induced by magnetic fluxes, electromagnetic forces, or high ground-fault currents causes deterioration. Although DC and transient voltages may cause aging, it is normally AC-voltage-induced effects that cause the most severe damage. In addition, broken bars or shorting rings in squirrel-cage induction motor rotors and synchronous machine rotor damper windings cause rapid thermal degradation of their lamination insulation and may be considered an electrical cause of problems. Degradation from electrical aging can occur in all types of stator and rotor laminated cores.

10.2.1 General Process

There are several electrical processes that can lead to deterioration. Excessive fluxes in specific areas of stator cores induce elevated eddy current losses, which can cause overheating. Such overheating is most commonly found in large synchronous machines. Failure of stator or rotor winding insulation in or near the end of the core leads to the flow of high ground-fault currents that can cause localized core damage. High currents that flow through the rotor laminations, between bars, in squirrel-cage induction motor and synchronous machine damper windings, as a result of broken bars or shorting rings, cause rapid thermal aging of the laminations and deterioration in their mechanical strength. Since induction motors have relatively short air gaps, high electromagnetic forces of attraction between the stator and rotor can cause the two to make contact, causing core insulation surface shorting.

10.2.2 Root Causes

The root causes of electrically induced core insulation degradation can be subdivided into the categories of overheating due to over- or underexcitation, winding ground faults, and stator-to-rotor rubs due to unbalanced magnetic pull effects. These are discussed separately below:

Stator Core End Overheating Due to Underexcitation. The main air gap flux in synchronous machines is in the radial direction. This flux is responsible for generating the voltage in the stator winding. In addition, synchronous machines have significant leakage fluxes in the end region, especially when the rotor winding is underexcited. These fringing fields are produced by currents in the stator and rotor end-windings and by the discontinuities at the stator and rotor core surfaces. The axial component of this field generates circulating currents within the segments of the end region stator laminations, generating some electrical losses and, thus, heat.

The eddy currents due to the axial magnetic field cause stray losses in the end regions. The axial magnetic field is sensitive to changes in load and power factor. During leading power factor (under excitation) operation, this field can be quite high in large machines, especially if they have direct-water-cooled windings. This can degrade the interlaminar insulation as follows:

- Higher temperatures occur, which may reduce the dielectric strength of the interlaminar insulation over time and also give rise to other stresses due to expansion and relative motion between components. Based on tests performed over limited time, suppliers have claimed that interlaminar insulation has the capability to withstand temperatures as high as 500°C. However, in some machines, especially if they were built before 1970, this limit is much lower. This compares with the Class B winding insulation temperature limit of 130°C used on a typical large generator of that era. However, the long-term effect of operating near the rated temperature capability of the insulation could reduce its life.

- The circulating currents in the laminations can result in relatively high voltages being developed between adjacent core laminations. Under extreme conditions, this voltage may be an order of magnitude higher than normal. It has been shown that minor defects in the interlaminar insulation may provide a path for circulating currents, causing further, perhaps serious, local deterioration.

The combined effect of the above two mechanisms in conjunction with other existing stresses can damage interlaminar insulation in the stator core end regions near the bore. This increases circulating currents between laminations, causing temperature rise, local weakening, and tooth chatter and breakage. Several thermocouples may be installed on generators in this area during manufacture. A rising trend or a sudden increase in temperature recorded by these sensors can provide an early warning of the problem.

Overheating of Back-of-Stator Core Due to Overexcitation. In order to keep the physical size of large two- and four-pole synchronous machines within reasonable limits, it is necessary to excite the stator core at a fairly high magnetic flux density. Laminated steel, as well as the lamination insulation, is, therefore, selected to avoid high core losses. The core lamination steel is either grain oriented or specially processed to give low core loss (Section 6.19). The lamination insulation is selected for its low dielectric permitivity and good insulation properties under high stress. High temperatures due to increased core loss can result from overexciting the field winding, thus producing higher than normal magnetic flux.

The large volume of core behind the slot is more prone to overheating due to increased flux compared to the tooth area. The core behind the slot has relatively less ventilation than the teeth. Consequently, the core losses can quickly raise the temperature of the back of core. The temperature increase is particularly steep as the iron begins to saturate.

Once the temperature has been elevated, the chances of breakdown of the lamination insulation are increased. Such a breakdown would give rise to interlaminar faults and increased eddy currents, which can cause even higher temperatures. The higher temperature can also cause mechanical stresses, resulting in distortion and vibration. When combined, these effects can eventually lead to fusing of laminations, melting of iron, and core failure.

Stator Winding Ground Faults in Core Slots. The energy and heat produced by stator- or wound-rotor winding ground faults in or just outside the slot region are often sufficient to melt and fuse the core laminations at the core surface. If this core damage is not repaired when the failed coil or bar is replaced, the new coil could also fail to ground as a result of the heat generated by the shorted laminations. It is, therefore, important to perform tests described in Chapter 14 to check the condition of core insulation in the vicinity of ground fault damage before installing a new bar or coil.

Stator Core Faults from Through-Bolt Insulation Damage. In some medium to large motor and generator designs, the stator core pressure is maintained by bolts that pass though axial holes in the stator core laminations and endplates and have nuts and washers installed on either end. Core pressure is maintained by keeping these nuts tight. These through bolts have to be insulated from the core with tube insulation to prevent core insulation shorting. That is, if retaining nuts become loose or the bolts stretch, the insulation can fail from core movement. If this happens, core lamination shorting and core burning may occur.

Broken Rotor Bars or Short-Circuit Rings. Bars or short-circuit rings in a squirrel cage rotor winding or salient pole synchronous machine damper winding may crack or fracture to create an open circuit; then, rotor core insulation damage and shorting will occur. This results from the current flowing through the core laminations, between bars, because this is a lower-resistance path that that through the broken winding (Figure 10.1). The heat generated by this abnormal current flow is sufficient to cause insulation damage, eddy current flow, and

Figure 10.1. Core lamination burning due to broken rotor bars in an SCI rotor.

core melting. If a broken short-circuit ring is the cause, the resulting damage will occur at the end of the core nearest the break. Core damage due to broken bars will occur in the vicinity of these breaks. In addition, if bar breaks are outside the core, centrifugal forces may push the end of the broken bar through the weakened bridge at the top of the rotor core slot so that it rubs on the stator winding. If this happens, a stator winding ground fault will occur.

Stator-to-Rotor Rubs Due to Unbalanced Magnet Pull. Electromagnetic forces of attraction (unbalanced magnetic pull) between rotor and stator occur in all rotating machines due to uneven airgaps. The direction of these forces is toward the smallest airgap and their magnitude increases as the airgap gets smaller. If the rotor is not stiff enough, such forces can be high enough to cause the rotor to "pull over" and rub the stator. This causes the same type of damage that results from bearing failures. Small to medium size induction motors with relatively small airgaps and single-circuit stator winding connections are most susceptible to this problem since the unbalanced magnetic pull forces in these machines is relatively high. This problem can also result from the rotor not being properly centered in the stator bore during motor assembly, thus creating a much smaller airgap at one location. Salient pole pullover can also occur in hydrogenerators if the airgap becomes uneven due to movement of the stator or rotor. Airgap monitoring in hydrogenerators (Section 13.8) and current signature analysis for SCI motors (Section 13.7) can warn of this condition.

10.2.3 Common Symptoms

Stator Core End Overheating Due to Underexcitation. This results in interlamination shorts, discoloration, missing ventilation spacers, and, perhaps, broken teeth in the core end regions. If temperature sensors are fitted, a gradually rising temperature or a sudden increase in temperature may give an early warning.

Overheating of Back of the Stator Core Due to Overexcitation. Higher vibration and higher stator temperature are often the more common symptoms. However, these may not be immediately evident. It is not practical to install the number of thermocouples required to monitor the complete core of a large turbine generator. Inspection of the core through vent ducts using boroscopes and fiber-optic probes during overhaul is a good precaution. Cracked welds in the inner structure of the generator housing may be an indirect symptom.

Stator Winding Ground Faults in Core Slots. These result in shorting and melting of the laminations in the area of a coil ground fault.

Stator Core Through-Bolt Insulation Damage. Signs of core overheating are seen at core through stud locations.

Broken Rotor Bars or Short-Circuit Rings. Core insulation damage and melting may be associated with broken bars, short-circuit rings, or joints between the two. If a bar or short-circuit ring fractures, it may be bent radially outward by centrifugal forces to cause a stator winding fault. The earliest indication of a fractured rotor winding may be arcing, increased noise, and increased vibration. Current signature analysis, as described in Section 13.7, can provide early detection of such problems.

Stator-to-Rotor Rubs Due to Unbalanced Magnet Pull. This results in smearing of the stator core bore and rotor core outside diameter at or near their centers.

For hydrogen-cooled turbogenerators, condition monitors (Section 13.2) will indicate if any of these conditions are occurring.

10.3   MECHANICAL DEGRADATION

The most common causes of mechanical degradation in cores are inadequate core pressure applied in manufacture, core pressure reduction in service due to relaxation of the core support structure, core vibration, back-of-core looseness, and mechanical damage causing smearing of the core surface at the bore. Degradation due to core looseness is predominantly found in large generators and motors with a segmented core construction, as described in Chapter 1. Core insulation damage due to vibration is most commonly found in large two-pole turbine generators. Mechanical damage to the core bore due to impact can occur in any type of machine.

10.3.1   General Process

When the laminations in a stator core become loose they can move relative to one another under the influence of mechanical vibration and/or electromagnetic forces, and the insulation on them degrades due to abrasion. If not detected in time, all of the lamination insulation in the areas of core looseness is removed and lamination shorting occurs. The eddy currents that flow as a result of this shorting create excessive heat that can eventually lead to core melting and lamination fracture. Core looseness can also result from vibration caused by the natural frequency of the core and frame being too close to the main twice-power frequency (100 or 120 Hz), the electromagnetic core excitation frequency that occurs in all AC machines.

The stator cores in large machines with segmented laminations are built on axial keybars (two per lamination) welded to the frame. A dovetail fit between each core lamination segment and the keybar provides radial support for the core. If the fit between the core laminations and keybars is, or becomes loose, then arcing between the two and shorting of the core laminations at the back of the stator core will occur.

If the stator and rotor of a rotating machine make contact during operation or rotor removal, core surface smearing and shorting will likely occur. Also, in any machine, loose components entering the airgap will cause localized shorting of the stator core insulation and that of the rotor if it also has a laminated core or poles.

10.3.2   Root Causes

The root causes of mechanically induced core insulation degradation can be subdivided into the categories of core fretting, relaxation, and failure; core vibration; loose fit between back of core and keybars, causing overheating and burning; bearing failures in induction motors; and metallic debris in the airgaps of synchronous machines.

Stator Core Relaxation, Fretting, and Failure—Turbine Generators. Core design and manufacturing problems contribute to this type of core deterioration. The following gives some general information, which may not be applicable to particular situations.

- Excessive use of resilient materials during manufacture may contribute to relaxation during service. Some examples of such materials are varnish in core insulation, rubberized segments for thermocouples, and Nomex™ sheets used as electrical separators. Avoiding excessive resilience is particularly important in longer cores associated with higher ratings.

- Core pressure is another important factor to consider during manufacture. The applied and retained core pressure should be sufficiently high to ensure even distribution of the clamping force throughout the core laminations and to avoid slackness and distortion. This becomes more critical as the length of the core increases with the rating. If the core support structure relaxes in service, then the core laminations become loose. The most common location of such looseness is at the stator bore since this is farthest from where the core pressure is applied (see Chapter 1).

If the laminations at the core bore are loose at the end of the core, the following sequence of degradation occurs if this problem is not detected and quickly corrected:

1. The core insulation is abraded due to lamination relative movement under the influence of axial electromagnetic forces from end leakage fluxes.
2. Core lamination shorting and overheating starts to occur when the core insulation is removed by abrasion.
3. Eventually, pieces of lamination teeth will break off due to fatigue failures, and vent spacers may also break free. As indicated in Section 10.3.2 such debris can cause core insulation damage in other locations.

Stator Core Vibration—Turbine Generators. Some of the causes of high stator vibration in service are:

- Inadequate support of the core in the stator frame, creating an assembly resonant frequency close to twice the power supply frequency (100 or 120 Hz)
- Unbalanced phase loading
- Inadequate stator end-winding support (causing vibrations that are reflected back to the core).

Even without these factors, a certain amount of vibration exists due to the 100 or 120 Hz "ovalizing" force caused by the magnetic field. The displacement of a loose core caused by these forces may result in relative motion between laminations and fretting of the lamination insulation to the point of breakdown.

Stator Core Fretting, Relaxation, and Failure—Hydrogenerators. Since a hydrogenerator core has a large diameter and short depth behind the winding slot, it is relatively flexible and the frame is the main support. The magnetic forces between the rotor poles and stator core will, therefore, tend to produce displacements in the core.

In general, the displacements and resulting vibration are small in multipole generators. The traveling wave produced by the magnetic force has a number of nodes equal to twice the number of poles. This results in smaller displacement. The only exception is in the case of fractional slot windings (that is, windings having a noninteger number of slots, poles, or phases), where larger displacement is possible due to wavelengths that can be longer than the pole pitch.

The above mechanisms can cause high temperature and melting anywhere in the core.

However, the bore section is particularly susceptible. It is physically weaker than the section behind the slot and carries a higher flux density. As these mechanisms weaken the bond between the laminations, the teeth are likely to chatter and break off. Secondary damage to the stator and the rotor can occur if the debris finds its way into the airgap.

Back-of-Stator Core Overheating and Burning.

Overheating and burning of the back of a stator core can be caused by a loose connection between the core laminations and the stator frame. Axial keybars welded to the frame are used for piling the core laminations. A dovetail fit between each coreplate segment and the keybar provides radial support for the core. To allow assembly or stacking on the keybars, a clearance is necessary in the dovetail fit. If excessive, this clearance will permit relative motion, intermittent contact, circulating currents, and overheating during operation.

Leakage flux at the back of the core induces currents in the keybars. These currents flow to ground through the stator frame without causing any harm, provided there is no path for them to flow through the core. This is the case if the punchings are in good contact (positive grounding) with the keybars or if they are completely isolated electrically (insulated keybars). However, should the clearance become excessive at some point, any core vibration could cause intermittent contact at that point and cracking of keybar insulation, if this is present. This can lead to arcing, overheating, and core melting in the local area. Small amounts of melting are difficult to detect due to inaccessibility and the impracticability of monitoring a large area with thermocouples. An early indication of this problem is an upward trend in frame vibration due to the increase in dovetail clearances.

Should a number of intermittent contacts develop on the keybars, the possibility of current transfer between keybars would increase. The currents would begin to circulate through the low-resistance path offered by the laminations. The resultant overheating could escalate into failure of interlaminar insulation and an increase in circulating currents and temperature, leading to fusing of laminations, melting, and, ultimately, core failure due to widespread melting of the lamination.

This problem can be greatly reduced by interconnecting all the keybars at each end of the core by means of welded copper straps to form a ring for carrying the circulating currents. This is better done during manufacture since retrofitting can be a major task.

Stator-to-Rotor Rubs Due to\ Bearing Failures—Induction Machines.

These types of failures mostly occur in induction motors, which have relatively small airgaps. Bearing failures can allow the rotor to move toward the stator and rub on its core bore surface. If this occurs, smearing of both the stator core at its bore and the rotor core outside diameter will occur. In addition, the rotor-to-stator rubbing may generate sufficient heat to cause rapid thermal aging of the stator winding insulation and a consequent ground fault. Such rubs will cause shorting of the core insulation on both the stator and rotor.

The severity of core insulation shorting from such rubs and the need for repairs can be assessed from the tests described in Chapter 14.

Loose Metal Components Entering the Airgap.

Bolts and other metallic components that break free from machine internal components can enter the airgap. Such objects are "sucked" into the airgap by cooling air or gas flow and magnetic attraction, if they are made from magnetic steel. If this occurs, stator core gouging or smearing and consequent insulation surface shorting will occur. Since synchronous machines have much larger airgaps than induction types, they are more susceptible to such damage.

10.3.3   Symptoms

Core Relaxation, Fretting, and Failure—Turbine Generators. High or increasing core vibration is an early symptom. If no corrective action is taken, a winding fault may result due to abrasion of the stator groundwall insulation. A visual inspection of the core will reveal core dusting and, perhaps, core overheating in areas where the insulation has broken down. If there has been hydrogen seal oil leaking into a turbine generator, this will combine with the dust from core insulation abrasion to produce a greasy substance. Broken teeth, missing ventilation spacers, and loose end-windings may be other indications.

Core Vibration—Turbine Generators. High or increasing core vibration is an early symptom. If no corrective action is taken, a winding fault may result due to abrasion of the stator groundwall insulation. A visual inspection of the core will reveal core dusting and perhaps core overheating in areas where the insulation has broken down. If there has been hydrogen seal oil leaking into a turbine generator, this will combine with the dust from core insulation abrasion to produce a greasy substance. Broken teeth, missing ventilation spacers, and loose end-windings may be other indications.

Core Fretting, Relaxation, and Failure—Hydrogenerators. High or increasing core vibration, which may vary with generator temperature, is a symptom. If no corrective action is taken, winding faults may result. A visual inspection will reveal dusting of the core insulation and, perhaps, core overheating, melting, and broken or cracked teeth.

Back-of-Core Overheating and Burning. This results in increasing core vibration, with signs of arcing between the core and the key bars that support it. A visual inspection may show cracked insulation if keybars are insulated.

Stator-to-Rotor Rubs Due to Bearing Failures—Induction Machines. These result in smearing of the stator core bore and rotor outside diameter at the failed bearing end of the motor.

Loose Metal Components Entering the Airgap—Synchronous Machines. This can cause localized impact core smearing at the stator bore with evidence of one or more loose metallic components inside the machine.

10.4   FAILURES DUE TO MANUFACTURING DEFECTS

A number of different core insulation defects can be introduced during manufacture and these may not be detected if adequate quality assurance (QA) checks are not performed. Many of theses defects can be detected and eliminated if core testing as described in Chapter 14 is performed as part of the QA program during core manufacture or refurbishing.

10.4.1   General Process

Interlaminar shorts can be introduced during core manufacture or refurbishing. The main causes of such shorts are:

- The introduction of defects, such as burrs, during lamination manufacture
- Inadequate cleaning of the lamination surfaces before core insulation application

- Poor quality insulating materials
- Core surface smearing due to stator-to-rotor rubs during factory assembly/disassembly
- Excessive filing of core slots to smooth rough lamination surfaces

As described above, shorts between laminations, especially those in stator cores, result in the flow of high eddy currents that in turn cause excessive heating. The area affected by these shorts will "grow" as the core insulation in the vicinity also thermally ages and fails. Eventually, if these faults go undetected, portions of the core will melt. The heating principle that causes this melting is similar to that of an arc furnace. If this occurs at or near a slot, rapid thermal aging (i.e., burning) of the stator groundwall insulation may occur until, eventually, the winding fails to ground.

10.4.2 Root Causes

The main causes of insulation shorting introduced during core manufacture are as follows:

- Lack of adhesion between the insulation and the steel, causing the insulation to flake off. This results in weak insulation or shorts if adjacent areas of laminations are bare. Such problems generally occur because the lamination surfaces were not adequately cleaned before the insulation was applied.
- Poor edge deburring of the laminations prior to the application of the insulation, or blunt dies used for blanking and punching preinsulated core steel, will result in shorts being created when the core is built and pressed; i.e., the sharp edges on the laminations will cut through the insulation and create metal-to-metal contact.
- Hard rubs during rotor installation or removal during manufacturing, repair, or maintenance, causing the lamination surface at the stator bore and rotor outside diameter (in laminated rotors) to be shorted by smearing.
- Poor core building procedures and/or lamination manufacture can create high spots in the slot regions that may damage coil groundwall insulation. Such high spots can be removed by careful filing but, if not controlled, this can cause smearing and shorting of the core laminations in the slot region.

10.4.3 Symptoms

- High core temperatures (which may be difficult to detect in service)
- Melting of the core at the airgap surfaces or around slots having poor edge deburring, or which have been filed
- Overheating as well as melting of either the core outer surfaces or core body where a poor bond exists between the core insulation and steel
- If rubs during rotor installation are the cause of failures, they are most likely to be in the 5 to 7 o'clock location, especially in large turbogenerators where "skid plates" are used for rotor insertion and removal.

10.5 CORE REPAIRS

The repair methods described below have been well proven within the rotating electrical machine industry. In all cases, the condition of the core insulation before and after repairs should be assessed by one of the techniques described in Chapter 14 as well as visual inspection. If core insulation condition tests are not performed after repairs, there is a danger that core and, perhaps, consequential insulated winding failures will occur at some time in the future.

In all repair methods where core unstacking and/or slot damage is involved, it should be understood that the winding in the core has to be removed and replaced. The only exception to this may be in large generators with Roebel windings, in which the top and bottom bars can be separately removed and reinstalled after the core repair.

10.5.1 Loose Cores

The following repair procedures are based on the premise that the core looseness has been detected before significant insulation damage has occurred.

Core Looseness at the Stator Bore. This can be verified by the "knife test" described in Chapter 14 and be repaired by a procedure called "stemming." Stemming or "shimming" involves manufacturing thin (about 1 to 2 mm thick) sheets of epoxy–glass laminate, which are shaped to the profile of the stator core teeth [10.2]. Each of these stemming pieces is roughened and grooved on both sides to create a reservoir for penetrating epoxy. In some cases, holes are drilled through the stemming pieces to increase the capacity of the epoxy reservoir. The stemming pieces are then coated with a low-viscosity "weeping" epoxy and driven into a loose area of core to tighten it. Once this epoxy cures, it bonds the stemming pieces in place. It is also a good idea to brush the epoxy onto the stator bore in the area of looseness to help restore the core lamination insulation.

Core Looseness Away from Stator Bore. There is no simple remedy for this problem unless the core has through bolts, in which case it should be possible to tighten the nuts at the end of these and, hence, correct the core looseness.

For machines that do not have through bolts, core looseness can only be corrected by removing the windings and repressing the core assembly. If the core looseness is due to an inadequate core end support structure, then this must be replaced with one having an improved design before repressing. If the looseness is due to resilient materials in the core stack, then the core will have to be unstacked to remove these before repressing.

Loose Stator Core at Key Bar Supports. There is no easy repair for this problem, which results from excessive core-to-keybar clearances, established during core design and assembly. The back-of-core damage from arcing between the core and keybars can be greatly reduced by interconnecting all the keybars at each end of the core by means of welded copper straps to form a ring for carrying the circulating currents. This is better done during manufacture since retrofitting can be a major task. If this does not arrest the problem, then the most economic solution is to unstack the core, replace the keybars with new ones that give a tighter fit, and reassemble the core. Any core insulation damage can be repaired during this process.

If the core looseness resulted from a resonant frequency problem, a more radical approach is required since the root cause is design related. In most cases, a new core and, perhaps, a new frame design will be required to eliminate looseness from this cause.

10.5.2 Core Insulation Shorting

The following repair methods are based on the assumption that insulation shorting has been verified by visual inspection and by one of the tests described in Chapter 14. Once repairs have been completed, the core should be retested to verify that the lamination shorting has been eliminated.

Shorting Due to a Loose Core at Stator Bore. Shorting of insulation near the core bore is commonly found if core looseness has gone undetected for some time. This is most often found near the core ends and can be corrected by generously applying weeping epoxy and/or inserting pieces of mica between laminations every 2 to 5 mm of each tooth within the damage area. This should, of course, be done before any core stemming as described in Section 10.5.1 is performed.

Shorting Due to Surface Smearing. If the smearing is very localized, acid etching or high-speed grinding can remove it. Acid etching is perhaps the better of the two methods since it removes the thin insulation shorting layer very slowly to reestablish core lamination separation.

Core surface smearing over large areas is more difficult to correct. In smaller machines, it may be possible to eliminate stator bore smearing by installing the stator in a boring mill and removing the smearing by machining the surface with a sharp tool. The dangers of this approach are damage to the stator windings and, in the case of induction motors, degradation in performance due to increased airgap size. However, the costs associated with this type of repair are much lower since the stator winding does not have to be removed to allow the core to be unstacked to restore insulation integrity. In the case of laminated rotors, the smearing can be removed by installing them in a lathe and machining the core outside diameter. Again, there is a limit as to how much material can be removed if an induction motor rotor is involved since this again increases the airgap and weakens the core bridges at the top of the rotor bar slots.

In some instances, the only way to correct core surface smearing is to unstack the core, remove any remaining burrs, replace damaged core insulation, and rebuild the core.

Shorting Away from the Stator Bore. In most cases, the only solution to this problem is to unstack the core, reinsulate the laminations, replace any with significant damage, and rebuild the core. This is necessary because it is virtually impossible to restore damaged insulation in the body of the core. One technique that sometimes works for small to medium size stators is to VPI the damaged core with thin epoxy or polyester resin that is capable of penetrating the very small spaces between the laminations in the core body. After such treatments, the core must be baked in an oven to cure the resin.

10.5.3 Core Damage Due to Winding Electrical Faults

Ground faults in insulated windings and open circuits in uninsulated squirrel-cage windings can cause severe localized core insulation damage and core steel melting. Such damage occurs in or at the end of a core slot.

The method of repair is dependent on the size of the machine and the location of the damage, as follows.

Unstacking and Rebuilding Part or All of the Core. This repair method is used for:

- Stator and rotor cores with semienclosed slots, when the damaged area is away from the end of the slot.
- Cores in small to medium size machines with open slots, when there is damage to the groove that holds winding slot wedge in place

- Squirrel-cage rotor cores that have been damaged by core burning resulting from broken rotor bars

Once the core is unstacked, the damaged laminations can either have their insulation restored or new laminations can be acquired to replace them. If the original laminations are reused, they are reorientated during the core rebuilding process so that the damaged sections are distributed throughout a large number of slots. This eliminates significant irregularities in the slot or wedge grooves.

High-Speed Grinding. High-speed grinding is effective for removing local core damage in cores with open slots (if away from the wedge groove) and at the ends of cores with semi-enclosed slots. The core surface shorts are ground away to provide separation between the laminations. The ground area should then be treated with weeping epoxy and the depression where the grinding was done filled in with thixotropic (high-viscosity) epoxy or other type of filler to restore a smooth surface and prevent future winding mechanical damage.

10.5.4   False Tooth

This method of repair is used for large turbine generator cores in which severe localized tooth lamination damage has occurred to the extent that a significant number of laminations have broken off [10.2]. All of the tooth sections of the laminations in the damaged area are cut off. Radial securing dowels are epoxied into the root of the removed tooth section. A false tooth, made up of epoxy/glass material, shaped to the same contours as the removed section of tooth and containing holes to allow it to fit over the dowels, is then installed. The new tooth is bonded to the dowels by epoxy. If the false tooth covers more than one core packet, vent ducts are cut into it to allow cooling of the core below it.

10.5.5   Cracked Through-Bolt Insulation

Such damage can be repaired as follows:

1. Removing the bolts, one at a time
2. Removing the damaged insulation
3. Inserting new insulating tubes
4. Reinstalling the bolts
5. Fitting the washers and nuts to either end of the bolt
6. Torquing the nuts to the manufacturer's recommended value.

REFERENCES

10.1   ASTM A 976-97, "Standard Classification of Insulating Coatings by Composition, Relative Insulating Ability and Application."

10.2   G. Klempner "Ontario Hydro Experience with Failures in Large Generators due to Loose Stator Core Iron," paper presented at EPRI—Utility Generator Predictive Maintenance and Refurbishment Conference, December 1–3 1998, Phoenix, Arizona.

CHAPTER 11

GENERAL PRINCIPLES OF TESTING AND MONITORING

This and the next two chapters describe on-line and off-line tests and monitors that can be used on rotor and stator windings, as well as stator and rotor cores. This chapter discusses several issues that are common to all of the tests and monitors.

11.1 PURPOSE OF TESTING AND MONITORING

As will be seen in this and the following chapters, there are almost 40 different tests and monitors that can be used to diagnose motor and generator winding condition. One could spend a considerable amount of money, not to mention cause the machine to be out of service for long periods of time, if one did all of these tests. Therefore, before the individual tests and monitors are discussed, we review the reasons why testing is done and discuss concepts for selecting the tests and monitors. We believe that there are at least four distinct reasons for testing and monitoring. These are discussed in the following four subsections.

11.1.1 Assessing Winding Condition and Remaining Winding Life

Determining the winding condition or estimating the expected remaining useful life of a winding is a common reason cited for doing tests and monitoring. For example, a plant manager wants to know if a motor that is more than 15 years old can continue to operate for the foreseeable future, or if a new motor or rewind is needed. The manager may also ask how long the existing motor will run before failure occurs. Testing helps answer these questions. Unfortunately, testing and monitoring alone will not give a definitive answer. One reason is that the end of winding life depends, as discussed in Section 7.1, on *when* a transient occurs in the power system, or *when* an operator makes a mistake.

In addition, determining remaining life is also difficult because most of the tests measure a symptom, not the root cause of the failure, as we will see in later chapters. A medical anal-

Electrical Insulation for Rotating Machines. By Stone, Boulter, Culbert, and Dhirani
ISBN 0-471-44506-1 © 2004 Institute of Electrical and Electronics Engineers

ogy is useful here. If one has a cold, a symptom of the cold is that you sneeze. By measuring the sneeze rate (sneezes per hour), one can determine (1) if you have a cold, and (2), to some degree, how bad the cold is (more sneezes per hour may be a rough indicator of severity). However, if you are going to die from the cold (yes, this is unlikely) it will depend on the virus that is attacking your body (the root cause), and if the body is able to fend off the attack. Since most of the tests in the following chapters measure the symptoms and not the root cause, remaining life cannot be ascertained. Aside from hipot and surge testing, none of the other tests can determine if a rewind or new machine is needed. Some human intervention to correlate results from different tests and to perform an economic evaluation is needed to make this determination, as discussed in Section 7.3

Therefore, assessing winding condition must have a more limited objective. This could be to determine if winding deterioration has occurred. In some cases, the severity of the failure process can be determined and, perhaps, one can then estimate the "risk of failure." Risk of failure means the probability that the motor or generator will fail if a transient or operator error occurs.

11.1.2 Prioritizing Maintenance

Usually, a plant does not have just one motor or generator; it has many. In addition, the plant staff probably has neither the time nor resources to do maintenance on all of its machines. Thus, testing and monitoring can help maintenance personnel determine which machines are most in need of an outage for maintenance purposes. In other words, testing and monitoring can help maintenance personnel determine on a comparative basis the machines with the greatest risk of failure. This can be combined with the repair cost and the impact of in-service failure on plant production to prioritize which machines must be maintained. This is discussed in greater detail in Chapter 16, where testing is incorporated as part of a company's maintenance strategy.

11.1.3 Commissioning and Warranty Testing

When a motor or generator is purchased, or a rewind performed, the contract between the vendor and the user normally defines that the windings must meet certain minimum test requirements (Chapter 15). These tests are often done by the supplier before the machine leaves the factory or in the first year of operation (or whatever the warranty period is). Examples include hipot tests and power factor tests. Some tests are automatically required by referencing standards (e.g., NEMA MG 1 or IEC 60034) as part of the purchase specification, but additional tests need to be specified by the purchaser. The tests are needed to ensure that the windings meet the specifications.

11.1.4 Determining Root Cause of Failure

If a machine fails in service or as a result of a hipot test, then it is necessary to discover why the winding failed. The reasons for finding the root cause of a failure include:

- If a repair or a rewind is performed on a machine in which failure occurred before it was expected, the question arises: is the same failure process likely to happen again once the machine is returned to service? If one understands the root cause, it is possible that minor changes to the winding can be made during the repair, or a repair can be made that will extend its operating time. For example, if stator turn insulation failure due to voltage surges has occurred, one can install surge capacitors or other filters to extend the rewind life, or turn insulation can be upgraded.

- If there are multiple machines of the same design, then, by ascertaining the failure mechanism involved in a failure of one of these machines, one can determine if the other motors or generators are subject to the same failure mode.

If an in-service failure occurs, there is often tremendous pressure imposed on the plant and repair shop personnel to return the machine to service as soon as possible so that rarely can a full battery of tests be done. However, even performing some quick tests and inspections may permit several possible failure processes to be excluded. This can prevent future aggravation and wasted effort in making changes that may ultimately not be needed.

11.2 OFF-LINE TESTING VERSUS ON-LINE MONITORING

All the rotor and stator tests in Chapters 12 and 14 are performed when the machine is not operating. Many of the tests can be done from the machine terminals, avoiding full or partial disassembly of the machine. However, some of the tests can only be done with the machine at least partly disassembled. Thus, all of the off-line tests, by definition, require at least a short outage (or turnaround, as it is sometimes called in industry).

In contrast, on-line monitoring refers to "tests" that are done during operation of the motor or generator. Therefore, no outage is needed, although, for some monitors, the operating condition of the machine is changed to extract the greatest amount of diagnostic information. In this book, "tests" will normally refer to off-line tests, whereas "monitoring" will normally be applied to diagnostic information acquired with the machine on-line.

In general on-line monitoring is preferred because:

- No outage is needed to determine winding condition, at least for the failure mechanisms the monitoring can detect.
- Usually, the cost of acquiring the diagnostic data is cheaper than for off-line tests since, generally, one person can collect the data and it usually takes just a few minutes. In modern monitoring, data collection is often automated. In contrast, off-line tests may require a few people to isolate the machine, get the test equipment to the machine, hook it up, and run the test.
- It facilitates predictive maintenance (Chapter 16), since one can determine which machines are in need of off-line testing and repairs without taking the machines out of service.
- The stresses that occur in service are present. For example, it is difficult for an off-line test to properly simulate the AC stress distribution that occurs in service. An AC voltage applied to simulate the normal phase-to-ground stress in the slot will result in a 60% lower voltage between the line-end coils in different phases in the end-winding than occurs in service.

The disadvantages of on-line monitoring are:

- Usually, there is a higher capital cost, since sensors and sometimes monitoring instruments must be installed on each machine, in contrast with off-line testing, in which one instrument can be shared for tests on a number of machines.

- Not all failure processes can be detected with existing on-line monitoring. Thus, unexpected failures can go undetected if all diagnostic information comes only from on-line monitors.

As discussed further in Chapter 16, a judicious mix of off-line tests and inspections, together with on-line monitoring is needed to implement predictive maintenance. The mix will change from plant to plant, and even from machine to machine within the same plant, as one considers the importance of the machine to the plant and other economic criteria.

11.3 ROLE OF VISUAL INSPECTIONS

For each of the failure mechanisms in Chapters 8 to 10, a section was devoted to "symptoms." In addition to diagnostic tests, visual symptoms were often described. Our opinion is that a visual inspection of the winding is usually the most powerful tool for assessing the winding condition and determine the root cause of a developing problem. However, there are several problems with visual inspections:

- They require an outage, and at least some (and often major) disassembly of the motor or generator.
- They require expertise. For example, a thin white line at the end of a coil may mean nothing to a casual observer but, to an expert, this may be definitive evidence of the root cause of a problem. This visual evidence enables determining relative winding condition and the repairs needed. Unfortunately, experts who have many years of experience looking at machine windings are rare, especially among users.
- The disassembly required for a proper inspection may cause another problem when the machine is put back together. For example, a bolt may be left loose in the machine, or a cooling channel may be inadvertently left blocked.

Thus, inspections, wherever possible, should be triggered by an event such as an out-of-phase synchronization or a poor result from a test or monitor. The need for an expert to do the inspection should not be underestimated. Kerszenbaum has written books to help nonexperts inspect motors and generators [11.1, 11.2]. In addition, EPRI has funded the development of an expert system program to guide nonexperts through an inspection process [11.3]. In companies without an in-house expert, users can hire consultants from OEMs and large repair organizations. Also, individual consultants with extensive experience who formerly worked for OEMs or large users are often available.

11.4 EXPERT SYSTEMS TO CONVERT DATA INTO INFORMATION

All of the monitors discussed in Chapter 13, together with the tests in Chapters 12 and 14, produce "data." That is, temperatures, vibration levels, partial discharge magnitudes, insulation resistance readings, etc. are produced. These measurements in themselves convey little information about the winding condition to casual observers. Unless one has been schooled in interpreting the data, the readings themselves are meaningless. Since interpretation expertise is needed for the wide variety of tests and monitoring procedures, few people have the time or inclination to become "experts."

In the 1990s, new software technologies were developed that enabled the creation of "expert system" programs. Expert system computer programs attempt to recreate the reasoning processes that an expert uses to interpret test results, as well as other relevant information. If successful, an expert system lets nonexperts convert data into information, with accuracy approaching that of an expert. Expert systems have a secondary advantage in that they capture and permanently store the knowledge of experts, so that this knowledge is retained for long-term use.

Both on-line and off-line expert systems have been developed for interpreting the condition of rotor and stator windings. These are treated separately in the following subsections.

11.4.1 Off-Line Expert Systems

An off-line system is primarily intended solely for use by motor and generator maintenance personnel. The first and only off-line expert system for machine windings, called MICAA, was developed by Lloyd et al. [11.3]. The program calculates the risk of failure of stator windings, rotor windings, or cores. The risk of failure is calculated for most types of motors and generators, based on the following information:

- Known problems with the type of machine being analyzed
- Operating conditions (running hours, number of starts, and maximum operating temperatures)
- On-line monitoring data (for example, ozone levels, current signature analysis, vibration levels, partial discharge, etc.)
- Off-line test data
- The trend in on- and off-line data over time
- Visual inspections

The program uses Mycin reasoning to calculate the probability of any particular failure mechanism occurring, based on the above data [11.3]. The program then estimates risk of failure for each component. Mycin reasoning was developed by the medical community to diagnose diseases from a wide variety of tests and symptoms. Figure 11.1 shows a typical diagnosis produced by MICAA. Given a current diagnosis, MICAA also guides the user on what further tests or inspections are needed to improve the confidence in the diagnosis.

To obtain a reasonable diagnosis that has a high probability of being correct, the program requires information. Since the trend in a test result or monitoring procedure is usually a powerful indicator of a developing problem, historical data needs to be input. All this data can be tedious to collect and input. However, once this is done, the data is forever retrievable by everyone in the plant, and adding new test results is relatively easy. Although it is unlikely that the off-line expert system can produce a winding condition assessment as accurately as a true human expert, the program is able to identify the relatively few motors and generators in a plant that may be at risk of failure, and for which the expense of hiring an expert consultant may be worthwhile.

The off-line expert system is widely used by the North American utility industry.

11.4.2 On-Line Expert Systems

Westinghouse introduced the first on-line expert system that was capable of diagnosing the condition of generator windings [11.4]. Since then, several American and European manu-

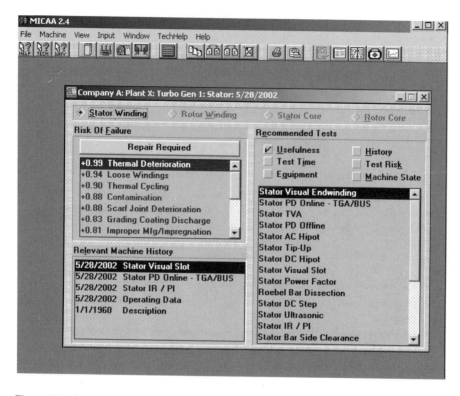

Figure 11.1. Output screen from MICAA. The program estimates the risk of rotor and stator winding failure, and identifies the most likely failure processes.

facturers of large turbine generators have produced their own versions. Recently, an on-line expert system was introduced for hydrogenerators [11.5].

The output of on-line expert systems is intended for both operators and maintenance personnel. If a rapidly developing problem is occurring, these systems identify the most likely cause of the problem, and if a failure may result in a short time, an alarm is raised to alert the operator. Most programs also produce advice for the operator on possible actions that may be taken to avert failure. For longer-term problems, for which failure may take weeks to years, the expert system output is usually intended for maintenance personnel.

Most of the input that drives the diagnosis comes from on-line monitors. Since turbine generators often have a wide variety of monitors, the diagnosis tends to be more credible on such machines. The more information that is available, the greater the accuracy. The commercial on-line expert systems use information from:

- Stator slot, core, cooling system, and rotor temperature readings
- Hydrogen and water cooling system pressure differentials
- Output current and voltage readings
- Bearing and frame vibration measurements
- Partial discharge levels
- Rotor flux readings

In general, off-line test results are not available to the expert system. However, most programs have a tremendous amount of customized information embedded within them regarding the design of the generator and the most likely failure processes and symptoms. This high degree of customization tends to make the initial purchase price of the software expensive. In an attempt to overcome this limitation, one program has an auxiliary expert system to make it easier and less costly to customize the software for a particular installation [11.6]. Unfortunately, the high commercial price for installation and maintenance of the software has greatly limited its application.

REFERENCES

11.1   I. Kerszenbaum, *Inspection of Large Synchronous Machines,* IEEE Press, 1997.

11.2   I. Kerszenbaum and G. Klempner, *Operation, Maintenance, and Troubleshooting of Large Turbine Driven Generators,* IEEE Press, 2003.

11.3   B. A. Lloyd, G. C. Stone, J. Stein, "Development of an Expert Monitoring System to Assess Machine Insulation Condition," in *Proceedings of IEEE Electrical Insulation Conference,* October 1991, pp. 33–37.

11.4   A. J. Gonzales, R. L. Osborne, and C. T. Kemper, "On-Line Diagnosis of Turbine Generators Using Artificial Intelligence," *IEEE Trans. EC,* 68–74, June 1986.

11.5   R. D. Milroy and B. A. Lloyd, "Development of an Advanced On-Line Expert Monitoring System for Hydrogenerators," paper presented at Doble Client Conference, Boston, April 1997.

11.6   G. S. Klempner, A. Kornfeld, and B. Lloyd, "The Generator Expert Monitoring System," paper presented at EPRI Conference on Utility Motor and Generator Predictive Maintenance," December 1991.

CHAPTER 12

OFF-LINE ROTOR AND STATOR WINDING TESTS

This chapter describes the main tests that are commercially available for assessing the condition of rotor and stator windings. All of the tests require the motor or generator to be removed from service, for at least a short time. Tremendous advancements in testing technology are being made, due to better electronics and the advent of computers with sophisticated data analysis software. However, only tests with a demonstrated usefulness at the time of printing this book are included. This does not mean that some newer tests are not useful; it just means that they have been in use for so short a time, that independent verification of claims has yet to be made.

The purpose of each test will be described, together with the types of machines and/or windings it is useful for. The theory of the test will also be described where it is not obvious. In addition, each test will be compared to other similar tests. Practical information is given on how to apply the test, including the state the winding must be in to do the test and the normal time it takes to do the test. Finally, a practical guide is given on interpreting the results. This interpretation will reflect the experience of test users rather than test vendors or the very general interpretation that may be published in industry standards.

To aid in test selection, Tables 12.1 and 12.2 show a summary of the most common off-line tests for stator and rotor windings, respectively.

12.1 INSULATION RESISTANCE AND POLARIZATION INDEX

This is probably the most widely used diagnostic test for motor and generator windings. It can be applied to all machines and windings, with the exception of the squirrel cage induction motor rotor winding, which does not have any insulation to test. This test successfully locates pollution and contamination problems in windings. In older insulation systems, the test can also detect thermal deterioration. Insulation resistance (IR) and polarization index

Electrical Insulation for Rotating Machines. By Stone, Boulter, Culbert, and Dhirani
ISBN 0-471-44506-1 © 2004 Institute of Electrical and Electronics Engineers

Table 12.1. Common Off-Line Stator Winding Tests

Name	Description	Performance Difficulty	Effectiveness	Relevant Standards
Insulation resistance (IR)	Apply DC voltage for 1 min to measure leakage current	Easy	Only finds contamination or serious defects	IEEE 43
Polarization index (PI)	Ratio of 1 min and 10 min IR	Easy	Only finds contamination or serious defects	IEEE 43
DC high potential	Apply high DC voltage for 1 min	Easy	Only finds serious defects	IEEE 95
AC high potential	Apply high AC voltage for 1 min	Moderate, due to large transformer needed	More effective than DC high potential	NEMA MG1 or IEC 60034
Capacitance	Apply low or high voltage to measure winding capacitance to ground	Moderate	Moderately effective to find thermal or water leak problems	
Dissipation (power) factor	Apply low or high voltage to measure insulation loss	Moderate	Moderately effective to find thermal or water leak problems	IEEE 286 or IEC 60894
Power factor tip-up	Differences in insulation loss from high to low voltage	Moderate	Effective to find widespread thermal or contamination problems in FW*	IEEE 286 or IEC 60894
Off-line partial discharge	Directly detect PD pulse voltages at rated voltage	Difficult	Finds most problems except end-winding vibration; for FW only	IEEE 1434
Surge comparison	Apply simulated voltage surge	Difficult to determine if puncture occurred in FW	Effective for finding turn insulation problems in RW* and multiturn FW	IEEE 522
Blackout	Apply high ac voltage and look for discharges with lights out	Moderate	Effective for contamination problems in end-winding	IEEE 1434
Wedge tightness	"Hammer" wedges to see if loose	Moderate	Effective to find loose windings in FW	
Side clearance	Insert "feeler gauges" down side of slot	Easy, after wedges removed	Effective to find loose windings in FW	

*FW is form-wound winding; RW is random-wound winding.

Table 12.2. Common Off-Line Rotor Winding Tests

Name	Description	Performance Difficulty	Types	Effectiveness
Insulation resistance (IR)	Apply DC voltage for 1 min to measure leakage current	Easy	All	Only finds contamination or serious defects
Polarization index (PI)	Ratio of 10 min to 1 min IR	Easy	All	Only finds contamination or serious defects
DC high potential	Apply high-voltage DC for 1 min	Easy	All	Only finds serious defects
AC high potential	Apply high AC voltage for 1 min	Moderate, due to large transformer needed	All	More effective than DC high potential
Open circuit	Measure generator output voltage as a function of field current to find short-circuited turns	Moderate	All	Effective only for generators; needs measurement when rotor OK
Impedance test	Apply 50 (60) Hz current and measure V/I at different speeds to find turn short circuits	Moderate	All rotors with slip rings	Effective
Pole drop	Apply 50 (60) Hz current and measure voltage drop across each pole to find poles with turn short circuits	Easy	SPR*	Only finds short circuits that are present when rotor stopped
Surge test	Find turn and ground faults by measuring discontinuities in surge impedance	Difficult	RR*	Effective if close to dead short circuit

*RR is round rotor; SPR is salient pole rotor.

(PI) tests have been in use for more than 70 years. Both tests are performed with the same instrument, and are usually done at the same time.

12.1.1   Purpose and Theory

The IR test measures the resistance of the electrical insulation between the copper conductors and the core of the stator or the rotor. Ideally, this resistance should be infinite since, after all, the purpose of the insulation is to block current flow between the copper and the core. In practice, the IR is not infinitely high. Usually, the lower the insulation resistance, the more likely it is that there is a problem with the insulation.

The PI test is a variation of the IR test. PI is the ratio of the IR measured after voltage has been applied for 10 minutes (R_{10}) to the IR measured after just one minute (R_1), i.e.:

$$PI = \frac{R_{10}}{R_1} \tag{12.1}$$

As discussed in Section 12.1.3, a low PI indicates that a winding may be contaminated or soaked with water.

The IEEE 43-2000 guide gives an extensive discussion of the theory of the IR/PI tests as applied to rotating machines [12.1]. There is no equivalent IEC procedure for the IR/PI test. In the test, a relatively high DC voltage is applied between the winding copper and the stator or rotor core (usually via the machine frame). The current flowing in the circuit is then measured. The insulation resistance at time t is then:

$$R_t = \frac{V}{I_t} \tag{12.2}$$

which is just Ohm's law. V is the applied DC voltage from the tester and I_t is the total current measured after t minutes. The reference to the time of current measurement is needed since the current is usually not constant.

There are at least four currents that may flow when a DC voltage is applied to the winding. They include:

1. A capacitive current. When a DC voltage is applied to a capacitor, a high charging current first flows, which then decays exponentially. The size of the capacitor and the internal resistance of the voltage supply, typically a few hundred kilohm, set the current decay rate. A form-wound stator coil may have a geometric capacitance of about 10 nF between the copper and the core. A large hydrogenerator may have a capacitance of 1 μF. Thus, this current effectively decays to zero in less than 10 seconds. Since this capacitive current contains little diagnostic information, the initial insulation resistance is measured once the capacitive current has decayed close to zero. This time has been set as 1 minute to ensure that this current does not distort the insulation resistance.

2. A conduction current. This current is due to electrons or ions that migrate across the insulation bulk, between the copper and the core. This is a galvanic current through the groundwall. Such a current can flow if the groundwall has absorbed moisture, which can happen in the older thermoplastic insulation systems, or if a modern insulation has been soaked in water for many days or weeks. This current also flows if there are cracks, cuts, or pinholes in the ground insulation (or magnet wire insulation in ran-

dom-wound machines), and some contamination is present to allow current to flow. This current is constant with time, and ideally should be zero. With modern insulation, this current usually is zero (as long as there are no cuts, etc.) since electrons and ions cannot penetrate through modern epoxy–mica or film insulation. If this current is significant, then the winding insulation has a problem.

3. A leakage surface current. This is a constant DC current that flows over the surface of the insulation. It is caused by partly conductive contamination (oil or moisture mixed with dust, dirt, fly ash, chemicals, etc.) on the surface of the windings. Ideally, this leakage current is zero. However, if this current is large, it is likely that contamination-induced deterioration (electrical tracking) can occur (Sections 8.8, 9.1.4, and 9.2.3). This current can be large in round rotor windings where the copper conductors are bare and the insulation is just slot liners.

4. The absorption current. This is a current that is hard to conceptualize. The current is due to precessing or reorientation of certain types of polar molecules in the applied DC electric field. Many practical insulating materials contain polar molecules that have an internal electric field due to the distribution of electrons within the molecule. For example, water molecules are very polar. When an electric field is applied across water, the H_2O molecules all align, just as magnetic domains become aligned in a magnetic field. The energy required to align the molecules comes from the current in the DC voltage supply. Once the molecules are all aligned, the current stops. This current is the polarization current, which is one component of the absorption current. There are many polar molecules in asphalt, mica, polyester, and epoxy. Experience shows that after a DC electric field is applied to such materials, the absorption current is first relatively high, and decays to zero after about 10 minutes. In all practical respects, the absorption current behaves like an RC circuit with a long time constant. The absorption current, like the capacitive current, is neither good nor bad. It is merely a property of the insulation materials. In addition to molecular realignment, absorption currents may arise in high-voltage laminated insulation (such as in high-voltage stator groundwalls), due to electron trapping at interfaces.

The total current I_t is the sum of all these current components. Unfortunately, we cannot directly measure any of these component currents.

The currents that are of interest, as far as a winding condition assessment is concerned, are the leakage and conduction currents. If just R_1 is measured (after 1 minute), the absorption current is still nonzero. However, if the total current is low enough, then R_1 may still be considered satisfactory (see Section 12.1.3). Unfortunately, just measuring R_1 has proved to be unreliable, since it is not trendable over time. The reason is that IR is strongly dependant on temperature. A 10°C increase in temperature can reduce R_1 by 5 to 10 times. Worse, the effect of temperature is different for each insulation material and type of contamination. Although some "temperature correction" graphs and formulae are available to enable users to "correct" R_1 to a common 40°C temperature, they are extremely unreliable. The result is that every time R_1 is measured at different temperatures, one gets a completely different R_1 corrected to 40°C. This makes it impossible to define a scientifically acceptable corrected R_1 from R_1 measurements over a wide range of temperatures. It also makes trending R_1 almost useless, unless one can be sure the measurement temperature is always the same.

The polarization index (PI) was developed to make interpretation less sensitive to temperature. Equation 12.1 shows that the PI is a ratio of the IR at two different times. If we assume that R_{10} and R_1 were measured with the winding at the same temperature, which is usually very reasonable to assume, then the "temperature correction factor" will be the same for both

R_1 and R_{10}, and will be ratioed out. Thus, the PI is relatively insensitive to temperature. Furthermore, the PI effectively allows us to use the absorption current as a "yardstick" to see if the leakage and conduction currents are excessive. If these latter currents are much larger than the absorption current, the ratio in Equation (12.1) will be about 1. Experience shows that if the PI is about 1, then the leakage and conduction currents are large enough that electrical tracking will occur. Conversely, if the leakage and conduction currents are low compared to the absorption current after 1 minute, then the PI will be greater than 2, and experience indicates that electrical tracking problems are unlikely. Thus, if we can see the decay in the total current in the interval between 1 minute and 10 minutes, then this decay must be due to the absorption current (since the leakage and conduction currents are constant with time), with the implication that the leakage and conduction currents are minor.

Some users define ratios other than the 1- and 10-minute ratio in Equation (12.1). For example, 10 seconds/1 minute or 30 seconds/5 minutes are other times that the IR is measured after to define a ratio. Such tests can be performed more quickly than the normal PI. The problem with these other ratios is that there is little agreement on what constitutes an acceptable ratio. Thus, unless one has an extensive database, interpretation is more difficult.

12.1.2 Test Method

The IR and PI can be measured with a high-voltage DC supply and a sensitive ammeter. The DC supply must have a well-regulated voltage; otherwise a steady-state capacitive charging current will flow (= $C \, dV/dt$). The ammeter must measure currents smaller than a nanoamp. There are several special-purpose "megohmeters" available commercially. Sometimes these are known as Megger Testers, after the name of the instrument first developed for this purpose (Megger is a trade name of AVO). A megohmeter incorporates a regulated DC supply and an ammeter that is calibrated in megohms. Modern instruments can apply voltages up to and exceeding 10 kV DC, and measure resistances higher than 100 GΩ (100,000 MΩ).

For stator windings, the test is best done right at the machine terminals, preferably one phase at a time, with other equipment such as cables and instrument transformers disconnected. This is relatively easy in generators, however, these test conditions can be difficult to achieve in motors. Motor leads in the motor terminal box would have to be untaped and disconnected, which is often tedious. Also, the neutral ends of each phase are usually not accessible; thus, all three phases are tested at the same time. These difficulties have lead many maintenance personnel to test the motor stator winding from the switchgear (motor control center or MCC). In this situation, the insulation resistance of the cable insulation (and especially the cable terminations) is in parallel with the stator winding. If failing readings are obtained, the IR/PI test must be repeated at the motor terminal box, with the cables disconnected from the motor. However, if the readings are good, then it implies that both the cables and the stator winding have good results.

If the motor is equipped with surge capacitors, then these *must* be disconnected prior to the IR/PI test. This can be done by disconnecting the high-voltage leads to the capacitor, or by temporarily isolating the low-voltage side of the capacitors from ground. Surge capacitors usually have a 10-megohm or so discharge resistor within the capacitor case. This resistor will yield a low IR and a PI of 1 if not removed.

For direct-water-cooled stator windings, R_1 and the PI will be significantly reduced if the stator is measured with the water in place. Ions in the water will move under the influence of the DC electric field. In addition, electrochemical reactions may occur in the water, releasing hydrogen, with the high DC voltage that is applied. Thus, it is advisable to do the test only after the water has been drained and a vacuum drawn for several days to dry out the cooling water channels. This considerably complicates the test on direct-water-cooled stators; therefore, many users no longer do this test on such machines.

For rotor windings, the test is done at the slip rings. If a brushless type of exciter is tested, then the winding must first be disconnected from the rotating diode rectifier to allow this test to be performed.

The IR/PI test will depend strongly on the humidity. If the winding temperature is below the dew point, there is no way that R_1 or PI can be "corrected" for the humidity. If the results are poor, then the test must be repeated with the winding temperature above the dew point. It will probably be necessary to heat the winding in some fashion for hours or days to dry off the moisture that has condensed on it.

The 2000 revision of IEEE 43 suggests that the test voltages be higher than recommended in the past, because tests at higher voltages are more likely to find major defects such as cuts through the insulation in the end-windings. Note that the test voltages are still well below the rated peak line-to-ground voltages of the windings. Thus the IR/PI test is not a "hipot" test (see Section 12.2). Table 12.3 shows the test voltages suggested in IEEE 43.

After each IR and PI test, the winding should be grounded for at least four times as long as the voltage was applied, i.e., 40 minutes. Premature removal of the ground will cause a high voltage to reappear, due to the time it takes for the molecules to again become random in orientation, and for the space charge to dissipate. Thus, a shock hazard exists. In addition, repeat IR/PI tests will be in error if the winding is not grounded for a sufficiently long time.

12.1.3 Interpretation

What constitutes a "good reading" and a "bad reading" depends on the nature of the insulation system and the component (stator or rotor) being tested. Until 2000, the minimum R_1 and the acceptable range for PI was the same for all types of stator winding insulation. However, it has been recognized that the modern insulation materials in random-wound and form-wound stators have essentially no conduction current (as long as there are no cracks or pinholes). Thus, it is possible for a clean, dry, form-wound stator winding to have an R_1 that is essentially infinite—greater than 100 GΩ. With an R_1 of infinity, calculations of a realistic PI are dubious. Such high R_1's are not likely in systems made before the 1970s. Consequently, the maintenance person needs to establish the type of insulation used in the winding, or at least the approximate age of the winding, before interpreting IR and PI results.

Table 12.3. Guidelines for DC Voltages to be Applied During IR/PI

Winding rated voltage (V)[a]	Insulation resistance test direct voltage (V)
<100	500
1000–2500	500–1000
2501–5000	1000–2500
5001–12 000	2500–5000
>12 000	5000–10 000

[a]Rated line-to-line voltage for three-phase AC machines, line-to-ground voltage for single-phase AC machines, and rated direct voltage for DC machines or field windings (from IEEE 43-2000).

Generally, if the PI is less than 1, the winding is wet or contaminated. If PI is greater than 2, it is clean and dry.

Table 12.4 summarizes how to interpret IR results in rotor and stator windings. The distinction between older and modern insulation systems was set at 1970, although this is somewhat arbitrary. Of note in this table are the following:

- If R_1 is below the indicated minimum, the implication is that the winding should not be subjected to a hipot test, or be returned to service, since failure may occur. Of course, if historical experience indicates that low a R_1 is always obtained on a particular winding, then the machine can probably be returned to service with little risk of failure.
- The minimum R_1 is the value corrected to 40°C. Unfortunately, that correction is unlikely to be valid.
- The minimum acceptable R_1 is much lower for old stators than new stators, and it depends on voltage class. For modern stators, the minimum acceptable R_1 only depends on whether it is a form-wound or random-wound stator.
- For modern form-wound stators, if a very high R_1 is measured (say greater than 5 GΩ), then PI is not likely to indicate anything about the winding. Thus, one can save time by aborting the test after the first minute of testing.
- If the IR or PI is below the minimum in a modern stator winding, it is only an indication that the winding is contaminated or soaked with water.
- If a high PI result is obtained on an older stator winding, then there is a possibility that the insulation has suffered thermal deterioration. This occurs because thermal deterioration fundamentally changes the nature of the insulation and, thus, the absorption currents that flow. The insulation has changed in an asphaltic mica winding if the asphalt has been heated enough to flow out of the groundwall.
- Rotor windings generally have a lower PI and R_1 than stator windings, because the insulation is thinner and there is much more surface area. In round rotor and salient pole strip-on-edge windings, there is often exposed copper, making leakage currents much higher.

In general, the IR/PI test is an excellent means of finding windings that are contaminated or soaked with moisture. Of course, the test is also good at detecting major flaws where the insulation is cracked or has been cut through. In form-wound stators with thermoplastic insulation systems, the test can also detect thermal deterioration. Unfortunately, there is no evidence that thermal deterioration or problems such as loose coils in the slot can be found in modern windings [12.1].

Table 12.4. Recommended Minimum Insulation Resistance Values at 40°C (All Values in MΩ)

Minimum insulation resistance after 1 minute	Test specimen
kV+1	For most windings made before about 1970, all rotor windings, and others not described below
100	For most stator windings built after about 1970 (form wound coils)
5	For most machines with random-wound stator coils and form wound coils rated below 1 kV

Notes. (1) kV is the rated machine terminal-to-terminal voltage, rms kV. (2) From IEEE 43-2000.

12.2 DC HIPOT TEST

The DC hipot test is an overpotential test that is applied to stator and rotor windings of all types (with the exception of SCI rotors). Hipot is a short form for high potential. In this test, a DC voltage, substantially higher than that which occurs in normal operation, is applied to the winding. The basic idea is that if the winding does not fail as a result of the high voltage, the winding is not likely to fail anytime soon due to insulation aging when it is returned to service. If a winding fails the DC hipot test, then repairs or a rewind are mandatory, since the groundwall insulation has been punctured. Stator windings are much more likely to be subjected to a DC hipot than rotors.

12.2.1 Purpose and Theory

The purpose of this test is to determine if there are any major flaws in the groundwall insulation before a winding enters service (commissioning or acceptance hipot test) or during service (maintenance hipot test). The principle is that if there is a major flaw in the insulation, a high enough voltage applied to the winding will cause insulation breakdown at the flaw. By IEC 60034 and NEMA MG1 standards, all new windings (original or rewound) must pass a hipot test prior to being accepted by the customer (Section 15.2).

Of course the main problem with hipot testing (both DC and AC, see Section 12.4) is that the winding may fail. If failure does occur, then either the insulation that punctured must be replaced, the coil with the puncture is removed from the circuit, or the coil or even the complete winding is replaced. These are all expensive alternatives, and all involve a delay in placing the machine in service.

Since a hipot test can be destructive and delay a return to service, many plants do not perform a maintenance DC hipot. The rationale for this is that the hipot test may cause a failure that would not occur for a long time in service, resulting in rewinding or significant repairs before they are really needed. This is true. However, the proponents of hipot testing argue that for many critical machines, an in-service failure (that could have been prevented if a hipot test were done) can result in a greater disruption to plant output than a hipot failure. For example, the in-service failure of a critical pump motor in a petroleum refinery can stop production for days or weeks, and cost as much as a million dollars per day. Also, an in-service fault can sometimes cause consequential damage such as stator core damage, a fire, or coils being ejected from the slot, resulting in much higher repair costs. Thus, whether a DC hipot is performed as a maintenance test depends on how critical the machine is to the plant, the availability of spares, and the philosophy of plant management being to avoid unexpected plant shutdowns.

There are differences between a DC and an AC hipot test. These are discussed in Section 12.4.

12.2.2 Test Method

There are several different methods for performing a DC hipot test. Most are reviewed in IEEE Standard 95 [12.2]. Some of the variations reduce the risk of a failure during the test, and some also give information of a diagnostic nature.

For all types of DC maintenance hipot test methods, the critical decision to be made concerns the maximum test voltage. For form-wound stator windings, IEEE 95 gives guidance. It suggests that the maintenance hipot should be as high as 75% of the acceptance hipot level. NEMA MG1 and IEC 60034 stipulate that the DC acceptance hipot be 1.7 times the AC hipot acceptance level of $(2E + 1)$ kV, where E is the rated rms phase-to-phase voltage in kV of the stator winding. Many users have adopted a DC maintenance hipot level of about $2E$ [12.8]; that is, a 4.1 kV winding would be tested at about 8 kV DC. This level was originally suggested since it approximates the highest likely overvoltage that can occur in the stator if a phase-to-ground fault occurs in the power system. Consequently, a maintenance hipot test just reproduces, in a controlled, off-line fashion, the overvoltage a stator can see in service. The idea here is that if the winding can survive this hipot test, it is unlikely to fail in service due to a voltage surge created by a power system fault.

In general, it is better to isolate the phases from one another in a stator winding hipot test. The hipot is applied to each phase in turn, with the other two phases grounded. All winding-temperature detectors should be grounded before this test is performed. Energizing each phase separately ensures that there is electrical stress between coils in the end-winding, so flaws in the end-winding are more likely to be detected. In motor stators, it is usually not possible to isolate the phases; therefore, the DC hipot test energizes all three phases at the same time. Since there is no potential difference between coils in the end-winding, end-winding problems that are distant from the core are not detected in the three-phase test.

If the DC hipot test is performed on a direct water-cooled winding, the water should be drained from the stator. The high-voltage DC may cause hydrogen to be created in the Teflon hoses at the end of each bar.

It is prudent to always perform an IR/PI test prior to a DC hipot test. If the winding is wet or contaminated, which will be discovered by the IR/PI test, then one should not perform the DC hipot test until the winding is dried and/or cleaned. Performing a hipot test on a wet or contaminated winding may unnecessarily puncture the insulation, greatly increasing the amount of effort needed to repair the winding.

In contrast with an AC hipot test (Section 12.4), the DC hipot test does not age the winding insulation, since partial discharges occur very infrequently under DC voltage. Thus, if the winding passes the DC hipot test, then the insulation has not been deteriorated in any way by the test. However, one should be aware that if the DC hipot test is done from the switchgear, and if the power cables have been soaked in water for years, then the DC hipot might age and even cause the power cables to fail. This occurs because power cables rated 2300 V and above often fail by a mechanism called "water treeing." A DC potential accelerates water treeing. If the cables have always been kept dry, then DC hipot testing should pose no risk to the cables.

Hipot testing poses a risk of electric shock to the test personnel. Even after the winding has been grounded for many hours after a DC hipot test, if the ground is removed, residual polarization can cause the copper conductors to jump up to a high DC voltage. Appropriate safety procedures are needed [12.2].

There are several alternative DC hipot test methods, as described below.

Conventional DC Hipot Test. In the conventional maintenance DC hipot test, a suitable high-voltage DC power supply (available from many suppliers) is connected to the winding, either at the switchgear or at the machine terminals. The DC voltage is quickly raised to the test voltage and held for either 1 minute or 5 minutes. After this time, the voltage is quickly lowered and the winding is grounded. If the insulation is sound, there will be no high-current surge, and the power supply circuit breakers will not trip. If the power supply breaker trips, then it is likely a puncture has occurred, since the insulation resistance will have instantaneously dropped to zero, which causes an "infinite" current to flow (by Ohm's law), and the power supply cannot deliver this "infinite" current. Circuit breaker tripping is an indication that the winding has failed and winding repairs or replacement are required. The conventional test contains little diagnostic value, although one can measure the DC current after the 1- or 5-minute application of the test voltage. If one trends the leakage current over the years, then an increasing trend is an indication that contamination is occurring.

Step-Stress Hipot Test. A variation is to use the same power supply as described previously, and gradually increase the voltage in either equal or unequal steps. For example, the DC voltage can be increased in 1 kV steps, with each voltage level being held for 1 minute before it is increased again. One then measures the DC current after the end of each step (since by this time the capacitive current will have dropped to zero), and plots it on a graph of current versus DC voltage. Ideally, the plot will be a line with a gentle upward curve. However, sometimes the current increases abruptly above a certain voltage. This may be a warning that the insulation is close to puncturing. If the tester acts rapidly, the test can be aborted (voltage turned off) before a complete puncture occurs. Experience shows that warning is likely if the flaw is in the end-winding, but little or no warning is given if the flaw is within the slot. By carefully applying this test, a hipot failure may be avoided. However, if the voltage at which the current instability was detected is below operating voltage, there is a high risk in returning the winding to service without repairs.

In the 1950s a variation of the step-stress test was developed that replaced the equal time steps with steps of variable duration. This test is often called the Schleif test, after its developer [12.3]. The intention is to make the plot of DC current versus applied DC voltage a straight line by using steps that linearize the absorption currents discussed in Section 12.1.1. This makes any abrupt increase in current easier to identify, further increasing the probability of detecting a flaw before puncture occurs. More details on applying this test are in [12.2]. The Schleif test was developed at the time when asphaltic mica groundwall insulation was present in most machines.

DC Ramp Hipot Test. A third variation of the DC hipot is called the ramp test. In this case, the DC voltage is smoothly and linearly increased at a constant rate, usually 1 or 2 kV/minute. Thus, there are no discrete steps in voltage or current. The current versus voltage plot is automatically graphed and displayed. By increasing the voltage as a constant ramp, the capacitive current is a constant current ($I = C \cdot dV/dt$, with $V = rt$; r being a constant, and t being time), which can be easily ignored, unlike in the stepped-stress test. The primary advantage of the ramp test is that it is by far the most sensitive way to detect that a current instability is occurring, since the capacitive charging current is not changing with time. Consequently, the ramp test is the method most likely to enable the user to avoid a puncture [12.4, 12.5]. As described in [12.5], the test may also enable the detection of such conditions as groundwall delamination. Adwel introduced a commercial DC ramp test supply in 2003.

12.2.3 Interpretation

Fundamentally, the DC hipot test is not a diagnostic test that gives a relative indication of the insulation condition. Rather it is a go–no go test; the winding is in good condition if it passes, and in severely deteriorated condition if it fails. However, the DC current measured at the time of the test can give some qualitative indication of condition, much like the IR test does. Specifically, if the current at any particular voltage increases monotonically over the years, it is an indication that the insulation resistance is decreasing, and the winding is gradually getting wetter or becoming more contaminated. However, caution is needed when trending the current over time. The current is very dependent on winding temperature and atmospheric humidity. Thus, in most cases the trend is erratic and impossible to interpret.

12.3 DC CONDUCTIVITY TEST

The DC conductivity test is a means of finding if the copper strands in the stator winding coils or associated ring busses are broken or cracked, if the conductors in insulated rotor windings are cracked or broken, or if the brazed or soldered connections in either type of winding are deteriorating. In addition, if the test is done from the switchgear, one can sometimes detect loose bolted cable connections.

12.3.1 Purpose and Theory

If the copper conductors in the rotor or stator winding crack or break, the DC resistance between the terminals of the winding will increase, due to the reduced copper cross section that must pass the current. There are several reasons for the resistance to increase:

- Stator winding end-winding vibration (Section 8.12) can fatigue crack copper strands.
- Operating events such as an out-of-phase synchronization can put such a high magnetic force on coils that the copper cracks or breaks at connection points.
- Copper connections between coils or to busses may be poorly brazed or soldered. This gives rise to local heating that oxidizes the connection, further increasing the temperature and increasing connection resistance.
- Stator winding leads may be poorly bolted or brazed to the power system bus or cables.
- There may be shorted turns in insulated rotor windings.

Cracked strands or high-resistance connections ultimately lead to failure due to thermal deterioration of the insulation (Sections 8.1, 9.1.1, and 9.2.1).

It is best to use DC rather than AC current to measure the resistance, except for detecting shorted turns in insulated rotor windings (Section 12.20). An AC measurement will be sensitive both to the resistance and the inductive reactance of the winding. The reactance can be modified by changes in the magnetic permeability near the machine and, thus, by rotor position; the amount of enclosure disassembly (if any); or even if the machine is adjacent to large objects made of magnetic steel. Consequently, unless one can guarantee the magnetic environment from test to test, AC measurements make it hard to trend the conductor resistance.

12.3.2 Test Method

The DC resistance (or conductance) is measured by passing a DC current through a winding and measuring the voltage across the winding. The resistance is the voltage divided by the current. The resistances of a stator or rotor winding are very low, ranging from just less than 1 ohm for small machines to less than a milliohm for a large machine. To accurately measure the resistance at such low values requires either a four terminal measurement or a bridge type of measurement, for example, a Kelvin bridge. There are many commercial instruments on the market to measure such low resistances. The measuring instrument should have an accuracy of at least 1% to detect winding problems.

For a stator winding, it is best to measure each phase individually between the phase terminal and the neutral terminal in the same phase. If a motor stator winding is measured, it is usually not possible to isolate the three phases. Instead, a measurement is made between the phase terminals, i.e., three measurements: A–B, B–C, and C–A. If resistance between phases is measured on motor stators from the switchgear (MCC), the resistance of the connections of the power cables to the stator is also measured. Since the connections between the power cables and the motor leads are often troublesome, this is very desirable. However, if the power cable length is long, there will be less sensitivity to incipient conductor cracks in the stator winding. Rotor winding resistance can be measured between the slip rings.

The DC resistance is strongly affected by the temperature of the stator winding; that is, as the winding temperature increases, so does the resistance. Thus, to trend the resistance in a winding over the years, temperature correction is needed. To correct a copper resistance R_T measured at temperature T to the resistance R_{20} at 20°C:

$$R_{20} = \frac{R_T}{1 + (T - 20)/255.5} \tag{12.3}$$

Humidity has little effect on the winding DC resistance.

12.3.3 Interpretation

Resistance measurements are easiest to interpret by comparisons between phases, comparisons to other identical machines, or by trending the measurement on the same winding over time.

For a stator winding, the resistance of each phase (or resistance of the phase-to-phase winding in a motor) should be compared. All three resistances should be within 1% (for form windings) and 3% (for random windings) of each other (or the accuracy of the measurement instrument). If one of the phase resistances is high, then it may be an indication that a strand is broken or that there is a bad connection on that phase, etc.

If there are several identical machines with identical rotor and stator windings, then all windings should have a resistance that is within about ±5% of one another. If a winding has higher resistance than the others, then perhaps it has a winding problem. Note that a rewound winding is not likely to have exactly the same resistance as the original winding, but resistance balance between phases should meet the criteria indicated above.

Trending the resistance of a winding over the years is perhaps the most useful way to find a developing problem such as cracking conductors or deteriorating joints. To make the trend meaningful, the resistances must be corrected to the same temperature, using Equation 12.3. If the corrected resistance increases by more than 1%, then problems may be occurring. It is prudent to use the same measurement instrument over the years to achieve the best sensitivity. Frequent calibration of the instrument is also needed.

As could be expected, it can be difficult to find cracked strands or one bad joint if a stator or rotor winding has many parallel circuits. Also, high-resistance joints or cracks may appear very quickly. Therefore, a test every year or so is unlikely to catch rapidly developing problems.

12.4 AC HIPOT TEST

The AC hipot test is similar to the DC hipot test (Section 12.2), with the exception that power-frequency (50 or 60 Hz) voltage is used. Sometimes 0.1 Hz AC is also employed. Both commissioning (acceptance) and maintenance AC hipot versions of the test are in use. This test is most commonly applied to form-wound stator windings. The maintenance AC hipot is rarely used in North America, but does find more widespread application in Asia and Europe.

12.4.1 Purpose and Theory

Most of the description given for the DC hipot test (Section 12.2) is relevant to the AC hipot test. Specifically, it is a go–no go test that ensures that major insulation flaws that are likely to cause an in-service fault in the near future can be detected in an off-line test.

The major differences between the DC and AC tests are the test voltage applied and the voltage distribution across the groundwall insulation. Both are linked. With DC voltage, the voltage dropped across insulation components within the groundwall and in the end-winding depends on the resistances (resistivity) of the components. Components with a lower resistance will have less voltage dropped across them. In contrast, the AC voltage dropped across each component in the groundwall or in the end-winding depends on the capacitance (dielectric constant) of each component. Thus, there tends to be a completely different electric stress distribution across the groundwall between AC and DC tests. In older insulation systems, particularly asphaltic mica systems, the differences between the AC and DC stress distributions were less pronounced because of the finite resistivity in older groundwalls due to the absorption of moisture. However, with modern epoxy–mica insulations, the resistivity is essentially infinite; thus, the DC voltage may all be dropped across a very thin layer of insulation. Consequently, significant flaws that might cause puncture may escape detection with a DC test but would be easily detected with an AC test because of the more uniform voltage distribution with AC stress.

For modern windings, the AC hipot test yields an electric stress distribution that is the same as that which occurs during normal operation. Consequently, the AC hipot test is more likely to find flaws that could result in an in-service stator failure if a phase-to-ground fault occurs in the power system, causing an overvoltage in the unfaulted phases. For this reason, the AC hipot is considered superior to the DC hipot, especially with modern thermoset insulation systems.

In the 1950s there was considerable research on the relationship between AC and DC hipot tests, specifically, the ratio of the DC to AC hipot voltages [12.6, 12.7]. Eventually, a consensus was reached that, under most conditions, the DC breakdown voltage is about 1.7 times higher than the AC rms breakdown voltage. This relationship has been standardized in IEEE 95. This research was based on older insulation systems and, unfortunately, is largely irrelevant in modern insulation systems, since, as described above, the voltage distribution is completely different under AC and DC. There have, however, been a few studies of the relationship between AC and DC breakdown in modern groundwall insulation systems. One of the largest of these studies pointed out that the ratio of DC to AC breakdown voltage on average was 4.3 in epoxy–mica insulation [12.8]. The 1.7 factor, then, no longer seems to be valid, but since the variability is so large, no replacement ratio has been proposed. Thus, we are left with 1.7.

NEMA MG1 and IEC 60034 define the AC acceptance hipot level as $(2E + 1)$ kV, where E is the rated rms phase-to-phase voltage of the stator. IEEE 56 recommends the AC maintenance hipot be 1.25 to 1.5E [12.9]. The AC hipot for a 4.0 kV motor that has seen service is about 6 kV rms, applied between the copper conductor and the stator core.

There normally is no diagnostic information obtained from the AC hipot test. A winding usually either passes the test or it fails because of a puncture. Although the current needed to energize the winding can be measured, this current contains little practical information, unless the winding is saturated with moisture or the insulation has not been cured.

The AC hipot will age the insulation. In most cases, the hipot voltage is sufficiently high that significant partial discharge activity will occur. These partial discharges will tend to degrade the organic components in the groundwall, thus reducing the remaining life. However, if one uses Equation 2.2 in Section 2.1.2, it is apparent that a 1-minute AC hipot test at $1.5E$ is equivalent to about 235 hours or 10 days at normal operating voltage. Therefore, the life is not significantly reduced by a hipot test if the expected life is about 30 years.

12.4.2 Test Method

The key element in an AC hipot test is the AC transformer needed to energize the capacitance of the winding. A 13.8 kV generator stator winding with a capacitance of 1 μF requires a charging current of 8 A at 60 Hz for a $1.5E$ maintenance hipot test. A minimum transformer rating is over 150 kVA. This is a substantial transformer and is definitely not very portable as compared to a DC hipot set. The AC hipot set is also much more expensive than the DC supply. It is because of the size and expense of the AC hipot supply that an AC hipot is rarely performed as a maintenance test.

An alternative to the power-frequency transformer is a "VLF" supply. VLF refers to very low frequency. Such supplies operate at 0.1 Hz. With this low a frequency, the capacitive charging current is 1/500 or 1/600 of the current needed at power frequencies. The current for the winding above would then be only about 13 mA. The VLF supply is then rated at only 275 VA, which is considerably smaller and more practical to move about in a plant. Modern VLF supplies are relatively cost-effective. Research indicates that the voltage distribution across the groundwall insulation is essentially the same as at power frequency, i.e., the capacitances govern the voltage distribution [12.10]. Thus, the VLF AC hipot has all the advantages of the power frequency AC hipot described above. A higher voltage is used for the VLF hipot than for the power frequency version [12.11]. In spite of the advantages of the VLF test, it still seems to be rarely used for maintenance AC hipot tests.

As with the DC hipot, it is best to test each phase separately, rather than all three phases at the same time, since this will allow more effective detection of any significant flaws in the end-winding. The test can be performed at the stator winding terminals or from the switchgear. AC or VLF testing does not age cables that are in good condition. As with the DC hipot test, considerable caution is needed when performing the test, since there is a personnel hazard.

12.4.3 Interpretation

A winding either passes or fails the AC or VLF hipot tests. There is no other diagnostic information provided. If the winding fails, as determined by the power supply circuit breaker tripping or an observed insulation puncture, then repairs or coil or winding replacement are required.

12.5 CAPACITANCE TEST

Measurement of the winding capacitance can sometimes indicate problems such as thermal deterioration or saturation of the insulation by moisture within the bulk of the insulation. This test is most useful on smaller random- and form-wound motor stators, or very large direct-water-cooled generator stators that may have water leaks.

Capacitance measurements are also made during manufacturing to determine (a) when the resin has impregnated a coil, bar, or global VPI stator; and (b) when the resin has cured.

12.5.1 Purpose and Theory

Some deterioration processes involve changes to the winding that fundamentally alter the nature of the insulation. For example, if a form-wound stator deteriorates due to long-term overheating, the groundwall insulation layers delaminate (Section 8.1). Some of the epoxy, polyester, or asphalt has either vaporized (giving the "burned insulation" smell) or, in the case of asphalt, it has flowed out of the groundwall. The result is that the groundwall now contains some gas, usually air. The dielectric constant of air is lower than the dielectric constant of all solid insulation materials. Specifically, the dielectric constant of gas is 1, whereas the dielectric constant of epoxy–mica is about 4. As the percentage of gas within the groundwall increases as a result of thermal deterioration, the average dielectric constant decreases. If the coil in a slot is approximated by a parallel plate capacitor then, from Equation 1.5 in Section 1.4.4, the capacitance will decrease, since the dielectric constant has decreased.

Similarly, if the insulation bulk has been saturated with water, then the capacitance will increase over time. The water may have soaked the insulation because the machine was flooded. In direct-water-cooled windings, the water can come from cracks in the water-filled copper tubes within the stator bar (Section 8.13). Since water has a dielectric constant of 80, the presence of water increases the average dielectric constant of the groundwall, increasing the capacitance.

If the end-winding of a stator is polluted with a partly conductive contaminant, then the ground potential of the stator core partly extends over the end-winding. This effectively increases the surface area, A, of the capacitor "plate" (Equation 1.5). Thus, winding contamination will increase the capacitance also.

By trending the change in winding capacitance over time, one can infer if thermal deterioration or problems with moisture or contamination are occurring or not. If the capacitance is unchanged over the years, then little deterioration is occurring. If the capacitance is decreasing, then the winding is likely to have experienced thermal deterioration. If the capacitance is increasing, then, perhaps, the winding has absorbed moisture from the environment, a water leak has occurred in the winding, or electrical tracking is present. A single measurement of the capacitance has no diagnostic value.

12.5.2 Test Method

The capacitance tests discussed in this section are generally performed at low voltage with commercial capacitance bridges. (Section 12.6 discusses capacitance tests performed at high voltage.) Since the amount of gas or moisture within the groundwall is usually a small percentage of the normal insulation, the change in capacitance over the years is also very small, even for very significant deterioration. Thus, the measurement device should have a precision of better than 0.1%. Capacitance bridges can easily achieve this precision.

Unfortunately, inexpensive "capacitance meters" are not suited for this application. In general, such meters have a true precision of only about 1%, which is inadequate to detect small changes in capacitance due to thermal deterioration. More importantly, most of these meters work on the principle of applying a voltage pulse through a known resistance to the load capacitance, and measuring the decay time of the capacitive charging current. The capacitance can then be inferred from the charging current time constant. As discussed in Section 12.1.1, stator winding insulation not only has a capacitive charging time constant, but the absorption current also has an apparent charging time constant. The absorption current confuses the capacitance meter, producing a capacitance reading that is much larger than it actually is.

The capacitance of the entire phase or winding can be measured in a "global" measurement. This version of the test determines overall insulation condition. In addition, for the specific problem of stator winding water leaks, the "local" capacitance of a portion of a stator bar is measured. With measurements on all bars, a "capacitance map" is created.

Winding Capacitance Test. This test is best performed at the stator winding terminals. If the test is done from the switchgear, then the cable capacitance between the switchgear and the stator may dominate the stator winding capacitance. This will make any change in the stator winding capacitance harder to discern. The longer the power cable, the less sensitive the test becomes. With small motors, if the power cables are longer than about 100 m, it is likely that the test results will not be meaningful. Similarly, any surge capacitors connected to the machine must be disconnected to achieve sensitive measurements.

On a generator, the capacitance of each phase can be measured since each phase can be isolated from the others by disconnecting the neutral connections. For most motors, the phases are connected at the neutral, and this can only be broken with great difficulty. Thus, no matter what phase is measured, the capacitance of all three phases is measured.

It is important to use a calibrated capacitance bridge for these measurements since the results are usually trended over many years, and small changes in capacitance are important. Any drift in the bridge may be mistaken for a problem in the winding. Also, it is prudent to use the same instrument over the years for the measurements.

Capacitance Map. When it became apparent that the water leak problem in direct-cooled stators was an important cause of failure, utilities developed the capacitance mapping technique [12.12]. In this method, the stator winding is grounded. A small metal plate is then placed over an accessible portion of the end-winding, usually near the connections since this is where the water is first likely to collect. The capacitance is then measured between the plate and the (grounded) copper conductor within the stator bar, using a capacitance bridge. The plate is then moved to another bar, and the capacitance to the plate in this bar is then measured. Similarly, the capacitance of the plate in every bar, at both ends of the bar and, possibly, at different axial positions of the bar in the end-winding is measured. The result is a map of capacitance versus slot number and axial position. The test is more useful if the dissipation factor (Section 12.8) is measured at the same time as the capacitance [12.12]. This is easily done since capacitance bridges measure the capacitance and the dissipation factor at the same time.

A statistical analysis of the capacitances is then performed. Specifically, the mean and standard deviation of all the capacitances of all bars in a given axial location are calculated. Recent improvements in the method have been aimed at making the measurements more repeatable. These include using a guard electrode to reduce the effect of stray capacitance. Also, a ratioing technique has been developed in an effort to reduce the effect of variable groundwall insulation thickness from bar to bar [12.13].

12.5.3   Interpretation

For the capacitance measurement on complete phases or windings, the key for interpretation is the trend. A significant amount of thermal deterioration will result in only a 1% drop in capacitance over the years. If the winding has been soaked with water, then a significant amount of deterioration will result in about a 5% increase in capacitance. If thermal deterioration or water leak failure processes are occurring at only a few locations, then the test is probably not sensitive enough. If the entire winding is affected, then the capacitance test is more likely to detect it.

The trend is less important with the capacitance map test to find bars with water leaks. Instead, one compares the capacitance measured on all the bars in the winding. The bars that have a capacitance (and dissipation factor) higher than the mean plus 3 or 4 standard deviations are then likely to have water in the insulation. Small variations are to be expected because of the variable thickness of the insulation (refer to Equation 1.5).

Coil manufacturers often use the capacitance test to monitor the impregnation and curing process in coil, bar, and global VPI stator manufacturing. Epoxy and polyester impregnating resins have a very high dielectric constant when in the liquid state. As they cure, the dielectric constant asymptotically reaches its low value (about 4) as it cures to a solid. When a coil is first impregnated with the liquid resin, its capacitance increases as the resin replaces the air between the mica-paper tape layers. It reaches a high steady-state value when complete impregnation is achieved. During the cure cycle, the capacitance starts to decrease as it cures. With experience, manufacturers can be assured of better impregnation and can define the optimum cure time by monitoring the initial increase and then decrease of the capacitance.

12.6 CAPACITANCE TIP-UP TEST

The capacitance tip-up test is a variation of the capacitance measurement on complete windings described in the previous section. This test is an indirect partial discharge test, and is closely related to the power factor tip-up test in Section 12.9. This test is only relevant for form-wound stator windings rated 2300 V and above. The test is not widely used in the Americas or Asia, but has been applied in Africa and Europe.

12.6.1 Purpose and Theory

Thermal deterioration, load cycling, and poor impregnation methods can result in air pockets or voids within the groundwall insulation of form-wound stators. If the winding is energized at a sufficiently high voltage, a partial discharge will occur within these voids, ionizing the gas for several milliseconds (Section 1.4.4). The ionized gas has sufficiently high conductivity that the void is "shorted out", i.e., it is much like a metal sphere within the void. Since part of the distance between the copper and the core is "shorted out" by the PD, the effective thickness of the insulation is reduced, which, by Equation 1.5, increases the capacitance. One void "shorted out" by the PD will have no measurable effect on the capacitance of the entire winding; however, if there are thousands of voids, all undergoing PD, then there will be a noticeable increase in the capacitance.

PD only occurs if there are voids and the electric stress within the voids exceeds 3 kV/mm (at 100 kPa). Thus, PD only occurs at high voltage. Measurement of the capacitance at high voltage is not sufficient to detect the voids. Instead, by measuring the capacitance at high voltage and subtracting from this the capacitance at low voltage, the result is the increase in capacitance due to PD activity [12.14]. Stated differently, the capacitance at low voltage is the normal capacitance of the insulation (voids plus solid insulation). The capacitance at high voltage is the capacitance of the solid insulation alone, since the voids have been "shorted out" by the PD. By taking the difference, one can estimate the capacitance of the voids. Of course, the larger this void capacitance, the more deterioration within the groundwall and, presumably, the closer the winding is to failure.

In summary, the capacitance tip-up test measures the void content within the groundwall.

12.6.2 Test Method

There are no standardized procedures for performing this test. An AC voltage supply is needed that can energize the capacitance of the winding to rated voltage. This can be a conventional transformer, a resonant supply or even a VLF supply (Section 12.4.2). For a large stator at power frequency, 20 to 30 kVA may be needed. An instrument is also needed that is capable of measuring the capacitance at least at the rated line-to-ground operating voltage of the stator winding. To cover all machines, the capacitance instrument should be able to operate up to about 25 kV. In addition, the device needs an accuracy of better than 0.1%. There are several types of devices that can meet these requirements. The Schering bridge and the transformer ratio bridge are the most common [12.15].

The test is best done with the phases isolated from each other, with all terminal equipment such as cables disconnected, since this will increase sensitivity. One phase at a time is tested, with the other two phases grounded. The low-voltage capacitance (C_{lv}) is first measured, usually at about 0.2E, where E is the rated phase-to-phase voltage of the stator. Then the applied voltage is raised to about the rated line-to-ground voltage (about 0.58E) and the high voltage capacitance (C_{hv}) measured. The capacitance tip-up is:

$$\Delta C = \frac{C_{hv} - C_{lv}}{C_{lv}} \tag{12.4}$$

It is usually expressed in percent rather than in Farads.

12.6.3 Interpretation

The higher ΔC is, the more voids there are within the insulation. Modern epoxy–mica groundwalls should have a ΔC less than about 1%. Older asphaltic mica windings should have ΔC of less than 3 or 4%. If ΔC is higher than these values, then it is an indication that thermal or load cycling deterioration, or poor manufacturing problems are present.

There are some limitations to this method that affect interpretation. The first is that ΔC is a measure of the total void content. There is no indication from the measurement whether there are many thousand tiny voids or whether there are just a few big voids. Of course, the winding is more likely to fail from the few large voids than from many little defects. Thus, the test is more sensitive to the average condition of the winding than to the condition of the worst coil.

The second limitation comes from the effect of the silicon carbide stress layers (Section 1.4.5) on the capacitance measurements. Silicon carbide has a resistance that depends on the applied voltage—at low voltage the layer is fully insulating, whereas at high voltage it is much like a grounded metal plate. When the capacitance is measured at high voltage, the conducting silicon carbide effectively increases the area of the capacitor plates (Equation 1.5), increasing the capacitance. Thus it is possible for the increase in capacitance at high voltage to be caused by both the PD in the voids as well as by the increase in capacitance due to the silicon carbide coatings. (The influence of the silicon carbide coatings can be reduced in tests on individual coils, since the layer's effect on the measurement can be neutralized with a suitable guarding circuit. See Section 12.8.) The net effect is that the silicon carbide coating creates a "noise floor," which tends to diminish the effect of voids on the capacitance.

The best means of overcoming these limitations is to trend ΔC over the years. The initial tip-up test on a complete phase has little meaning, due to the silicon carbide coating effect. However, if ΔC increases monotonically from year to year, perhaps doubling in magnitude, then it is an indication of progressive delamination, possibly due to overheating or load cycling.

12.7 CAPACITIVE IMPEDANCE TEST FOR MOTOR STATORS

This is another variation of the capacitance test described in Section 12.5. In reality, the capacitive impedance test is just another way of measuring winding capacitance. We include it as a separate section here because a number of specialized instruments were introduced in the 1990s for measuring motor stator winding condition. The capacitive impedance is one of the measurements made by some of these specialized motor test instruments.

The relevance of measuring the trend in capacitance of a motor stator winding is discussed in Section 12.5. In that section, the measurement was done with an AC capacitance bridge. The capacitance can also be calculated by an accurate measurement of the current and voltage across a winding, enabling the calculation of the capacitive impedance $X_c = V/I$. Specifically, the capacitance C of a winding is

$$C = \frac{1}{2\pi f X_c} = \frac{I}{2\pi f V} \qquad (12.5)$$

where f is the frequency in Hz, and V and I are the measured voltage and current across the winding. The frequency can be 50 or 60 Hz, as in some brands of power-factor-measuring equipment, or it can be 10 to 100 kHz as in some of the modern specialized motor testing instruments. At the higher frequencies, higher currents are measured and the measurement will suffer less from interference due to power-frequency currents. As discussed in Section 12.5, the voltage and current measurements must have accuracy better than 0.1% to produce results that can be successfully interpreted.

In addition to the capacitive impedance, the specialized motor testing instruments may also measure one or more of the following:

- IR and PI (Section 12.1)
- Conductivity (Section 12.3)
- Inductive impedance (Section 12.13)

All these measurements are usually made at relatively low voltage, which can impair effectiveness, especially in the IR/PI test. However, a computer usually automates the instruments, and a summary is made of many important quantities that can be easily trended over time.

12.8 DISSIPATION (OR POWER) FACTOR TEST

Dissipation factor and power factor provide an indication of the dielectric losses within the insulation. Certain deterioration processes, such as thermal deterioration and moisture absorption, will increase these losses. Thus, trending of the dielectric loss over time is an indication of certain types of insulation problems. There are two main ways to measure the dielectric loss: (1) dissipation factor (tan δ) and (2) power factor. These tests are relevant for stator windings only and are usually applied only to form-wound stators. The test is also employed by some OEMs to determine when the resin in newly manufactured coils, bars, or global VPI stators is cured.

12.8.1   Purpose and Theory

Dielectric loss is a property of any insulating material. Ideally, the winding insulation will act as a pure capacitor, that is, it will only store energy, not dissipate it. In practice, the materials used for groundwall insulation will heat up a little when excited by AC voltage, i.e., they will dissipate some energy. The cause of the dissipation is primarily the movement of polar molecules under the AC electric stress. As discussed in Section 12.1.1 in connection with the absorption current, common rotating machine insulation materials contain polar molecules that tend to rotate or vibrate when a DC electric field is applied, giving rise to the absorption current. With power-frequency AC applied, the polar molecules oscillate back and forth as the applied voltage changes from positive to negative and vice versa 50 or 60 times a second. Since the polar molecules are oscillating in a solid medium, the friction created with adjacent molecules (or parts of molecules) gives rise to heat. The energy to cause this molecular heating comes from the applied electric field, i.e., from the power supply. The dielectric loss is a property of any particular insulation material and its chemical composition. It is not an indicator of insulation quality.

When the organic parts of an insulation system are exposed to a sufficiently high temperature or nuclear radiation, some of the chemical bonds within the polymer vibrate with enough intensity that they break. If an oxygen molecule is near the broken ends of the polymer chain, a new chemical reaction occurs, with the oxygen forming a bond with the broken ends. This process is called oxidation and gradually makes the insulation brittle (Section 8.1). The uptake of oxygen into the polymer tends to create additional polar groups within the insulation. The result is that, since there are more polar molecules, more molecules are available, which will oscillate when excited by an AC electric field, thus increasing the dielectric loss. An increase in dielectric loss in a winding over the years may indicate insulation aging due to overheating or radiation.

Similarly, if a winding has been soaked with water, the dielectric loss will increase. This occurs since the H_2O molecules are polar.

12.8.2   Test Method

There are two different ways to measure the dielectric loss. Both involve recognizing that the winding is essentially a capacitor with a small dielectric loss.

1. Dissipation Factor or tan δ. The dissipation factor (DF) is measured with a balanced bridge-type instrument, where a resistive–capacitive network is varied to give the same voltage and phase angle (tan δ) as measured across the stator winding (Figure 12.1). The DF is then calculated from the R and C elements in the bridge that give the null voltage. This method can easily achieve a 0.01% accuracy.

2. Power Factor. For materials with a relatively small dissipation factor, which is the case for most stator winding insulation, the power factor and the dissipation factor are about the same. The power factor is measured by accurately measuring the voltage (V)

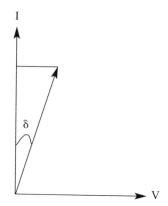

Figure 12.1. Calculation of phase angle for tan δ measurement.

applied between the copper and the core of a winding, and detecting the resulting current (I). At the same time, the power (W) to the winding is measured with an accurate wattmeter. The power factor PF is

$$PF = \frac{W}{VI}$$

As with DF, PF is usually expressed in percent. The PF test is less accurate in measuring the dielectric loss than the dissipation factor method, but tends to be less expensive, since it does not use a bridge-type instrument.

Both methods can be performed by applying low-voltage AC to the winding. However, as discussed in Section 12.9, there are advantages to measuring the dielectric loss at high voltage also. In most cases, both methods are accomplished by using power frequency AC voltage to energize the winding. However, if low-voltage power-frequency AC is used, there may be some induced power-frequency currents in the winding due to inductive or capacitive pick-up from other energized equipment. These induced signals can lead to false results. Thus, a frequency different from 50 or 60 Hz is often used. Alternatively, the measurements could be done at a high power-frequency voltage to minimize any interference from other energized equipment.

The specific technology used by each method is given in detail in [12.15–17]. The DF is usually measured on a capacitance bridge (Section 12.5). The DF can be converted to PF using

$$PF = \frac{DF}{(1 + DF^2)^{0.5}}$$

However, the numbers produced by both methods are essentially the same for most rotating-machine insulation systems.

As with the capacitance test, the dielectric loss is most sensitively measured at the stator terminals. Power cable can distort the readings. However, if less than 100 m of polyethylene cable, cross-linked polyethylene cable, or isolated phase bus (with air insulation) are connected to the terminals, reasonably accurate measurements of dielectric loss can be measured. This is because these three materials have very low (<0.01%) loss. Oil-paper cables and most rubber-insulated cables, have too high an inherent loss and will mask the loss readings from the stator winding.

It is best to measure each phase separately, since this increases sensitivity. However, in motors, this is usually not possible due to the neutral connection.

12.8.3 Interpretation

For maintenance tests on complete windings, the initial DF measured is irrelevant. Typical DFs are about 0.5% for modern epoxy and polyester impregnated insulation. The DF can be 3 to 5% for asphaltic mica windings. If the DF is measured on a regular basis, say every few years, and if it remains constant over time, then it is an indication that there is no thermal aging or gross contamination of the winding. If there is an increase over time, then it is an indication that insulation overheating is occurring or the winding is becoming more contaminated by moisture or partly conductive contaminants. A significant amount of deterioration has occurred if the DF has increased by 1% or more from the initial value.

If the capacitance and DF are measured at the same time, and if the trend in C decreases while DF is increasing, this is strongly indicative of general thermal deterioration. If both C and DF increase over time, then this is a powerful indicator that the winding is contaminated or has absorbed moisture.

Dissipation factor trend is indicative of the average condition of the insulation, since it measures the total dielectric loss in a winding. The additional loss for a single seriously overheated coil in a winding, when the rest of the coils are in good condition, will be small. Thus, this test cannot detect a few deteriorated coils.

Coil manufacturers sometimes use the dissipation factor test as a process monitor for the impregnation process. As the groundwall is impregnated, the DF will increase, since liquid resin has a higher DF than the air the resin is replacing. As the coil cures, the DF will decrease to its steady final level, since the DF of liquid resin is higher than the DF of cured epoxy or polyester.

12.9 POWER (DISSIPATION) FACTOR TIP-UP TEST

The power factor tip-up test, sometimes called the dissipation factor tip-up test, is an indirect way of determining if partial discharges are occurring in a high-voltage stator winding. Since PD is a symptom of many high-voltage winding insulation deterioration mechanisms, the tip-up test can indicate if many failure processes are occurring. The power factor tip-up is a complement to the capacitance tip-up (Section 12.6). This test is only relevant to form-wound stator coils rated at 2300 volts and above.

In addition to its use as an off-line maintenance test, the tip-up test is widely used by stator coil and bar manufacturers as a quality control test to ensure proper impregnation by epoxy and polyester during coil manufacture.

12.9.1 Purpose and Theory

As discussed in the previous section, all practical insulating materials have a dielectric loss, which can be measured with a power factor or dissipation factor test. At low voltages, the PF and DF are not dependent on voltage. However, as the AC voltage is increased across the insulation in a form-wound coil, and if voids are present within the groundwall, then at some voltage, partial discharges will occur. These discharges produce heat, light, and sound, all of which consume energy. This energy must be provided from the power supply. Consequently, in a delaminated coil, as the voltage increases and PD starts to occur, the DF and PF will increase above the normal level due to dielectric loss, since the PD constitutes an additional loss component in the insulation. The greater the increase in PF or DF, the more energy that is being consumed by the partial discharge.

In the tip-up test, the PF or DF is measured at a minimum of two voltage levels. The low-voltage PF, PF_{lv}, is an indicator of the normal dielectric losses in the insulation. This is usually measured at about 20% of the rated line-to-ground voltage of the stator. The voltage is then raised to the rated line-to-ground voltage and PF_{hv} is measured. The tip-up is then

$$\text{tip-up} = PF_{hv} - PF_{lv} \qquad (12.6)$$

The higher the tip-up, the greater is the energy consumed by PD. Some organizations will record the PF or DF at several different voltage levels and calculate several different tip-ups between different levels. By plotting the tip-up as a function of voltage, the voltage at which PD starts is sometimes measurable. It is not uncommon for coil manufacturers to measure the tip-up to full phase-to-phase voltage. If the PF or DF is measured in percent, then the tip-up is in percent. Since the tip-up on windings rated greater than 6 kV is usually significant, both DF and PF measurement methods yield about the same result. That is, the greater accuracy possible with a DF measurement will usually not yield superior tip-up results.

12.9.2   Test Method

Detailed test methods are described in References 12.16 and 12.17. Historically, the test was first applied to high-voltage stator bars and coils, to ensure that the insulation was completely impregnated or the layers of resin-rich tape were fully bonded and air eliminated by the compaction process. However, since the late 1950s, some generator operators have applied the test to complete windings to detect various aging mechanisms that produce PD. In tests on machines, it is important to test as few coils as possible at a time, since this will increase sensitivity (see the next Section). Thus, as a minimum, each phase should be tested separately while the other two phases are grounded. Preferably, the winding should be partitioned into parallels or coil groups to gain maximum sensitivity.

Measurement of the tip-up is complicated by the presence of silicon carbide stress control coatings (Section 1.4.5) on coils rated at 6 kV or above. At low voltage, the silicon carbide is essentially a very high resistance coating, and no current flows through it. Thus, there is no power loss in the coating. However, when tested at rated voltage, by design the silicon carbide coating will have a relativity low resistance. Capacitive charging currents flow through the insulation and then through the coating. The charging currents flowing through the resistance of the coating produce an I^2R loss in the coating. The DF or PF measuring device measures this loss. Since the loss is zero at low voltage and nonzero at operating voltage, the coating yields its own contribution to tip-up. It is not uncommon for the tip-up due to the stress relief coating to be 2 or 3%. This coating tip-up creates a noise floor. Very significant PD must be occurring in most windings for the PD loss to be seen above the silicon carbide tip-up.

When manufactures test individual coils and bars in the factory as a QA test, the tip-up contribution due to the stress relief coating can be negated. The most common way is to "guard" out the currents due to the silicon carbide by overlapping the coatings with grounded aluminum foil. Other methods are also possible [12.16, 12.17]. Unfortunately, it is not practical to guard out the coating tip-up in complete windings.

If the tip-up test is made on direct-water-cooled stator windings, and the water is present during the test, then the water may cause an additional tip-up. Reference 12.16 describes how this effect can be compensated for. Some manufacturers of high-voltage stator coils insert a metallic shield adjacent to the copper conductors in the straight portion of the coil. The purpose is to reduce the effect of high electric fields caused by sharp points on the copper or any misaligned strands. Such shields have been known to cause a negative tip-up, which further complicates interpretation.

12.9.3   Interpretation

As a maintenance tool for complete windings, the tip-up test is used for trending. The initial value of the tip-up on a phase is of little significance, because it will be dominated by the stress relief coating contribution to tip-up. However, if the tip-up is measured every few years and the tip-up starts increasing from the normal level, then it is likely that the winding has significant PD activity. To increase the tip-up above the normal level requires widespread PD. The most likely causes of this PD are thermal deterioration and load cycling.

The tip-up test is not likely to be sensitive to loose coils in the slot, semiconductive coating failure, or end-winding electrical tracking. In all these cases, the PD is at a relativity low repetition rate or the damage is confined to the relatively small portion of the winding, and, thus, the PD contribution to tip-up is relatively minor. Details on each of these mechanisms are found in Chapter 8.

For tests on newly manufactured stator bars or coils, absolute limits to the tip-up on any individual coil can be established based on the mean and standard deviation of previously manufactured coils made with the same process and materials. If a new coil has a tip-up greater than, perhaps, the mean plus 3 standard deviations, then it is likely that the coil groundwall has not been properly impregnated. Sensitive tests are achievable on individual coils since the stress control coatings can be guarded out. Coils that have a high tip-up are sometimes segregated and installed in the stator where they will operate at a low voltage, i.e., near the neutral end of the winding.

12.10   OFF-LINE PARTIAL DISCHARGE TEST

The off-line partial discharge test directly measures the pulse currents resulting from PD within a winding energized at rated line-to-ground voltage. Thus, any failure process that creates PD as a symptom can be detected with this method. The test is mainly relevant for form-wound stator windings rated at 2300 V and above. However, a variation of the test is also relevant for random-wound stators intended for use on PWM-type inverter-fed drives.

12.10.1   Purpose and Theory

Many of the stator winding failure processes described in Chapter 8 had PD as a direct cause or a symptom of the process. When a partial discharge pulse occurs, there is a very fast flow of electrons from one side of the gas filled void to the other side. Since the electrons are moving close to the speed of light across a small distance, the pulse has a very short duration, typically a few nanoseconds [12.18]. Since the electrons carry a charge, each individual discharge creates a current pulse ($i = dq/dt$). In addition to the electron current flow, there will be a flow of positive ions (created when the electrons are ionized from the gas molecules) in the opposite direction. However, the ions are much more massive than the electrons and, consequently, move much slower. Since the transition time of the ions across the gas gap is relativity long, the magnitude of the current pulse due to ions is very small and usually neglected.

Each PD pulse current originates in a specific part of a winding. The current will travel along the coil conductors, and since the surge impedance of a coil in a slot is approximately 30 ohms, a voltage pulse will also be created, according to Ohm's law. The current and voltage pulse flow away from the PD site, and some portion of the pulse current and voltage will travel to the stator winding terminals. A Fourier transform of a current pulse generates frequencies up to several hundred megahertz [12.18].

Any device sensitive to high frequencies can detect the PD pulse currents. In the off-line PD test, the most common means of detecting the PD currents is to use a high-voltage capacitor connected to the stator terminal. Typical capacitances are 80 pF to 1000 pF. The capacitor is a very high impedance to the high AC voltage (needed to energize the winding sufficiently to create the PD in any voids that may be present), while being a very low impedance to the high-frequency PD pulse currents. The output of the high-voltage capacitor drives a resistive or inductive–capacitive load (Figure 12.2). The PD pulse current that passes through the capacitor will create a voltage pulse across the resistor or inductive–capacitive network, which can be displayed on an oscilloscope, frequency spectrum analyzer, or other display device. Older oscilloscopes had trouble displaying the very short duration PD pulses on the screen. Thus, some types of detectors use an inductive–capacitive load since the PD current will then create an oscillating pulse at lower frequency, which can be easily viewed on oscilloscopes. The bandwidth of the detector is the frequency range of the high-voltage-detection capacitor in combination with the resistive or inductive–capacitive network load. Early detectors were sensitive to the 10 kHz, 100 kHz, or 1 MHz ranges. Modern detectors can be sensitive up to the several hundred megahertz range [12.18].

Every PD will create its own pulse. Some PD pulses are larger than others. As described in [12.19], in general, the magnitude of a particular PD pulse is proportional to the size of the void in which the PD occurred. Consequently, the bigger the detected PD pulse, the larger is the defect that caused the discharge. In contrast, smaller defects tend to produce smaller PD pulses. The attraction of the PD test is that one concentrates on the larger pulses and ignores the smaller pulses. In contrast to the capacitance or power factor tip-up tests, which are a measure of the total PD activity (or the total void content), the PD test enables the measure-

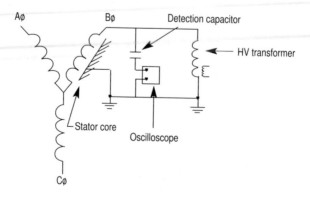

Figure 12.2. Measurement of PD pulse currents.

ment of the biggest defects. Since failure is likely to originate at the biggest defects and not at the smaller defects, the PD test can indicate the condition of the winding at its most deteriorated portion.

12.10.2 Test Method

Like the tip-up tests in Sections 12.6 and 12.9, the off-line PD test requires a power supply to energize the winding to at least rated phase-to-ground voltage. Thus, for large generator stators, a conventional or resonant transformer rated at 20 to 40 kVA may be needed. In addition, a low-noise 0.1 Hz (VLF) power supply could also be used (see Section 12.4.2).

It is best to perform the PD test at the machine terminals, energizing one phase at a time, with the other two phases grounded. PD tests can be measured from the switchgear, but it is then important to measure the PD in a frequency range less than about 1 MHz. As discussed in Section 13.4, power cables tend to strongly attenuate the higher-frequency components of the PD pulses as they travel from the stator winding to the detector at switchgear. Thus, unrealistically low PD signals will be measured at the switchgear if the detector primarily operates at frequencies greater than 1 MHz.

In the off-line PD test, it is common to gradually raise the applied voltage while monitoring the PD pulses on an oscilloscope screen. The voltage at which the PD is first detected is called the discharge inception voltage (DIV). The voltage then is raised to normal line-to-ground operating voltage. The winding should remain energized for 10 to 15 minutes at this voltage, and then the PD recorded (Figure 12.3), including the peak magnitude of the PD pulses (Q_m). The "soak" time is needed since the PD tends to be higher in the first few minutes after the voltage is applied. Space charge effects cause this, together with the build-up of gas pressure in the void due to deterioration caused by the PD. The voltage is then gradually lowered and the voltage at which the PD is no longer discernable is measured. This is the discharge extinction voltage (DEV). The DEV is usually lower than the DIV, and it is desirable to have the DIV and DEV as high as possible.

For machines rated at 6 kV or more, the maximum test voltage is normally the rated line-to-ground voltage. As will be discussed in Section 13.4, a test at this voltage will usually detect deterioration years before an in-service failure is likely. For machines rated at 2300–4100 V, a test at rated voltage may not produce significant PD, even in severely deteriorated stator insulation. This is because there may be insufficient electric stress within the defects to achieve the 3 kV/mm needed in atmospheric air to cause PD. Some users then per-

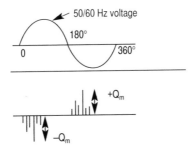

Figure 12.3. Oscilloscope display of PD pulses (lower trace) versus the power-frequency voltage waveform.

form the test at the rated line-to-line voltage; i.e., a 4 kV motor would have 4 kV applied between the copper and the ground. This is a small hipot test, and the stator owner should be aware that such an overvoltage may lead to failure.

Conventional PD analyzers cannot be used with IFDs or surge testers (Section 12.12). The risetimes produced by such voltage sources can be as short as 50 ns, which is only ten times longer than the PD pulses. These surges generate frequencies up to 5 MHz. At this frequency, the high-voltage capacitor in Figure 12.2 will have low impedance, resulting in a high-voltage surge being applied to the oscilloscope or other recording device. This may destroy the input. Specialized sensors have been developed that can extract the PD from the 1000 times higher voltage surge [12.20].

12.10.3 Interpretation

Unlike the tip-up test, which produces a single number representing the total PD activity, direct measurement of the PD produces several results. The key measure is the peak PD magnitude Q_m, i.e., the magnitude of the highest PD pulse. This can be measured in several units:

- picoCoulombs (pC) if a laboratory PD measurement device is used. pC is a measure of the apparent number of electrons that were involved in each discharge.
- millivolts (mV), where the PD magnitude is measured with an oscilloscope or electronic pulse magnitude analyzer (PMA). A PMA also counts the number of PD pulses of each magnitude range.
- milliamps (mA) if the PD pulses are measured with a high-frequency (ferrite core) current transformer and displayed on an oscilloscope.
- decibels (dBm) if a frequency spectrum analyzer records the pulses.

Unfortunately, there is no standardized measurement unit. In Europe, there is a tendency to use pC, in spite of various standards indicating that the pC calibration procedure is not intended for use in inductive apparatus [12.21-23]. In North America, mV and dB are more common.

The detected PD magnitude of a PD pulse within the winding but measured at the stator terminals depends on a large number of factors:

- The size of the defect. In general, physics indicates that the larger the volume of the defect, the larger will be the detected pulse [12.23].
- The capacitance of the winding. If the winding has a large capacitance, the impedance to ground at high frequencies will be very low. Thus, most of the PD pulse current is immediately shorted to ground, leaving little to be detected at the stator terminals.
- The inductance between the PD site and the PD detector. The pulse will be attenuated as it propagates through the winding to the terminal. In general, the further the PD site is from the PD detector, the lower will be the magnitude detected at the machine terminal.

These plus other effects make it difficult to define a "high" PD magnitude that indicates that a winding has seriously deteriorated [12.24]. The off-line PD test is thus a comparison test. One can determine which phase has the highest Q_m and, thus, which phase has the greatest deterioration. One can also compare several similar machines to see which has the highest PD or the lowest DIV or DEV. Finally, one can compare the PD from the same stator over time, i.e., trend the data. In general, if the PD doubles every 6 months, then the rate of deterioration is increasing [12.23].

Direct measurement of the PD pulses also enables one to measure how widespread the PD is. Studies show that as many as 10,000 PD pulses may occur per second in a stator winding. Although the physics of PD is beyond the scope of this book, it seems that a single defect only produces at most one or two PD pulses per half AC cycle. Thus, if only a few hundred PD pulses are occurring per second, then there are only a few PD sites in the winding and the deterioration is localized. If there are 10,000 PD pulses per second, then there are thousands of PD sites and the deterioration is widespread. The pulse count rate can be easily measured with a pulse magnitude analyzer, which is incorporated into most modern commercial PD analyzers.

If there is one dominant deterioration mechanism in a winding, the PD test can sometimes give the approximate location of the deterioration within the groundwall [12.23]. As shown in Figure 12.3, both positive and negative PD pulses are created. If the positive PD pulses (which, by definition, occur in approximately the negative half of the AC cycle) are larger than negative PD pulses, then it is likely that the PD is occurring on the surface of the coil (due to loose coils or defective semiconductive coatings). If the negative PD is predominant, then the PD is most likely occurring at the copper. If there is no polarity predominance, then the PD is likely to be between the groundwall insulation layers (due to delamination). Note that the polarity definition differs from place to place. In Europe, a negative PD pulse occurs on the negative portion of the AC cycle.

12.11 PARTIAL DISCHARGE PROBE TESTS

The direct or indirect partial discharge tests in Sections 12.6, 12.9, and 12.10 indicate that partial discharges are occurring somewhere in the winding, thus, some stator winding insulation deterioration has occurred. However, these tests do not give any indication of where in the winding the problem is occurring. PD probe tests can answer this question. The probe test is useful for stator windings rated at 2.3 kV and above. This test replaces or is a complement to the "blackout" test in Section 12.26.

12.11.1 Purpose and Theory

If high PD has been detected using any of the direct or indirect PD tests, then it is sometimes prudent to locate where the PD is occurring in the stator, so that the visual inspection of the winding can concentrate on the most deteriorated regions. Two special probes have been developed to help locate the PD sites. One probe detects the RF energy emitted by PD. The other detects the acoustic energy emitted.

The RF probe was first developed by Johnson in the late 1940s, and subsequently improved by the utility Tennessee Valley Authority (TVA) [12.25]. The probe is called either the corona probe or the TVA probe. The RF probe is essentially a modified am radio, usually tuned to about 5 MHz. The loop stick antenna is at the end of a rod. A radio detector, with no automatic gain control, is connected to the antenna. As the antenna comes close to a coil with high PD, the PD generates radio waves that the antenna detects. The bigger the PD, the stronger the signal, thus yielding a louder demodulated signal and a higher meter indication. The TVA probe is calibrated in "quasipeak" mA, which means that it tends to register the highest PD pulse detected.

The ultrasonic probe is a directional microphone that is pointed at the winding. When a partial discharge occurs on the winding surface, high-velocity electrons and ions move through the air to create a pressure wave, a tiny version of thunder from a lightning strike. This pressure wave creates an acoustic pulse that has its greatest intensity around 40 kHz. A directional microphone and associated frequency downscaler tuned to 40 kHz can make the ultrasonic noise audible to the human ear. As the microphone points to the PD sites, a louder indication will be heard from the device.

12.11.2 Test Method

For best access to the stator winding, the rotor is usually removed. The winding is then energized to the rated line-to-ground voltage, using a conventional or resonant transformer. This is usually the same power supply as used for the direct or indirect partial discharge tests. Most users of the probe tests energize one phase at a time and ground the other two phases. However, as will become apparent, this will increase the time it takes to perform the test threefold. The windings are usually energized for at least 10 to 15 minutes (some "soak" the winding for 1 hour) to allow the PD activity to stabilize. Extreme caution is needed when performing this test, since, by its nature, one has to get close to an energized winding.

With the winding energized, the RF probe antenna is placed over the slot near the end of the core and a helper records a reading. The antenna is usually in contact with the core and bridges the stator slot. The probe antenna is then moved to the next slot and the reading recorded. This is continued until the RF activity in all slots is measured. It is then common to measure the RF activity at the other end of the stator core. Sometimes, the activity in each slot in the middle of the core is also recorded. Some users also "wave" the probe antenna over the end-winding region, being careful to have the antenna no closer to the winding than about 2–3 cm. This can sometimes detect PD in the end-winding area. After one phase has been completed, the other phases are energized in turn, and the entire RF map repeated. On a large hydrogenerator, it can easily take 10–20 hours for two people to complete the test, because there are several hundred slots in the stator.

In general, the ultrasonic probe test is performed just after the RF probe test is completed on each phase. This test usually is much quicker, since the directional ultrasonic microphone is relatively quickly scanned over the end-windings and the slots. Slots or end-winding locations showing the greatest noise activity are noted.

12.11.3 Interpretation

Usually, slots having the highest RF probe or highest ultrasonic microphone readings will have the greatest PD and, hence, are more likely to be more deteriorated. Reference 12.23 gives a consensus indication of what constitutes a high reading for the TVA version of the RF probe test. If the indication in any slot is above about 20mA for a modern epoxy–mica insulation, then significant PD is occurring in that area. There are no agreed-upon limits for what constitutes a high ultrasonic reading; thus, one just notes that the loudest noise is probably from the most deteriorated location.

If both the RF and ultrasonic probes point to the same site as having high PD, then this is solid evidence that there is significant PD occurring on the surface of the coils. This PD activity should be visible. If the RF activity is high and the ultrasonic probe gives little indication, then the PD may be occurring within the groundwall insulation and, thus, may not be visible.

There should be no region in which the ultrasonic probe gives a high indication and the RF probe gives little or no reading. If this occurs, then the acoustic indication is probably spurious. The high acoustic reading probably is just detecting acoustic reflections. Cross coupling and resonance can also result in spurious high readings from the RF probe. If one phase is energized at a time, it is not uncommon for the RF probe to show high PD occurring in a slot that contains no energized coils!

Thus though the RF and ultrasonic probe tests are simple in theory, considerable skill is needed to ensure that slots with coils in good condition are not misidentified as being bad.

12.12 STATOR SURGE COMPARISON TEST

None of the tests discussed above directly measure the integrity of the turn insulation in form-wound or random-wound stator windings. The stator voltage surge test does this by applying a relatively high voltage surge between the turns. This test is a hipot test for the turn insulation and may fail the insulation, requiring a repair, coil replacement, or rewind. The test is valid for any random-wound or multiturn form-wound stator. The test is not relevant for half-turn Roebel bar windings, since there is no separate turn insulation (or more correctly, the turn insulation is the same as the ground insulation, and no realistic surge voltage is likely to puncture the groundwall).

12.12.1 Purpose and Theory

Section 8.7 discussed that motor turn-on was accompanied by a fast (i.e., short) risetime voltage surge caused by breaker or switch contactors coming together or separating. Similar voltage surges occur from IFDs and faults in the power system. These fast risetime surges result in a nonuniform voltage distribution across the turns in the stator winding. If the risetime is short enough, the surge voltage high enough, and the turn insulation weak enough, then the turn insulation punctures, rapidly leading to a stator ground fault (Section 1.4.2).

The surge test duplicates this action of an external surge. As such, this test is analogous to the AC and DC hipot tests: apply a high voltage to the turn insulation and see if it fails. The surge test is a destructive, go–no go test. If the turn insulation fails, then the assumption is that the stator would fail in service due to motor switch-on, IFD surges, or transients caused by power system faults. If the winding does not puncture, then the assumption is that the turn insulation will survive any likely surge occurring in service over the next few years. Thus, the main question is whether a maintenance surge test is performed or not, and this is a philosophical question identical to that posed for the DC hipot (Section 12.2).

The main difficulty with the surge test is determining when turn insulation puncture has occurred. In the DC or AC hipot test, groundwall puncture results in the insulation resistance plummeting to 0 Ω. This causes the power supply current to increase dramatically, opening the power supply circuit breaker. There is no question that puncture has occurred. A turn-to-turn puncture in a winding does not cause a huge increase in current from the power supply. In fact, if there are 50 turns between the phase terminal and neutral, the failure of one turn will only slightly reduce the inductive impedance of the winding, since the impedance of only one turn has been eliminated. Thus the other 49 turns can continue to impede current flow, and the circuit breaker does not trip.

In the surge test, turn failure is detected by means of the change in resonant frequency caused by shorting out one turn. The schematic of a simplified surge tester is shown in Figure 12.4. The inductor is the inductance of one phase of a stator winding, or in the motor stator where the neutral ends cannot be isolated, the inductance of (say) the A phase and B phase windings in series. A high-voltage capacitor within the surge tester is charged from a high-voltage DC supply via the winding inductance. Once the capacitor is charged to the desired voltage, the switch in Figure 12.4 is closed. The switch is a thyratron in older surge testers or

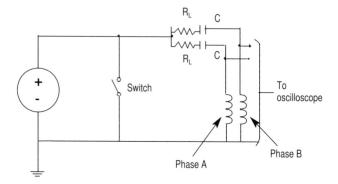

Figure 12.4. Schematic of a surge comparison tester applied to two of three phases of a stator.

an IGBT in modern sets. The energy stored in the capacitor then oscillates back and forth with the winding inductance. The resonant frequency (f) of either the voltage or current waveform is approximately

$$f = \frac{1}{2\pi\sqrt{LC}} \tag{12.8}$$

If there is no turn fault, there will be a fixed frequency of oscillation. If a turn fault occurs as a result of the short risetime surge imposed on the line end turns, together with weak turn insulation, the inductance of the winding will decrease and, thus, the resonant frequency will increase, according to Equation (12.8). Thus, one looks for the increase in frequency of the voltage surge on an oscilloscope screen as the voltage is gradually increased and as the winding moves from no turn shorts to having a turn short.

The increase in frequency is small, typically only a few percent. Such a small increase is difficult to detect. To aid in detecting the frequency shift, modern surge testers digitally capture the resonant waveform at low voltage, where the turn insulation is still intact. The surge voltage is gradually increased by raising the voltage that the capacitor charges to, and triggering the switch after the capacitor has charged up (usually the switch is automatically triggered once per second or once per 50 or 60 Hz cycle). If a change in the waveform is noted above a certain voltage, which can be detected by scaling the low-voltage stored waveform up to the current applied voltage, then turn insulation puncture has occurred.

Older surge test sets were called surge comparison testers. They consisted of two energy storage capacitors, which were connected to two phases (Figure 12.4). The waveform from each phase is monitored on an analog oscilloscope. The assumption is that the waveform is identical for the two phases. As the voltage is increased, if one of the waveforms changes (increases in frequency) then turn puncture occurred in the phase that changed. This approach has lost favor since it is possible for two phases to have slightly different inductances due to different circuit ring bus lengths, midwinding equalizer connections, or even due to rotor position (since it affects the permeability).

It is easy to detect turn insulation failure on individual coils, since the shorting of one turn will have a much larger impact on the total inductance of a coil, thus drastically changing the waveform. Machine manufacturers and rewind companies use individual coil surge testing to check the quality of the turn insulation. Such testing is best done after the coils are wound, wedged, and braced, since by then they have been exposed to all the mechanical handling and stresses associated with the winding process. However, as a quality check, most coil manufacturers do the surge test prior to inserting coils into the slot. Ground faults are easily detected by a surge test, since the waveform collapses. Surge testing is also useful to identify wrong connections in the winding.

12.12.2 Test Method

IEEE standard 522 provides the best description of both an acceptance and maintenance surge test [12.26] for form-wound stator windings. As an acceptance test, the surge is recommended to have a risetime of 100 ns and a maximum magnitude of 3.5 per unit, where 1 per unit is the peak line-to-ground rated voltage. For a maintenance test performed after the winding has seen service, the surge should have the same risetime, but reach only 2.6 per unit. As for the DC and AC hipot test voltages, these limits were set because they represent the worst surge that is most likely to occur in normal service. Voltages higher than these maximums should not be applied to the stator winding, otherwise there is a significant risk that good turn insulation will fail unnecessarily. Baker Instruments and Electrom are two manufacturers of surge test sets.

NEMA MG1 and IEC 60034 Part 15 set generally lower acceptance surge test requirements and do not specify maintenance surge tests. NEMA requires that IFD motors should withstand 100 ns risetime surges at 3.7 per unit. As discussed in Section 8.7, the longer the surge risetime, the lower is the voltage applied across the turn insulation. Thus, longer risetime surges are not as severe as short risetime surges.

The test should be applied at the machine terminals. If the test is performed from the switchgear, the long cable or bus will lengthen the surge risetime, decreasing the effectiveness of the test, since there will be less interturn voltage.

As discussed above, the surge voltage is gradually increased to the maximum recommended test voltage. If the waveform changes on the oscilloscope, then the turn insulation has likely been punctured. If the winding is form wound, the failed coil will have to be located and isolated. This is likely to be a phase end coil. If the lights are dimmed around the stator, it is sometimes possible to find the faulted coil by reapplying the surges and looking for a glow through the ground insulation from the puncture site. Otherwise, the coils must be separated from one another and tested separately until the faulted coil is found. This can be tedious. In a form-wound stator, the faulted coil must be replaced or the coil must be cut out of the circuit (Section 7.3). If the turn insulation fails in a random-wound stator, it is often possible to locate visually the faulted turn by dimming the lights and reapplying the surge. Often, the turn insulation in a random wound stator can be restored by cleaning and dipping the stator in varnish (dip/bake).

If a turn puncture has occurred, it is not acceptable to ignore it and return the stator to service. Once the first significant surge occurs in service, the punctured turn insulation will break down again, allowing power-frequency currents to flow, rapidly leading to groundwall failure.

12.12.3 Interpretation

The surge test is a go–no go test, and the stator either passes or fails. There is no real diagnostic information obtained. If one combines the surge test with a partial discharge test, then it may be possible to detect significant voids between the turns, before actual puncture occurs. This requires a special PD detector, since conventional PD detectors will be damaged by the high-voltage surges [12.20].

Although the surge test was stated above to be the only test that directly determines the condition of the turn insulation, there are some caveats. In random-wound stators, the turn insulation and the ground insulation are, to a large extent, the same. That is, the insulation film on the magnet wire serves as both turn and part of the ground insulation. Thus the IR/PI, capacitance, and dissipation factor tests discussed above will also indicate the condition of the turn insulation.

For form-wound stators, the turn insulation is again part of the groundwall insulation where the turn insulation is around the outside of the copper conductors (Figure 1.9). Thus, the diagnostic tests described above (particularly the capacitance, dissipation factor, and PD tests) also indirectly evaluate the condition of the turn insulation. However, the turn insulation between two adjacent turns is not tested with groundwall tests. The surge test specifically tests this "between-turn" insulation.

12.13   INDUCTIVE IMPEDANCE TEST

This test is a low-voltage version of the stator surge test and can be applied to any three-phase stator winding. This test is often included in the motor winding test instruments discussed in Section 12.7, to complement the IR/PI, conductance, and capacitance tests. The test set applies a high frequency, f (typically in the range of 1 kHz) and voltage, V (usually a few volts) to a pair of phase terminals and measures the AC current, I, that flows. The inductive impedance X, is

$$X = \frac{V}{I}$$

Since $X = 2\pi f L$, the higher the inductive impedance, the greater is the inductance of the winding. The inductance is measured three times, between A and B, between A and C, and between B and C phases. The inductance will depend on the number of turns between the two-phase terminals and the permeability of any surrounding steel.

Ideally, the three inductances should be equal, within ± 1%. If the three inductances are not equal, this could be for several reasons:

- One of the phases has a shorted turn. For example, if B phase has a shorted turn, the A–B and B–C inductances should be lower than the A–C inductance. If there are 50 turns between phase and neutral, then there are 100 turns between the two phases. Thus, a single turn short will result in 1% less inductance between the two phases.
- The rotor is in a slightly different position with respect to the coils in each phase. Since the rotor is made of magnetic material, it will affect the inductance via the permeability. Since the rotor is not a completely homogeneous magnetic material (if nothing else, there are slots cut into the surface for the nonmagnetic rotor bars), moving the rotor makes small differences in the inductances of different coils in the stator.
- Presence of steel end shields and other steel around the motor. Unless the motor is completely enclosed in steel, structural steel beams, other motors, or other close-by magnetic objects affect the inductance of the coils, since this affects the permeability.

Unfortunately, the last two effects can completely overwhelm the very small change in inductive impedance due to a shorted turn. This makes the test very difficult to interpret.

There is another limitation with this test. The purpose of the test is to detect the presence or absence of a turn short. Since only a few volts are applied to the winding, the volts per turn are only millivolts. Unless the short already exists, this low voltage is not likely to cause a puncture in the turn insulation. As discussed in Section 1.4.2, if a turn insulation failure occurs in service, it is very likely that the high circulating current that results from the short will rapidly melt the copper turn. This melted copper then burns a hole through the ground insulation, rapidly leading to a ground fault. In form-wound machines, the time between the onset of a turn fault and the ground fault is understood to be seconds or minutes. For a random-wound stator, the time may be longer due to the higher resistance that may occur between the shorted turns (although, if the resistance is high, the inductance measurement will not be affected, either). Thus, the inductive impedance test can only be useful if it can detect a turn fault before it progresses to a ground fault. Since this time is extremely short, it will be virtually impossible for this test to give any more than a few minutes warning of failure. Reference 12.27 discuses the problems with the inductive impedance test in greater detail.

The stator surge test (Section 12.12) is much more effective for determining the condition of the turn insulation, since it develops hundreds or thousands of volts across the turn insulation. If the insulation is weak, it will be punctured, changing the resonant frequency of the winding. The voltage from the inductive impedance test is below that of the normal in-service voltage. If the in-service voltage was not able to puncture the turn insulation, it is impossible for an even lower voltage to do this.

12.14 SEMICONDUCTIVE COATING CONTACT RESISTANCE TEST

The measurement of the resistance between the semiconductive coating and the grounded stator core can indicate if coils are loose in the slot or if the coating has deteriorated. This test is only useful in form-wound stator windings that have a semiconductive coating. Thus, the test is usually only applied to stators rated at 6 kV and above.

12.14.1 Purpose and Theory

Section 1.4.5 discussed the need for a partly conductive coating on the surface of high-voltage coils and bars, in the slot region. Specifically, this "semiconductive" coating prevents partial discharge (also known as slot discharge) between the coil surface and the stator core. Poorly manufactured coatings or abrasion of the coating due to coils and bars being loose in the slot will reduce the effectiveness of the coating and promote the onset of the slot discharges.

As the semiconductive coating deteriorates, its resistance increases. Also, if the coils are loose in the slot, there may be only a few points of contact between the semiconductive coating and the core, rather than many. In both these situations, the contact resistance between the coil surface coating and the core will increase. In this test, the electrical resistance between the coil surface and the core is directly measured.

12.14.2 Test Method

The test requires access to the stator winding coils/bars, just outside of the slot. This is easiest to achieve if the rotor has been removed. An electrical contact must be established to the semiconductive coating that extends 2 or 3 cm beyond the slot at each end of the core. If this area has been painted with an insulating varnish, then the varnish needs to be carefully scraped away on the coil surface closest to the bore. Contact is then made to the semiconductive coating with a short length of braided wire, making sure the wire does not contact the core. The resistance is then measured with a standard digital volt–ohm meter between the braid and a solid ground point on the stator core. The resistance is preferably measured at both ends of the core on each top coil in the stator (it is virtually impossible to make electrical contact to the bottom coil in a slot). If this is too time consuming, then as many coils as possible that are connected to the phase terminals should be measured.

If the rotor is out and some stator wedges can be removed, then the resistance from the exposed semiconductive coating on the top of the coil to the core can also be measured in several locations.

12.14.3 Interpretation

The lower the resistance, the better the contact between the coating and the core, and thus the coils are tighter in the slot. Coils showing high resistance should be visually examined to determine if the semiconductive coating is becoming lighter in color or if dusting is occurring due to coil looseness.

In general, all the coils should have a resistance less than about 2000 ohms. If the resistance is higher than about 5 kΩ in phase end coils, then it is likely that slot discharges will occur.

12.15 CONDUCTOR COOLANT TUBE RESISTANCE TEST

Section 1.1.3 mentioned that large turbine generator stators are sometimes directly cooled with hydrogen. The hydrogen flows from one end of the stator bar to the other in stainless steel tubes that are adjacent to the high-voltage copper conductors. Resistors are connected between the tubes and the copper. The resistance of these resistors is tested occasionally to ensure that they have not open-circuited.

12.15.1 Purpose and Test Method

Since the stainless steel tubes are conductive, they will pick up some voltage by capacitance coupling from the copper conductors in the stator bar. To ensure that high potential differences do not occur between the copper conductors and the tubes, resistors are connected between the coolant tubes and the copper conductors. If the end-winding starts to vibrate (Section 8.12), these resistors sometimes crack. If these resisters fail, sparking can occur, causing local overheating and possibly damaging the insulation. A resistance test can determine if the resistors are open or not.

With the generator off-line and the end-windings exposed, the resistors are located near the end of the bars. The resistance of each resistor is measured with an ohmmeter. If there is an open circuit (infinite resistance), then the resistor is defective and must be replaced. If several resistors are open-circuited, then it is likely that severe end-winding vibration is occurring.

12.16 STATOR WEDGE TAP TEST

This test determines if the stator wedges are loose and, thus, if the windings are likely to be vibrating in the slot. The test is relevant for conventional form-wound stators, i.e., non-global VPI stators. Global VPI stators are presumed to always have tight wedges, since they have been glued into place.

12.16.1   Purpose and Theory

The loose coil/bar problem was described in Section 8.4. One of the root causes of this failure process is stator wedging that was never installed tightly, or became loose over the years. If the wedges are tight, then it is less likely that the coils will be loose. The most direct means of assessing if the wedges are tight is to tap each individual wedge.

12.16.2   Test Method

The original way of measuring wedge looseness is to tap each end of each wedge sharply with the ball end of a ball peen hammer or similar tool. This version of the test can only be performed if the rotor has been removed.* If the wedge is loose, the hammer will make a "thud" sound as it contacts the wedge and vibration will be felt if a finger is placed along side the point of hammer impact. If the wedge is tight, it will "ping." Often, there is an uncertain state between loose and tight. This is clearly a subjective judgment, but someone with experience can easily grade the three levels of looseness. The looseness of every wedge in every slot is measured. A two-dimensional "wedge map" is created before the test. This has a row for every stator slot and a column for each wedge in a slot (or vice versa). The degree of looseness of each end of each wedge is marked on the map.

Several vendors, mostly large generator OEMs, have developed instruments that measure the looseness. Usually a calibrated "hammer" strikes the wedge and an accelerometer placed against the wedge measures the vibration of the wedge in response to the impact. If the vibration is highly damped, the wedge is tight. If the output is undamped (i.e., takes longer to decay to zero), then the wedge is loose. With suitable signal processing, the wedge can be easily graded into three levels of looseness. The device is moved from wedge to wedge and slot to slot to create the wedge map. The result is a more objective wedge map, although a very experienced person can probably be more accurate in defining the relative looseness of the wedges, especially if top ripple springs are present.

The wedge tap test can be automated by using a tractor or robot to carry the instrument along the slot. The tractor moves the instrument over each wedge, stops for the impact test, transmits the results to a computer, and then moves to the next wedge. This innovation certainly reduces the laboriousness of the test. In addition, since large steam turbine generators often have a space between the stator bore and the rotor surface of 5 cm or more, the robot can often complete the wedge map with the rotor in place. This same tractor or robot can also carry a tiny video camera for looking down the core ventilation ducts, to see if abrasion is occurring, and a stator core insulation tester such as the EL-CID type described in Chapter 14. If the only reason to remove a rotor is to perform the wedge tap test, the robot alternative can sometimes result in considerable savings. One deficiency of this type of tester is that it does not reliably check the tightness of end wedges, especially if the core is stepped at the ends. This is because the device cannot be adequately supported at this location. Most large generator OEMs can provide a robot test service.

12.16.3   Interpretation

When viewing the wedge map, a few loose wedges are of little concern. However, some consideration for general or local rewedging is warranted if

---

*In large hydrogenerators with salient pole rotors, the removal of the rotor can sometimes be avoided if one or two poles are removed instead, so that a person can be lowered into the space where the poles were. They can then perform the tap test on several slots. The rotor is then manually rotated so that the wedges in other slots can be tapped.

- Two or three adjacent wedges in the same slot are loose
- Any wedges at the end of the slots are loose
- More than 25% of the wedges in the core are loose
- 10% or more wedges are looser than on the previous test
- None of the wedges should be cracked

Note that loose wedges in themselves are not a problem (unless they are so loose they are migrating out of the slot). However, loose wedges tend to lead to loose coils or bars, especially if there are no side ripple springs or other conformable materials in the slot (Section 1.4.6). Also of importance is that if wedges are loose when the unit is out of service, they will also be loose at full load if the windings are made with modern epoxy–mica or polyester–mica insulation. Older thermoplastic windings expand as they heat up. Therefore, loose wedges in thermoplastic windings do not necessarily imply they will be loose at full power. Within reason, loose wedges are less of a problem with asphaltic mica windings.

12.17   SLOT SIDE CLEARANCE TEST

This is another test to compliment the tests in Sections 12.15 and 12.16 that can indicate how loose a coil is in a stator slot. The test is only valid for nonglobal VPI form-wound stators.

12.17.1   Purpose and Theory

This is a simple test to measure the gap that exists between the surface of a coil and the stator core. If there is too large a gap, then the coil or bar is likely to be free to vibrate in the slot and rubs against the stator core, reducing the groundwall insulation thickness (Section 8.4). The test is most applicable to stators that have a flat side packing system, and which do not use a conformable coating on the side of the coils/bars (Section 1.4.6).

12.17.2   Test Method

The rotor must first be removed and then wedges are removed in some slots. Usually the slots containing phase end coils in the top position in the slot should be examined. "Feeler gauges" (thin metal fingers with known thickness) are then slid down the side of the coil/bars between the coil and the core. The purpose is to find the thickest feeler gauge that can be easily slid between the coil and the core and moved back and forth in the axial direction for at least a few centimeters. The gauge should be inserted at both ends of the core, as well as at random positions along the slot.

12.17.3   Interpretation

If the winding is insulated with modern epoxy–mica or polyester–mica groundwall insulation and uses flat side packing, then the largest feeler gauge that should be able to fit between the coils and the core is 0.125 mm [12.28]. If larger gauges can be inserted in many spots and slots, then the winding is loose enough to warrant corrective action. For asphaltic mica windings and the early polyester (pre-1970) windings, it is not uncommon to have larger gaps and still not have the winding at risk. This is because the older materials expand when operating at full load and high temperature, thus reducing the size of the gap that may exist when the winding is cold. However, even with older groundwall insulation systems, the gap should be no larger than 0.5 mm.

12.18 STATOR SLOT RADIAL CLEARANCE TEST

The importance of a tight fit for stator conductor bars in the slot is underscored by yet another test to complement the stator wedge tap and the slot side clearance tests (Sections 12.16 and 12.17). Applicable to large form-wound stators, this test, also referred to as the "bar jacking test," determines the amount of radial clearance in the stator winding slot.

12.18.1 Purpose and Theory

Conductor bar support systems are designed to restrain the motion of the stator winding due to twice-per-cycle magnetic forces of several tons acting on each conductor bar. Use of conforming materials, tapered wedging systems, and ripple springs, as well as curing of conforming materials under pressure and elevated temperature are the principal methods used to eliminate radial clearance in the slots. Nevertheless, bars become loose in the slot, as described in Section 8.4. This test directly measures whether stator bars are free to move up and down in the slot.

Radial clearance may not be identified by the stator wedge tap test. The end wedges may still be tight with radial clearance present between the conductor bars or between the bottom bar and the bottom of the slot. It is, therefore, useful to conduct the radial clearance test to determine the risk of bar abrasion, particularly in the presence of oil.

12.18.2 Test Method

The test method involves the application of a known amount of radial force in the 2200 Nt (500 lb) range and measuring the deflection of the conductor bar in the same direction. The potential for damage to nearby generator components is significant when performing the clearance test. Special tooling and considerable expertise are required to safely conduct the test. It is, therefore, recommended that the OEM should perform this test.

The test can be performed with the rotor removed. However, in-situ inspection systems offered by most OEMs have been adapted to allow the performance of this test without removing the rotor. Extra care is required with in-situ inspection because of the potential for surface damage to the rotor retaining ring, which is used for supporting the jacking equipment, and because of the confined space involved.

12.18.3 Interpretation

Generally, "zero clearance" is the requirement for slot content design. Rewedging, including changes to the slot support system, may be required if the clearance does not meet this criterion. A minor amount of clearance may be permissible until the next outage if there is no oil contamination, the stator was recently rewedged, and the clearance is not widespread. However, the impact of all contributing factors has to be assessed, together with the results from visual inspection, wedge tap, and clearance tests, to determine the course of action.

12.19 STATOR END-WINDING RESONANCE TEST

This test objectively determines how loose the stator end-winding is, and if it is susceptible to high levels of end-winding vibration. The test is most useful for large two-pole or four-pole generators and motors. Such machines are most likely to suffer from the end-winding vibration deterioration process (Section 8.12).

12.19.1 Purpose and Theory

Magnetic forces tend to make the stator bars/coils vibrate at twice power frequency (100 or 120 Hz) as well as at twice the frequency of any other strong harmonics in the stator current. If the length and mass of the end-windings are such that they have a mechanical resonant frequency near twice power frequency, then it is very likely that the end-windings will vibrate no matter how well they are braced. As discussed in Section 8.12, this will eventually lead to groundwall insulation failure due to fatigue cracking or abrasion. One of the purposes of this test is to ensure that the end-windings do not have a natural vibration frequency at 100 Hz or 120 Hz, for 50 or 60 Hz AC, respectively.

Similar to the wedge tightness test, if the end-windings are hit with a hammer, the tightness of the end-winding bracing system can be established by how quickly the vibrations of the end-winding damp out. If the coils/bars just "ping," then they are tight. If they respond with a dull thud, then the blocking and bracing may be coming loose.

12.19.2 Test Method

The best tool for this test is the standard vibration analysis instrumentation used by NDE (nondestructive evaluation) specialists. This equipment includes

- A "calibrated hammer" that can impact the end-winding and measure the magnitude of the impact force with an accelerometer mounted on the hammer
- Detection accelerometers that are temporarily bonded to the coils/bars. At least two accelerometers are needed to measure the vibration in the circumferential and radial directions.
- An FFT type of spectrum analyzer that can respond to frequencies up to about 100 kHz to simultaneously capture the three accelerometer responses and produce the frequency spectrum analysis

Specialized personnel are needed for this testing. Most OEMs can provide the equipment and specialists.

The testing should be done at several locations at both ends of the stator. Performing the tests on coils/bars connected to the circuit ring buses is important, since the vibration pattern may differ from the rest of the coils/bars.

12.19.3 Interpretation

If the test reveals that there is a resonant frequency within about ± 3 Hz of twice the power frequency, then it is likely that severe end-winding vibration may eventually occur. The OEM should be contacted for possible remedies.

It is difficult to set an acceptable level for vibration damping. Instead, a fingerprint of the vibration spectra should be obtained when the stator is new and is presumed to have a tight blocking and bracing system. The test is then repeated in the same locations every 5 years or so. If there is any change in the resonant frequencies or the damping over the years, this will be an indication that the end-winding support system is becoming less effective.

12.20 ROTOR VOLTAGE DROP TEST

This test is used to determine if there are shorted turns in the salient pole windings of a synchronous machine rotor. It is quite effective for all rotor sizes and speeds and does not require any special test equipment. This test is not 100% reliable since some turn shorts can disappear when the centrifugal forces due to rotor rotation are not present.

12.20.1 Purpose and Theory

This test is designed to verify the presence of shorted turns in salient pole winding coils. It takes advantage of the fact that if an AC power supply is applied to a mainly inductive circuit, shorted turns will create a significant reduction in inductive impedance. Thus, if an AC voltage is applied between the terminals of a salient pole winding and the voltage across each coil in it is measured, coils with shorted turns will have a lower voltage drop, due to their reduced impedance, than those with no turn shorts.

12.20.2 Test Method

For small synchronous machines, it is more convenient to remove the rotor for this test. On the other hand, it is usually possible to perform this test with the rotor in place if access to the individual pole connections can be attained. This test also requires access to both ends of the rotor winding. This is readily available if the winding is connected to sliprings. On the other hand, if the machine has a brushless exciter with a rotating diode rectifier, the ends of the winding have to be disconnected from the rectifier to allow the test to be performed.

The test equipment required is simple, comprised of only a 120/240 V variac (variable autotransformer) plus common meters. The AC voltage is applied across all the poles via the variac, which is adjusted to a convenient voltage, and the voltage is measured across the terminals in each pole.

12.20.3 Interpretation

If the minimum voltages measured across any of the pole winding coils is 10% or less than the average of the voltage drops across all the poles, then it is unlikely that there are any shorted turns in the winding. The only proviso is that, in some cases, turn shorts are only present under the influence of the centrifugal forces from rotation, and at standstill these shorts are not present. If the minimum voltage measured across a coil is greater than 10% of the average value (that is, <0.9 of the average pole drop) then the presence of shorted turns is confirmed.

12.21 ROTOR RSO AND SURGE TESTS

Both of these off-line tests can be used for detecting shorted turns, ground faults and high-resistance connections in synchronous machine rotors. The data obtained from the recurrent surge oscilloscope (RSO) test can also be used to identify the location of the fault. These test procedures are primarily for turbine generator rotors, but can also be applied to salient pole rotors.

12.21.1 Purpose and Theory

Both tests rely on the fact that a healthy rotor winding, when viewed from its terminals, is electrically symmetrical and a winding with a turn fault, ground fault, or high-resistance connection is not.

RSO Test. This operates on the basis that if low voltage (<100 V peak), identical high-frequency electrical pulses are injected at both ends of a healthy winding, their travel time through the winding will be identical when they reach the other end. If a turn short or ground fault exists in a rotor winding, the impedance seen by each set of pulses will be different and some of the pulse energy will be reflected backed to the end of the winding, thus changing the input pulse waveforms in a way that is dependent on the distance to the fault. Thus, the fault will produce different waveforms at each end of the winding unless the fault is exactly halfway through it. This technique is a variant of time domain reflectometry. A high-resistance connection also creates different winding impedances to produce a similar effect.

This technique can also be applied to rotors with sliprings while they are being rotated. Thus, it can detect turn shorts or ground faults that disappear when the rotor is at standstill.

An experienced operator is required to perform this test since interpretation of the results can be difficult.

Surge Test. This is based on the same principles as the stator winding surge comparison test described in Section 12.12. If a turn short, ground fault, or high-resistance connection exists in a rotor winding, a voltage surge applied to the winding will cause oscillations at a frequency that depends on the presence of shorted turns. Thus, if high-frequency surges are injected at one end and then at the other end of the winding, the resulting waveforms will be seen to be different. The peak voltages used for such tests are much lower than that for stator windings since round-rotor and salient pole rotor windings have much lower operating voltages, typically well below 500 V DC.

12.21.2   Test Method

RSO Test. A special instrument, sometimes called a reflectometer, is required to perform the RSO test. If the machine has a brushless exciter, its field winding must be disconnected from the rotating diode rectifier to allow the test instrument to be directly connected to it.

The test instrument alternately or simultaneously injects identical, fast-rising, high-frequency, voltage pulses with maximum peak magnitude of less than 100 V at each end of the winding. The potential at each injection point is then recorded as a function of time using an oscilloscope. In the absence of a fault, identical records should be obtained for the two injection points due to the symmetry of the winding. Features found on one trace and not on the other are, therefore, indicative of a winding fault. The time at which the irregularity occurs can be used to locate the fault.

Surge Test. The same test equipment that is used for stator winding surge testing can normally be used for round-rotor and salient pole rotor windings. As for the RSO test, the rotor windings with brushless excitation must first be disconnected from the rotating diode rectifier. The rotor body must then be grounded for this test and it is advisable to perform a one-minute, 500 V IR test first to ensure that there are no dead shorts between the winding and ground. Providing the IR test results give a value of at least 1 megohm, the winding can be surge tested. A peak voltage in the region of 1000 V is normally used and the voltage surges are applied to one end of the winding. The resulting waveform is stored electronically. The test is then repeated from the other end of the winding and the two stored waveforms are overlaid and compared for differences.

12.21.3   Interpretation

As already indicated, some experience is required to interpret the results of both of these tests, especially when a turn fault is indicated. It should be noted that for the standstill RSO and surge tests, some faults might not be detected. This is because the removal of the centrifugal forces present during operation may significantly increase the resistance of the fault to the extent that it may not be detected. For both tests, it is advisable to have a set of reference baseline test results taken when the on-the-rotor winding was known to have no faults in it. This aids in the interpretation of the results.

RSO Test. The RSO test will detect ground faults having a fault resistance of less than 500 ohms. Since standard generator protection systems can usually detect such faults, the value of the RSO is mainly as a confirmatory test for ground faults.

The RSO test will detect interturn faults if the fault has a resistance of less than about one ohm. Faults that are significant during operation but are less severe when the rotor is at standstill may not be detected since they are likely to have a resistance of more than 10 ohms.

Surge Test. If the two recorded traces are identical, then no fault is present. On the other hand, if there are significant differences in their frequency, the type of fault present can be determined, by an experience operator, from the nature of these differences.

12.22   ROTOR GROWLER TEST

This is a simple test that has been used for many years to detect open-circuited bars in squirrel cage rotors [12.31]. It can be particularly useful for diecast rotor windings in which none of the bars are visible. There are two types of growlers. One consists of a U-shaped laminated core with a multiturn 120 V AC coil wound on to it. This device is placed across the teeth on either side of a bar and moved axially from one end of the rotor to the other. This process is repeated for each suspect bar in the rotor. The other type also has a multiturn 120 V AC coil wound onto an open V-shaped laminated core into which the rotor outside diameter will fit. In this type, each bar is scanned by turning the rotor while it is in the core "V." There is a limit to the size of rotor that can be tested by this latter type.

12.22.1   Purpose and Theory

Growlers are used to detect rotor bars that are open-circuited inside the core and therefore not visible. The growler induces a current in healthy bars by creating a flux, which encircles them; i.e., the flux generated by the winding in the growler passes around a circuit consisting of it own and the motor rotor laminated cores. For the U-shaped growler, a magnetic strip, such as a hacksaw blade, or a metal plate integral with the growler is placed on top of the bar. This metal strip will rattle under the influence of the magnetic force from the induced core flux if there is a break in the bar. Also, if the bar is loose in the slot, it will be heard to rattle. The open V-type of growler usually has an ammeter and resistor in series with its excitation coil to limit the current from the power supply. The ammeter reading will drop significantly when an open bar passes through the growler "V."

12.22.2   Test Method

For this test, the rotor must be removed from its stator and be supported at either end. The supports used must be such that the rotor can be rotated by hand. For the simple U-shaped growler, each bar is traversed axially with the growler and those that produce vibration in the metal strip will have open-circuits in them. With the open V-type growler, the rotor is placed in it and rotated slowly. Any bars that indicate a significantly lower ammeter reading as they pass through the center of the "V" will have open-circuits in them.

12.22.3   Interpretation

The growler will reliably detect completely open-circuited bars as indicated above. It will not, however, detect open circuits that close up with the rotor at standstill and cold.

12.23   ROTOR FLUORESCENT DYE PENETRANT TEST

This test is used to detect the presence of hairline cracks in squirrel cage rotor windings.

12.23.1   Purpose and Theory

Hairline cracks in fabricated winding rotor bars, shorting rings, and the brazed or welded joints are often difficult to detect with the naked eye. This statement also applies to diecast winding shorting rings and their integral fan blades. If a fluorescent penetrant dye is sprayed onto areas with suspected cracks and an ultraviolet light beam is aimed at the dye-treated surfaces, cracks will glow in the presence of this light.

12.23.2   Test Method and Interpretation

Spray a fluorescent penetrant dye from an aerosol can on to the parts of the rotor winding suspected to have cracks in them. Then shine an ultraviolet light on the area. If cracks are present, the entire length of the crack will glow under the light.

12.24   ROTOR RATED FLUX TEST

This test can be used to check for cracks or breaks in squirrel cage rotor windings, to indicate the severity of rotor lamination surface smearing due to stator rubs, and to detect loose rotor cores [12.31].

12.24.1   Purpose and Theory

If an AC current is passed through the shaft of a squirrel cage rotor it will induce a flux in its core which, in turn, will generate currents in its winding. If the rotor core is mounted on shaft spider arms, the induced flux and current can be generated by a winding made from a number of turns of cable connected to an ac power source. If the excitation flux is selected to produce enough ampere-turns to induce rated core flux, the rotor outside diameter surface will become hot in areas where the lamination insulation is shorted and the currents induced in the winding will create core hotspots at rotor bar cracks and open circuits. If the rotor core is loose it will vibrate under the influence of the magnetic flux induced in it. Also, if iron shavings or magnetic paper are placed over the rotor surface, a clear representation of each bar covered by these should be seen. If any discontinuity or misalignment appears, then there is a broken bar present.

12.24.2   Test Method

AC flux is induced in the rotor core and currents in the rotor squirrel cage winding by applying a low voltage across the ends of the rotor, or by a core loop test as described in Section

14.2. Safety precautions are very important during this test. Iron shavings or magnetic paper is placed over the rotor and a clear representation of the bar should be seen. If any discontinuity or misalignment appears, then there is a broken bar present. If there is evidence of broken bars, warm the rotor winding by increasing the applied voltage. This will produce hot spots at the location of a cracked rotor bar, which should be detected and recorded with an infrared scanning device. If the rotor core is loose it will vibrate during this test.

12.24.3   Interpretation

If hot spots on the rotor surface are greater than 10°C above the ambient core temperature, then significant surface lamination insulation shorting is present and/or there are broken rotor bars. The majority of cracked or broken bars occur at or near the connections between the bars and short-circuit rings. Sometimes, the fault area may spark due to the voltage differential across the crack. This test can also to be used to determine the effectiveness of rotor core and winding repairs.

12.25   ROTOR SINGLE-PHASE ROTATION TEST

This is an often-used test for broken rotor bars, short-circuit rings, and the joints between them in squirrel cage induction rotors. The test can be performed without motor disassembly.

12.25.1   Purpose and Theory

If a single-phase 50/60 Hz power supply is connected across two of the three stator winding leads of a squirrel cage induction motor, the resulting current flow will create a nonrotating sinusoidal flux in the airgap. Breaks in the rotor cage winding will create a nonuniform impedance in the rotor winding. If the rotor with cage winding breaks is rotated through the airgap flux created by a single-phase power supply, the current drawn will fluctuate significantly due to the variation in rotor impedance.

12.25.2   Test Method

This test is done with the rotor within the stator bore and able to rotate. An instrumented, single-phase AC power supply is connected to two stator winding leads to apply a voltage of 10 to 25% of the rated line-to-line value. The rotor is slowly turned while monitoring the stator winding current. *It is important to note that this test will produce rapid heating in both the stator and rotor windings; therefore, the test time should be kept to a minimum.*

12.25.3   Interpretation

If current fluctuations exceed 5%, this indicates broken rotor bars and/or short-circuit rings. It should, however, be noted that this test may not detect cracks or breaks that close up when the rotor is not at normal operating speed and temperature.

12.26   STATOR BLACKOUT TEST

This test helps to locate the sources of surface partial discharge activity in form-wound stators rated at 6 kV and above. It complements the probe tests described in Section 12.11.

12.26.1 Purpose and Theory

End-winding partial discharge is a symptom or cause of several problems such as grading-coating failure, pollution, and inadequate end-winding spacing (Sections 8.6, 8.8, and 8.11). In addition to electrical and acoustic signals, PD also creates an optical output. Some of the light output is in the visible frequency range, but most is in the ultraviolet frequency range. If a stator winding is energized one phase at a time, there will be some electrical stress in the end-winding. If any of the above deterioration mechanisms are occurring, then PD may occur, which can be observed by the light emission.

12.26.2 Test Method

Each phase of the stator is energized to rated line-to-ground voltage, with the other two phases grounded. This test is often done at the same time as the off-line PD test (Section 12.10), with the same AC supply. If a higher voltage is used, for example applying the rated line-to-line voltage between phase and ground, then this would be an AC hipot test and a coil may be punctured.

The voltage is applied for about 15 minutes to stabilize the PD activity (see Section 12.10). The test is best performed at night with all the lights turned out near the machine under test. If this is not possible, a temporary shroud is constructed around the machine to block the ambient light. One or more people then stand near to the machine (or within the shroud) and the voltage is applied. While the PD is stabilizing, the observer's eyes will become accustomed to the dark. The number of points of light, if any, is then counted. Using a flashlight that is briefly turned on, the points of light are located on the end-winding. Note that extreme care is needed when performing this test, since personnel are close to an energized winding. Since personnel are in the dark, they may not see the high-voltage power supply or details of the stator.

To reduce the danger of this test to personnel, ultraviolet imaging cameras have been developed [12.29]. The camera amplifies the intensity of the ultraviolet light emitted by the PD, while strongly attenuating the visible light. The output is displayed on an imaging screen or directly on a video screen. The main advantage of this device it that the UV camera makes it possible to view the PD without turning off the lights in the plant. In addition, it is much easier to pinpoint the PD locations. Note that the UV camera tends to be more sensitive than the eye, even after long periods of acclimatization in the dark, and, thus, changing from one method to the other may produce nontrendable results. With appropriate lenses, the camera can be used from a position well away from the energized stator winding.

12.26.3 Interpretation

If the test is performed at rated line-to-ground stress, ideally there should be no light emitted by the PD. Windings that have operated in service will usually have a few surface PD sites. In each test, the approximate number of points of light should be estimated. The test should be repeated every few years. The winding is deteriorating if the number of points emitting light increase by more than five times between tests.

The location of the light should be noted and, in particular, if the light is coming from

1. The interface between the semiconductive and silicon carbide coatings
2. The end of the silicon carbide coating furthest from the slot (about 12 cm out from the core)

3. In the space between coils that are in different phases and which appear to be close

4. At blocking or bracing spots

The first two problems are associated with semiconductive or grading paint deterioration or insufficient silicon carbide coating length, the third problem is due to poor manufacturing (Section 8.11), and the last is due to electrical tracking caused by pollution. Note that with a test at rated line-to-ground voltage, the last two problems may not show up at this voltage, since the stress is lower than that which occurs in service. If the test is performed with the phase-to-phase voltage applied line-to-ground, there will almost certainly be PD at the silicon carbide coating, which should be ignored, since it is stressed much higher than would occur in service.

12.27 STATOR PRESSURE AND VACUUM DECAY TESTS

This is a pair of tests that are performed on direct-water-cooled stator windings. The tests detect if there are any cracks in the water tubes (copper or stainless steel) that are within the stator bar. The test can be performed on individual bars, or complete windings.

12.27.1 Purpose and Theory

End-winding vibration and poor brazing methods can lead to small cracks in the water-filled tubes that run the length of stator bars for cooling of large-generator stator windings (Sections 8.12 and 8.13). The water leaks can degrade the groundwall insulation, leading to failure. Two tests are available to detect if small cracks in the coolant tubes have occurred. One is the pressure decay test and the other is the vacuum decay test. The basic idea is that if the coolant system cannot maintain a vacuum after the air has been evacuated, or if a static pressure is applied and the pressure gradually reduces, then there is probably a leak in the system. If a leak is detected, then there are methods using tracer gases that are sometimes helpful in locating the leak. If leaks are confirmed, a capacitance test (Section 12.5) may confirm if the water leak has degraded the insulation.

12.27.2 Test Methods and Interpretation

Pressure Decay Test. This is a standard test to detect leaks from enclosed systems, and it is usual to perform this test after any dismantling of the generator. The stator winding coolant system is completely dried out internally by blowing air, and then by using a vacuum pump. Then the system is pressurized with dry air, nitrogen, or helium to the level of the design hydrogen operating pressure. The generator enclosure itself is at normal atmospheric air pressure. Since the pressure is highest within the cooling channels, the pressure is in the same direction that would cause the water to leak. The location of the leaks can sometimes be identified if a suitable "bubble" solution is applied at the potential leak sites.

The test is relatively insensitive to small leaks and is much affected by the changes in the environment temperature and barometric pressure. The pressure will drive the water in the leak direction and may aggravate the insulation damage. In a typical test at 300 kPa, a volume of 0.03 m³ (1 cubic foot) must leak out of the generator to produce a change of 10 kPa (1.5 psi). Therefore, the test has to be done over about 24 hours and very accurate pressure gauges are needed to obtain sensitive results. Ideally, there should be no drop in the pressure, after correction for winding temperature and atmospheric pressure. A 10 kPa drop in pressure over 24 hours is significant.

Vacuum Decay. In this test, a high vacuum is applied to the stator water coolant vacuum system fitted with a vacuum gauge and measurements of the decay in the vacuum are made at about 0.0002 kPa. Prior to the test, the coolant system must be thoroughly dried by first blowing dry air through the system and then by pulling a rough vacuum for 2 or 3 days. The test is very sensitive and can be applied without any access to the winding itself. Accurate results can be obtained in 1 to 2 hours as compared to 24 hours required for a pressure decay test [12.30]. The test is relatively insensitive to changes in ambient temperature and atmospheric pressure. Tests on individual bars may be done to locate any leaks.

The test may have to be repeated several times, because any water vapor will raise the vacuum pressure, indicating a test failure. The pressure differential is from the insulation into the coolant system and, thus, will not drive any water into the insulation. This test usually precedes the pressure decay test. Tests on individual bars may also be done to locate any leaks.

If there is no increase in pressure in 2 hours, then there are no leaks.

Leak Location. If the bubble test does not locate the bar(s) with the leaks and the axial locations of the leaks, then tracer gases can often be used. The winding is pressurized with a tracer gas, usually helium, and the potential leak sites are carefully inspected with hand-held detectors that are sensitive to the tracer gas. The test may take over a day to perform, since the helium detector must be within about 2 cm of the leak.

Rotor removal is needed to measure the entire length of the stator winding. Sometimes, the location of the leak at the stator bar surface can be a significant distance from the crack in the copper.

REFERENCES

12.1   IEEE 43-2000, "IEEE Recommended Practice for Testing Insulation Resistance of Rotating Machinery."

12.2   IEEE 95-1977, "IEEE Recommended Practice for Insulation Testing of Large AC Rotating Machinery with High Direct Voltage."

12.3   Schleif, F. R. and Engvall, L. R. "Experience in Analysis of DC Insulation Tests for Maintenance Programming," in *AIEE Transactions (Power Apparatus and Systems),* vol. 78, 1959, pp. 156–161.

12.4   W. McDermid and B. G. Solomon, "Significance of Defects Found During High Direct Voltage Ramp Tests," in *Proceedings of IEEE Electrical Insulation Conference,* Cincinnati, October 1999, pp. 631–6.

12.5   L. M. Rux, "High-Voltage DC Tests for Evaluating Stator Winding Insulation: Uniform Step, Graded Step, and Ramped Test Methods," in *Proceedings of 1997 Conference on Electrical Insulation and Dielectric Phenomena,* October, 1997, Minneapolis, pp. 258–262.

12.6   C. M. Foust and B. V. Bhimani, "Predicting Insulation Failures with Direct Voltage," *AIEE Trans. PAS, 76,* 1120–1130, 1957.

12.7   K. M. Stevens and J. S. Johnson, "Destructive AC and DC Tests on Two Large Turbine Generators of the Southern California Edison Co.," *AIEE PAS, 73,* 1115–1122, 1954,.

12.8   B. K. Gupta, G. C. Stone, and J. Stein, "Use of AC and DC Hipot Tests to Assess Condition of Stator Winding Insulation," in *Proceedings of IEEE Electrical Insulation Conference,* Chicago, September 1995, pp. 605–608.

12.9 IEEE 56-1977, "IEEE Guide for Insulation Maintenance of Large Alternating Current Rotating Machinery."

12.10 R. Miller and I. A. Black, "PD Energy Measurements on Electrical Machines When Energized at Frequencies Between 0. 1 Hz and Power Frequency," *IEEE Trans. EI*, 127–135, June 1979.

12.11 IEEE 433-1973, "IEEE Recommended Practice for Insulation Testing of Large AC Rotating Machinery with High Voltage at Very Low Frequency."

12.12 B. Milano and R. Arbour, "Diagnosing High-Potential Test Failures in Large Water-Cooled Hydrogenerators," in *Proceedings of IEEE Electrical/Electronics Insulation Conference*, Chicago, IL, September 1989, pp. 228–235.

12.13 V. Green, "Stator Bar Wetness Detector," Electric Power Research Institute Report TR-108733, September 1997.

12.14 A. D. W. Woolmerans and H. J. Geldenhuys, "Routine Diagnostic Measurements on High Voltage Machines," in *Proceedings of IEEE International Conference on Properties and Applications of Dielectric Materials*, Tokyo, July 1991, pp. 777–781.

12.15 R. Bartnikas, *Engineering Dielectrics, Vol. IIB, Electrical Properties of Solid Insulating Materials: Measurement Techniques*, ASTM Publication STP 926, 1987.

12.16 IEEE Standard 286-2000, "Recommended Practice for Measurement of Power Factor Tip-Up of Electric Machinery Stator Coil Insulation."

12.17 IEC Standard 60894 (1987), "Guide for the Test Procedure for the Measurement of Loss Tangent on Coils and Bars for Machine Windings."

12.18 S. A. Boggs and G. C. Stone, "Fundamental Limitations in the Measurement of Corona and Partial Discharge," *IEEE Transactions on Electrical Insulation*, 143–150, April 1982.

12.19 R. Bartnikas and E. J. McMahon, *Engineering Dielectrics, Vol. 1, Corona Measurement and Interpretation*, ASTM Publication STP 669, 1979.

12.20 S. R. Campbell and G. C. Stone, "Examples of Stator Winding Partial Discharges Due to Invertor Drives," in *Proceedings of IEEE International Symposium on Electrical Insulation*, April 2000, Anaheim CA, pp. 231–234.

12.21 ASTM D1868-1986, "Standard Methods for Detection and Measurement of Partial Discharge (Corona) Pulses in Evaluation of Insulation Systems."

12.22 IEC 60270, "High Voltage Test Techniques. Partial Discharge Measurements."

12.23 IEEE 1434-2000, "IEEE Trial Use Guide to the Measurement of Partial Discharges in Rotating Machinery."

12.24 G. C. Stone, "Calibration of PD Measurements for Motor and Generator Windings—Why It Can't Be Done," *IEEE Electrical Insulation Magazine*, 9–12, Jan 1998.

12.25 L. E. Smith, "A Peak Pulse Ammeter–Voltmeter Suitable for Ionization (Corona) Measurements in Electrical Equipment," paper presented at Doble Client Conference, Paper 37AIC, April 1970.

12.26 IEEE 522-1992, "Guide for Testing Turn-to-Turn Insulation on Form Wound Stator Coils for Alternating Current Rotating Electrical Machines."

12.27 R. L. Nailen, "Are Those New Motor Maintenance Tests Really That Great?," *Electrical Apparatus Magazine*, 31–35, January 2000.

12.28 J. F. Lyles, T. E. Goodeve, and H. Sedding, "Parameters Required to Maximize a Thermoset Hydro-electric Stator Winding Life—Parts 1 and 2," *IEEE Trans. EC*, 620–635, September 1994.

12.29 K. Forsyth, U.S. Patent 5,886,344, "Corona Detector with Narrow-Band Optical Filter."

12.30 A. M. Iversen, "GE TIL 1098 Update," paper presented at Electric Power Research Institute Conference on Maintaining Integrity of Water Cooled Generator Stator Windings, Tampa FL, November 1996.

12.31 R. L. Nailen, "Is Finding Broken Rotor Bars Easy?," *Electrical Apparatus Magazine*, 26–30, April 1998.

CHAPTER 13

IN-SERVICE MONITORING OF STATOR AND ROTOR WINDINGS

Chapter 12 discussed 27 different tests that can be performed on rotor and stator windings to assess insulation condition. All these tests are done with the motor or generator shut down and sometimes disassembled. This chapter describes a number of different monitors that determine insulation condition (and induction motor rotor winding integrity) during normal operation. The advantages and disadvantages of in-service monitoring versus off-line testing were reviewed in Section 11.2.

The monitors in this chapter measure thermal, chemical, mechanical, and electrical phenomena, and together can detect most of the important stator and rotor winding failure processes that are likely to occur. For the first time in history, it is now possible to have a very high likelihood of detecting most (but not all) of the likely problems that can lead to winding failure, without ever having to remove the machine from service. Thus, if the capital resources are available to implement all the appropriate monitors on a machine, times between shutdowns for visual inspections can be greatly increased. All the monitors identified in this chapter are commercially available. Other works that review this subject are books by Tavner and Penman [13.1] and Gonzales [13.2]. An extensive bibliography on on-line motor monitoring is in [13.3].

Section 11.4 discusses software to convert the data produced from on-line monitors into practical information that can be used by machine operators and maintenance personnel.

In general, except for alarms presented to operators indicating that exceptionally high temperatures or very small airgaps have been detected in a machine, the output from these monitors is intended for use by maintenance personnel. Most of the monitors detect problems years before failure (i.e., a ground fault) will occur. Since plant operators tend to only focus on the very short term, it is best *not* to provide monitoring output data for them to analyze. In fact, most operators, if given access to many of the monitors described below, will tend to say they do not work, since they do not consider any change in a monitor output as leading to immediate failure.

Electrical Insulation for Rotating Machines. By Stone, Boulter, Culbert, and Dhirani
ISBN 0-471-44506-1 © 2004 Institute of Electrical and Electronics Engineers

13.1 THERMAL MONITORING

Insulation failure caused by gradual deterioration of the insulation by long-term operation at high temperatures was extensively discussed in Chapters 8 and 9. Thermal deterioration is one of the leading causes of failure in air-cooled machines, especially hydrogenerators, gas turbine generators, and motors of all sizes. The deterioration rate depends on the operating temperature of the insulation plus the duration at that temperature. The higher the temperature and/or the longer the operation at that temperature, the more deteriorated the insulation will be. Thus, it is not surprising that virtually all stator windings, above a few hundred kW or HP, are equipped with embedded temperature sensors. In most cases, these sensors are used during initial machine acceptance testing and, once in operation, are connected to alarms to warn of very high temperature excursions. However, maintenance personnel and system engineers can diagnose that certain failure processes are occurring by monitoring and trending winding temperature information.

This section reviews how to make better use of the existing temperature monitors in motors and generators, to extract diagnostic information. Other on-line temperature monitoring such as infrared thermography is also discussed. Temperature monitoring, especially where temperature sensors are already in place, is probably the most cost effective and easiest monitoring to perform.

13.1.1 Stator Winding Point Sensors

Small-machine stator windings rarely have temperature sensors within the winding. The most they may have is a thermal cutout switch, which turns off the machine if the winding operates above a specified temperature. Some small motor and generator stator windings may be fitted with a thermistor to give an alarm if a specific temperature is reached. However, modern stator windings in larger motors and generators normally have at least a few temperature sensors, which can be continuously monitored. These sensors measure the temperature at specific points. They can be at a variety of locations:

- Embedded within the stator winding. In random-wound machines these are inserted between coils. In form-wound machines, they are between the top and bottom coils in a slot (Figure 13.1). In these locations, the sensor is most sensitive to the copper temperature, rather than the stator core temperature or the cooling air temperature. The sensors will be 5 to 20 °C cooler than the copper in an indirectly cooled machine. The thicker the groundwall insulation, the greater will be the temperature difference. Consequently, older windings and stators operating at higher voltages have the greatest temperature difference between the copper and the sensor. In most air-cooled motors and generators, 3 to 12 slots may contain sensors. These slots are randomly selected, except in 2.3 to 4.1 kV stators, in which the sensors are installed in slots containing coils operating at a low voltage. (Recall that such stators usually do not have semiconductive coatings on the coils and, thus, an effort is made to not expose the sensors to high voltages.) The axial position of the sensor depends on the cooling system. The sensors should be installed at the end (or middle) of the slot, where the temperature should be the hottest.
- Cooling water or gas channels. In direct-water- or hydrogen-cooled stator windings, temperature sensors are installed in the ground side of the water coolant hoses, on the end of the stator where the water has passed through the stator bar. Similarly, sensors are placed near the hydrogen gas outlet duct at the end of the direct-hydrogen-cooled

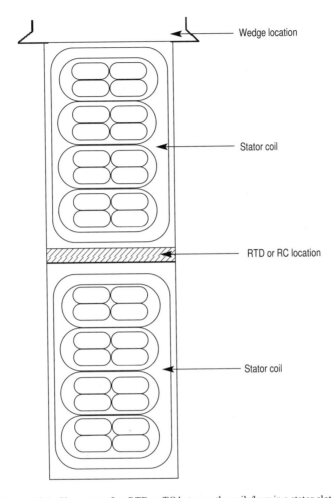

Figure 13.1. Placement of an RTD or TC between the coils/bars in a stator slot.

stator bars. Usually there is a sensor for each bar in a water-cooled winding. On direct-hydrogen-cooled stator bars, one sensor per parallel is usually used.

- Stator core, to measure the core temperature.
- Stator frame, although this is only loosely correlated to the winding temperature.
- At the air or hydrogen gas inlets and outlets of water coolers in totally enclosed machines to measure the temperature of the coolant leaving and going back into the machine.

There are many types of temperature sensors employed in machines. The most common is the resistance temperature detector (RTD). This sensor takes advantage of the fact that the electrical resistance of a conductor is directly proportional to the temperature. RTDs contain a thin strip of metal, many centimeters long, with a nominal resistance of 10 or 100 ohms. A fixed current is passed though the metal strip and the voltage across the strip is measured. This allows calculation of the resistance. Since the temperature dependence of the resistance is known, the temperature can be calculated. RTDs used in stators usually have three leads. A modified Wheatstone bridge measures the resistance, from which the temperature can be calculated. To reduce interference and thus obtain better readings, the leads from superior RTDs will be shielded. RTDs tend to be installed between the top and bottom coils in slots.

The thermocouple (TC) makes use of the property that when two dissimilar metals are welded together at a single point, and the metal leads are brought away from the heat source, a voltage will be induced that is proportional to the temperature at the weld. The output of the TC is two wires composed of the two different metals (copper and constantan are the metals often used for machine TCs). As with RTDs, noise immunity is better if the two leads are shielded. TCs tend to be used for measuring water- and hydrogen-cooling channel and core temperatures in large turbine generators.

13.1.2 Rotor Winding Sensors

The rotor winding temperature sensor is usually the winding itself. For synchronous machine rotor windings that employ slip rings, the voltage and current applied to the slip rings can be accurately measured. This allows the calculation of the rotor winding resistance from Ohm's law. As discussed in Section 12.3.2 and Equation (12.3), the resistance of the rotor winding will vary with temperature. With an initial calibration, the average rotor winding temperature can be inferred from the rotor current and voltage measurement. Most manufacturers can then provide a rotor temperature indication for synchronous machines with slip rings. This average temperature will not be indicative of the temperature of hot spots; for example, where there are local blockages of cooling gas channels. The indication is most useful in detecting rotor winding overloading or general malfunctions in the machine cooling system. As discussed in Section 13.6, an apparent but false reduction in rotor temperature will be detected if shorted rotor turns occur. Such shorted turns reduce the DC resistance of the rotor winding, which is interpreted as a phantom temperature reduction.

Temperature sensing of rotors with brushless excitation systems or induction motor rotors is not easily achieved. The problem is bringing the temperature signal off the rotor, in the absence of any slip rings. Also, the very high rotational forces imply that the temperature sensors must be very mechanically robust. Point temperature sensors for rotor windings have been developed for research or special commercial applications. These sensors tend to use a TC from which the output voltage is converted to a digital signal and broadcast off the rotor using a radio frequency transmitter [13.4]. These sensors are point temperature sensors. Careful thought is needed for their placement. They are rarely employed in normal motor or generator applications.

13.1.3 Data Acquisition and Interpretation

In the past, a machine equipped with temperature sensors may have had one or two of them connected to an alarm, to warn operators when very high temperatures were occurring in the machine. Although operators may have manually recorded temperatures on a periodic basis, rarely was this information used for diagnostic purposes.

The introduction of computer-based SCADA, computer-based protective relaying, and computer-based machine data acquisition systems facilitated the direct and continuous monitoring of all the temperature (as well as vibration, coolant flow, etc.) sensors in a machine.

More importantly, this data can be continuously archived in a computer database and stored with the ambient temperatures, voltage, current, and other relevant operating data taken at the same time.

Once the data is stored in a computer file, relatively simple programs can enable dynamic temperature alarming. That is, instead of the high-temperature alarm being set based on the maximum measured temperature rise for the hottest day of the year with the machine at full load, the alarm temperature is determined from minute to minute, based on ambient temperatures and the present stator/rotor current. With an initial calibration, and knowing that the winding temperatures are approximately proportional to the current squared, a temperature alarm can be raised even if the machine is operating at 10% load. This allows problems such as coolant channels that were inadvertently blocked during maintenance to be found at low load, and before major winding damage may result.

In addition, from a maintenance planning point of view, the database can be explored to determine the trend in winding temperatures over the years under constant operating load and ambient conditions. If there is a consistent increase in temperature under identical operating and ambient conditions over the years, then it may indicate one of the following problems:

- Gradual blockage of the heat exchangers or coolant ducts in the rotor or stator. If the winding temperature and coolant temperatures increase under constant operating conditions by more than a few degrees, then it may be wise to plan an inspection and cleaning. The use of the temperature monitoring in this way is a classic example of enabling the planning of maintenance, in this case cleaning, based on need, rather than an assumed time interval for the machine to become fouled.

- Another cause for a gradual temperature increase from sensors embedded in stator slots, especially if the coolant temperatures are relatively constant, is delamination of the groundwall insulation and/or strand shorts. As the insulation delaminates, the thermal impedance between the indirectly cooled stator conductors and the core will increase (see Section 1.4.3). Consequently, the copper temperature will rise, increasing the measured slot temperature.

- If there is an erratic variation in temperature of more than a few degrees over the years (under constant operating conditions), then one possible cause is voltage imbalances between the phases, which tend to vary as plant load requirements shift or the energy supplier has power quality problems (Section 8.1).

The key requirement for monitoring is the economic archiving of all the temperature and operating data. Most machine OEMs and protective relay manufacturers can easily provide the data to be off-loaded to a computer. Data acquisition systems (sometimes called scanners), both general purpose and specific to machines [13.5], are also widely available.

13.1.4 Thermography

Infrared imaging cameras are widely used in plants to find overheated electrical joints and cables, hot spots in boilers and process fluid pipes, and hundreds of other applications. Such cameras are sensitive to the infrared portion of the electromagnetic spectrum. Since hot objects give off infrared energy in proportion to its temperature, the normal temperature of the surface of apparatus can be easily assessed, and the location of any abnormally hot areas determined. Modern thermal imaging cameras contain computers to directly indicate the temperature of the surfaces, as well as to provide the ability to compare past images with the present image, enabling changes in the thermal image to be easily detected.

As applied to motors and generators, thermography has limited use, since key areas cannot be observed when the machine is in operation. However, gross problems such as severe cooling system blockages can be detected.

If only the machine surface can be observed, then comparison of the present image with past stored images with the machine operating under the same load and much the same ambient temperature will give a general indication if the machine is getting hotter. This is particularly true for totally enclosed fan-cooled (TEFC) enclosures for which all the heating losses in the stator and rotor are dissipated on the outer surface of the frame. The first image is not likely to be informative, and any temperature differences over the surface are probably a natural result of how the cooling system operates. But if the images are stored over the years, and if more recent images show most of the machine surface and outlet cooling air to be getting warmer, then this is most likely due to general blockage of the cooling system. This can occur if the machine is open-ventilated and the stator and rotor core ventilation ducts and/or filters have become clogged with dirt. In machines with heat exchangers, the cooling channels may have become obstructed. Because of natural changes in the machine surface, which can affect the measurement, temperature differences should be consistently about 5°C higher than in the past (under the same load and ambient conditions) to be considered significant. If uncorrected, thermal deterioration of the insulation will occur.

In comparison with past stored images of the entire machine, if one particular area is showing a greater temperature difference to the rest of the machine than it has in the past, there may be several causes:

- Local blocked cooling passages in the stator, or perhaps only one heat exchanger is becoming clogged
- A phase-to-phase voltage imbalance, resulting in one phase running hotter than the others

Thermography can also be used to ensure that the connections from the power system to the machine are well made. If a general increase in termination box temperature is occurring over the years (or after a maintenance outage), then the connections may not be properly torqued, or they may be oxidizing. However, the greatest sensitivity is obtained if the termination box can be opened and the thermal imaging camera can directly see the connections. If one connection is hotter than the other two, or if there is a significant temperature increase (more than 5°C under the same load and ambient conditions) from past readings, then the connections may be degrading. An off-line conductance test (Section 12.3) would confirm this.

13.2 CONDITION MONITORS AND TAGGING COMPOUNDS

Condition monitors, sometimes also called core monitors, are essentially smoke detectors. Since they are most frequently applied to large generators, they are called generator condition monitors (GCM); or core monitors since they were first developed to detect stator core problems. GCMs have been primarily applied to hydrogen-cooled machines to detect severely overheated insulation within the generator. Sometimes, condition monitors are used with "tagging compounds," which are special paints that emit characteristic chemicals when exposed to specified high temperatures. The compounds, when released, can trigger the condition monitor and provide a much better indication of the temperature and location of a hot spot.

Recently, GCMs have been applied to totally enclosed air-cooled machines to detect burning insulation.

13.2.1 Monitoring Principles

GCMs. The first condition monitors were developed in the 1960s specifically to detect burning of the stator core lamination insulation in large hydrogen-cooled turbine generators (Section 10.1). When overheating occurs inside the generator, any organic material affected, such as epoxy insulation, will thermally decompose to produce a great number of particulates ("smoke") the size of condensation nuclei (0.001 to 0.1 μm). These are readily detected by the GCM since, under normal operation, particulates of this size should not be present in the hydrogen gas [13.6]. The condition monitor is able to detect burning organic material, whether on the rotor, the stator winding, or in the stator core.

The condition monitor is installed in the recirculating hydrogen gas stream as close to the generator as possible to keep piping lengths short. This is essential since the rotor fan differential pressure, which is not substantial, is used to produce the flow of hydrogen through the GCM. Because GCMs are fully integrated with the hydrogen cooling system, with the consequent safety concerns, GCMs tend to be very expensive.

In the GCM, the hydrogen gas flows through an "ion chamber" where it is bombarded by a weak nuclear radiation source. This produces a large number of hydrogen ion pairs in which the negative ions are attracted to a collector electrode. The current produced is amplified and displayed as a percentage on the GCM front panel meter. Changes in gas flow, pressure, and temperature also affect the detected current flow so these must be maintained constant.

When particulates are introduced in the ion chamber with the hydrogen gas, the current will decrease. This is because some of the ions become attached to the particulates, greatly increasing the mass-to-charge ratio. Since the mobility of the particulates is less than the ionized hydrogen, fewer ions are collected per second, resulting in reduced current. Therefore, the reduction in current is proportional to the number of particulates produced and, hence, the extent of overheating. The output of the GCM is in percent of current flow. If the output is 100%, no particles are present and the organic insulation material in the machine is presumed to be healthy. If the current falls to 50%, some particles are present and possible burning of the insulation somewhere in the machine is indicated. The GCM output is usually continuously plotted on a graph of percent of current versus time, or transmitted to the plant computer for archiving.

Most modern GCMs have a preheated ion chamber to vaporize any oil mist, which can be present in the hydrogen atmosphere. This is done to eliminate nuisance alarms due to oil mist rather than particulate emission from overheating. The theory is that if the ion chamber is heated only sufficiently to gasify the oil, the temperature required (roughly 160°C) is not enough to gasify any particulates from actual overheating, and only true alarms will be received. Some manufacturers of GCMs claim that hot ion chambers desensitize the GCM by gasifying real particulates from overheating. It seems that desensitization may take place with hot ion chambers, but this can be compensated for by changes in the alarm levels.

The introduction of large air-cooled gas turbine generators and ever-larger hydrogenerators has led to the reengineering of the GCM for application in air-cooled machines [13.7]. The basic concept is the same as for hydrogen-cooled machines, with the exception that the system is installed in the air-cooling system in a totally enclosed fan-cooled machine. Instead of an ion chamber, which is not sensitive enough, a "cloud chamber" detection method is usually employed. The system has the potential to detect burning anywhere in the machine, unlike thermal monitoring (Section 13.1) that can only detect high temperatures where the sensor is located.

Tagging Compounds. Most insulation will not produce particulates (smoke) until they are at very high temperatures. In rotor and stator windings, if the groundwall insulation is at such a high temperature, it is likely that a ground fault will follow within minutes or hours. This is in contrast to stator cores, where the core lamination insulation can burn locally but this does not imply that the machine will fail even within the next few years, if at all (a few shorted laminations are not hazardous to the machine). Consequently, GCMs are usually not able to give much warning of overheating problems in the rotor and stator windings.

To overcome this limitation, tagging compounds were developed in the 1970s to be used in conjunction with GCMs. Tagging compounds are applied as a paint to any desired surface in a hydrogen-cooled or totally enclosed air-cooled machine. Usually, several different types of tagging compounds will be used in the same machine. When the surface heats up to a specified temperature, particles with unique chemical signatures are released into the cooling air or hydrogen [13.8]. These particles will normally trigger the GCM alarm, since they have the same particle size as smoke. Some types of GCMs have the ability to divert the hydrogen or air stream containing these particles through a filter, which can trap the particles for later analysis. When convenient, the filters are removed from the GCM and the particles subjected to chemical analysis, either with a gas chromatograph (GC) or a mass spectrometer (MS). The chemicals in the tagging compounds are easy to identify in the GC or MS. The idea is that if these chemicals are detected on the filter, the surface that was painted with them reached the temperature at which the chemicals are released. Thus, not only can the temperature of the winding be measured, but the location of the hot spot can be determined, as long as one knows where each type of tagging compound was painted.

The tagging compounds are usually chlorine compounds encapsulated within microspheres. When the microsphere exceeds a specified temperature, it bursts, releasing the chemicals. The critical bursting temperature is higher than normal for the winding, but much lower than the melting or ignition temperature of the insulation. Usually, there are many slightly different chemical compounds, each with its own unique chemical signature. Thus, if compound A is painted on the stator core, compound B on the rotor wedges, compound C on the rotor copper, and compound D on the stator end-winding, and compound C is detected in the GCM filter, then we know the copper conductors (and the adjacent insulation) reached a specified temperature.

Although effective, because a GCM is required and since the compounds usually have to be analyzed by the vendor, this temperature measurement approach is reasonably expensive and, thus, used only in very large generators where there is already some belief that overheating problems are possible. Many compounds have a fixed life. Since the chemicals can eventually diffuse out of the microsphere, if a high temperature occurs many years after compound installation, there may be insufficient compound left to trigger the GCM or allow analysis. How long the compound will be useful depends on the temperature of the surface the compound is painted on. If the surface is consistently near the "trigger" temperature, then the life will be shorter. The compounds should be replaced every 5 years or so. Unfortunately, the compounds can only be replaced during an outage, when the rotor is pulled.

13.3.2   Interpretation

Interpreting GCM readings, in some sense, is very straightforward. But since the typical output is an alarm to the operators, there have been many misunderstandings of GCM readings.

Traditionally GCMs have produced an alarm when the current dropped below about 40% of normal. If an alarm occurred and the cause of the alarm was in the stator core, often the alarm would apparently clear itself after a few hours. Many operators believed this was an indication of a "false alarm," that is, the GCM was not working properly. In reality, the GCM behaved as specified and detected the burning of the insulation. The operator was in error in not understanding that an alarm can naturally clear itself, and that a GCM alarm will often not result in a generator failure in the near future. Similar "false alarms" sometimes can occur if a surface within the machine is very hot and oil drops on the surface. The oil may "mist," that is, create many small droplets that can trigger a GCM alarm. This alarm also does not indicate that failure is imminent, but does indicate that there are hot surfaces within the machine.

The plant operator should only be concerned with the GCM output if there is a sustained, very severe drop in current (say, to less than 20%), which indicates that a sizeable amount of insulation is creating smoke. In all other situations, maintenance personnel should be responsible for collecting and analyzing GCM data.

Without the aid of tagging compounds, the GCM is most useful for finding hot spots in the stator core. If the lamination insulation degrades by any of the mechanisms described in Chapter 10, then shorted laminations will occur. These shorted laminations will allow circulating currents to flow. If sufficient laminations are shorted, the current flow is such that the steel laminations may weld together and become hot enough to burn the insulation in the local area, triggering the GCM. After a few hours, the insulation in the affected area will be all burned away, and the GCM returns to a normal reading. As the insulation in other areas of the core degrades, leading to shorts, other burning will occur. The result is that the GCM yields bursts of activity over days, months, and even years. Eventually, enough lamination insulation will be compromised that a failure occurs due to very high local core temperatures. Thus, repetitive bursts of GCM activity over a sustained period are an indication of core lamination insulation burning. For many core lamination problems, a burst can sometimes follow a situation in which the machine is overexcited.

If tagging compounds have been installed, then the particulate samples should be analyzed once a GCM alarm occurs. As discussed above, the chemical analysis, together with information on where each compound was installed, will indicate the location and the temperature.

13.3 OZONE

Ozone, or O_3, is a gas that is a by-product of partial discharge in air. PD in the air, that is, on the surface of the stator coils or bars, occurs as a result of loose windings, semiconductive or grading coating deterioration, and/or insufficient coil spacing in the end-winding (Sections 8.5, 8.6, and 8.11). Devices have been developed that can measure the ozone concentration and, thus, detect these problems during normal operation of the machine. Ozone monitoring is only relevant to air-cooled machines rated at 6 kV and above.

13.3.1 Monitoring Principles

Two main methods are available to measure the ozone concentration. The cheapest method uses inexpensive gas analysis tubes, which are sensitive to ozone. These tubes, available from chemical supply houses, contain a chemical that reacts with the ozone. When the tube is broken open, a chemical inside the tube changes color and the approximate ozone concentration in the surrounding air can be read. The tubes come in different ranges of ozone concentration. To do a test, the tube is broken open by an operator in the stream of the air exhaust from an open-ventilated motor or generator. The test should be repeated about once every six months.

Ozone tubes cannot be used with totally enclosed machines. Persons using these tubes should be reminded that ozone is a health hazard, and many jurisdictions insist that personnel not be continuously exposed to ozone with concentrations higher than 0.1 parts per million (ppm).

A second technique uses an electronic instrument that can measure the ozone concentration continuously [13.9]. A sensor is placed within the machine enclosure or in the exhaust air stream. The measuring instrument can be located remotely from the machine; thus, this method can be employed in both open-ventilated machines as well as totally enclosed machines. The instrument continuously measures the ozone concentration.

The ozone produced by a stator winding is affected by many factors, such as

- Operating voltage
- Air humidity
- Power factor in synchronous machines. As discussed in Section 13.4, slot discharges can be affected by the ratio of reactive to real power, since it affects the forces acting on the coils/bars.

Thus, the trend in ozone concentration is only meaningful if the readings are taken with the above factors held constant.

13.3.2 Interpretation

As with the electrical techniques to measure the partial discharge activity on stator windings, it is best to monitor the ozone in a machine over time. If there is a persistent increase in the ozone concentration, then other tests or a visual inspection should be planned. The ozone level depends on the magnitude of the surface PD, as well as the how many PD sites there are. Thus, like the tip-up tests (Sections 12.6 and 12.9), ozone measures the average surface PD activity. The monitor is not sensitive to one severe PD site.

In open-ventilated machines, the concentration is high if it is near or exceeds 1 ppm. Concentrations of 1 ppm are immediately identifiable by ozone's characteristic odor. Some open-ventilated machines have had ozone concentrations in excess of 4 ppm [13.10]. With such a high level, there is usually widespread PD activity on all the coils/bars operating at high voltage in the stator winding. If a high level of ozone is measured, the operator should ensure that it is not due to a high concentration in the general environment by repeating the test remote from the machine. The normal ozone background of up to 0.1 ppm or so sets a lower limit for detecting winding problems. Ozone occurs naturally after lightning storms. Also, copying and fax machines can increase the ozone background in a plant.

Ozone measurements within a totally enclosed machine will yield substantially higher levels of ozone. The ozone concentration not only depends on the rate of generation by the surface PD, but it also depends of the rate of ozone consumption. The latter depends on the surface area of materials that react with the ozone, such as uncoated steel and rubbers. This is highly variable from machine to machine. Thus, it is virtually impossible to state the level of ozone below which the winding is not suffering from excessive PD. The method for interpreting results is to note that if there is an increasing trend in ozone concentration over time (under the same machine operating conditions), then the rate of insulation aging is increasing. In totally enclosed machines, the rate can easily increase dramatically (10 to 100 times) in the first few years, but tends to saturate as equilibrium is achieved between creation and consumption of the ozone.

Semiconductive coating deterioration (Section 8.5) tends to produce the most ozone, even though this is a relatively slow failure process. Loose coils in the slot (Section 8.4) and inadequate end-winding spacing (Section 8.11) can produce moderate levels of ozone. Silicon carbide coating deterioration (Section 8.6), although creating very large PD, produces relatively little ozone, since the number of pulses and the number of possible PD sites are few. Consequently, higher levels of ozone do not imply a shorter time to failure.

Internal PD within the groundwall insulation does not produce measurable ozone. Thus, ozone monitoring is better for epoxy–mica windings than for the older asphaltic, shellac, or polyester–mica splitting types of winding.

13.4 ON-LINE PARTIAL DISCHARGE MONITOR

This monitor directly detects stator winding partial discharge electrical pulses during normal operation of the motor or generator. In most respects, it is very similar to the off-line PD test discussed in Section 12.10. On-line PD monitoring can be done on a periodic basis, typically every six months. In addition, continuous on-line PD monitoring can also be performed. Since partial discharges are a symptom, or cause, of about half the stator failure processes discussed in Chapter 8, on-line PD monitoring is a powerful tool for assessing the insulation condition in form-wound stators. More recently, on-line PD testing has also been applied to random-wound motors supplied by PWM-type inverter drives.

13.4.1 Monitoring Principles

As discussed in Section 12.10, every partial discharge creates a small current pulse that will propagate throughout the stator winding. Since the pulses are typically only a few nanoseconds in duration, using a Fourier transform, each pulse creates frequencies from DC to several hundred megahertz [13.11]. These electrical pulses are detected and processed in several different ways by the various PD monitoring systems.

In virtually all systems, the following elements are found:

1. Sensors, such as antennae, high-voltage capacitors on the machine terminals, and/or high-frequency current transformers at the machine neutral or on surge capacitor grounds are needed to detect the PD. These sensors are sensitive to the high-frequency signals from the PD, yet are insensitive to the power frequency voltage and its harmonics. Most sensors have an inherent wide range of frequency sensitivity, usually from a few kilohertz, and sometimes ranging up to 1 GHz.

2. Electronics convert the pulse signals from analog to digital form. By far the most common approach is to use pulse magnitude analyzers. More recently, pulse phase analyzers also digitally record where the PD pulses occur with respect to the power frequency AC cycle. In older systems, the signals were directly displayed on an oscilloscope or spectrum analyzer.

3. Signal processing techniques are used to reduce the information to manageable quantities and/or help discriminate PD signals in the winding from electrical noise to ensure more reliable interpretation. In addition, signal processing can perhaps be used to determine which type of insulation failure mechanism is occurring.

All of the existing PD measurement technologies available today rely on enhancing one or more of the above elements to implement on-line PD monitoring. The following first describes the sensors employed, and then describes the monitoring systems.

PD Sensors. The first sensor used for on-line PD measurement was a high-frequency current transformer (HFCT), sometimes called an RFCT [13.12]. The HFCT usually has a ferrite core, around which 10 to 100 turns are placed. The bandwidth of most commercial HFCTs that have a large enough opening is from about 100 kHz to 30 MHz, into a 50 ohm load. The HFCT, as originally used, was placed around the cable connecting the generator neutral point to the neutral grounding transformer or impedance. Each PD, which originates in the coils/bars operating at high voltage, propagates through the winding and generates a small current pulse in the neutral grounding current, which is detected by the HFCT [13.12, 13.13]. On motors equipped with surge capacitors (Section 1.4.2), the HFCT can also be placed on the ground lead from the surge capacitor to machine ground (Figure 13.2). Since the HFCT on surge capacitors is adjacent to the coils likely to have PD, much larger PD signals are detected in this location compared to the HFCT located at the neutral. HFCTs cannot be placed around power cables to the stator except in the smallest of machines since HFCTs do not have a large enough opening and the ferrite core of the HFCT will tend to saturate. Air core HFCTs (Rogowski coils) can be placed around the power cables but suffer from very low sensitivity since there is no ferrite core to concentrate the magnetic field from the PD current pulse.

The most popular sensors used in on-line PD monitoring are high-voltage capacitors installed on each phase terminal. As in the off-line PD test, the capacitor blocks the 50 or 60 Hz high voltage while serving as a low-impedance path for the high-frequency PD pulses. The most common capacitor rating is 80 pF; earlier users employed capacitors from 375 pF to 1000 pF, since these were the most common capacitors used with laboratory PD detectors. In general, the capacitors feed a 50 ohm load. Thus, the basic bandwidth of the detection system is formed from the high-pass filter of a capacitor in series with a resistor. An 80 pF capacitor detects PD pulse frequency content above 40 MHz. A 1000 pF capacitor detects signals above 3 MHz. In the absence of electrical noise, and assuming the capacitors are located near the coils/bars most likely to experience PD, either capacitor size can detect the PD. If noise is present, communication theory shows that the signal-to-noise ratio is larger with the 80 pF capacitor [13.11]. In addition, the smaller capacitance implies that a much thicker dielectric can be employed in the same volume of sensor, greatly reducing the risk of capacitor failure.

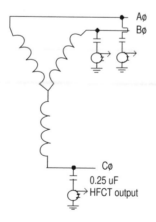

Figure 13.2. Placement of HFCT sensors to measure PD.

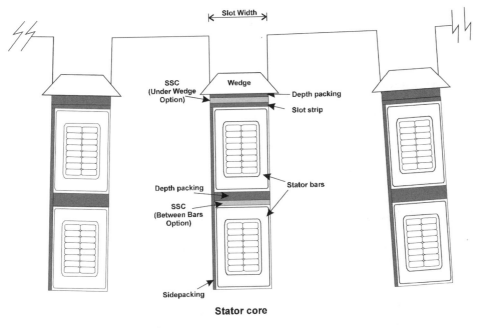

Figure 13.3. SSC location either under a wedge or between the top and bottom coils/bars.

Unlike the HFCT, the capacitor is connected to the high-voltage terminal, and it will cause a ground fault if it fails. Thus, capacitor reliability is crucial [13.14].

Various forms of antennae installed within the motor or generator enclosure are also employed to detect PD on-line.* The most popular "antenna" is called the stator slot coupler (SSC). It is a strip line directional coupling antenna that is usually installed under the wedge (Figure 13.3) in a stator slot containing coils or bars operating at high voltage [13.15]. The SSC directly detects the electromagnetic wave from PD that travels along the slot. The SSC has two outputs, so that it can directly distinguish between PD originating in the end-winding and PD originating in the slot, based on the direction the pulse is traveling. This type of sensor is primarily used in large, hydrogen-cooled turbine generators. SSCs have a bandwidth from 10 MHz to 1000 MHz.

Over the past decades, RTD and thermocouple temperature sensors have occasionally been used as PD sensors [13.16]. The PD is usually detected not by the temperature sensor but, rather, by the cable from the sensor, through antenna-like coupling. Unlike all the PD sensors discussed above, these sensors do not need to be installed in machines, since they are already in the stator slot in some machines. This is a major advantage. However, in spite of this, the sensors tend to fall out of favor a few years after each reintroduction by a different vendor. The primary reason is the difficulty in interpreting results. The RTDs/TCs are rarely in the slots containing high-voltage coils, especially in machines rated 4.1 kV

---

*Antennae installed outside of the machine enclosure will not detect PD, unless there are large openings in the enclosure. Faraday's law prevents the radiation of RF signals outside of a metallic enclosure.

and below.[†] Since the PD magnitude is extremely dependent on the voltage, the largest PD pulses usually only occur at the phase end. Consequently, the PD signal will propagate along the winding until a slot containing an RTD is encountered. Since the propagation path can be from a few coils to a dozen coils, the signal detected at the RTD can easily vary over an order of magnitude for the same initial PD. Thus, the detected magnitudes are meaningless unless one does a teardown of each machine and then uses pulse injection tests to establish the relative sensitivity of each RTD. As discussed later, the only means for interpretation is trending, which may take years of data to compile.

Conventional capacitors, HFCTs, or antennae cannot be used directly for on-line PD measurements of motors driven by IFDs. Since the frequency content of the IFD surges and the PD are similar, the PD can only be detected with very low resolution. A proprietary coupling method has been developed for this on-line PD monitoring application [13.17].

Electrical Interference. The main difficulty with on-line PD monitoring is coping with the electrical interference (noise) that is inevitably present. In the off-line PD test, there is little interference so, usually, no measures are needed to separate the noise from the stator winding PD. On-line monitoring is performed with the machine connected to the power system. There are many signals that originate in the power system that have the pulse-like nature of PD, including

- Corona from overhead transmission lines
- Poor electrical contacts on buses, which tend to spark
- Corona from electrostatic precipitators
- Sparking from slip rings in machines and power tools
- Shaft ground brush sparking
- Fast risetime transients from IFDs and computer power supplies

All the early on-line PD monitors used an oscilloscope or RF spectrum analyzer for displaying the signals from the PD sensors. With experience, a skilled test operator could subjectively separate the stator PD from the electrical interference based on frequency content, pulse shape, repetition rate, and/or phase position in the 50 or 60 Hz AC voltage cycle. With a suitably experienced person, this "subjective" PD monitor produced good results and is still the preferred method of many on-line PD monitoring service providers. When this test is done with an HFCT and a spectrum analyzer for display, it is called RF (for radio frequency) or EMI (for electromagnetic interference) monitoring.

For those desiring continuous PD monitoring, a subjective approach to separating noise from PD is not cost-effective and would likely lead to many false indications of stator winding problems. Furthermore, many machine users desire more objective data from even periodic monitoring, since expensive decisions are based on the data, and plant managers prefer something more substantial than a result that no one else could independently verify.

Two predominate methods have been developed to separate the PD from the interference. One requires the installation of at least two sensors per phase and measuring the relative time of arrival. The other is to look at each pulse and determine if the pulse shape is characteristic of PD or noise. The following presents information on how each method is implemented for different types of machines.

[†]As discussed in Section 1.4.5, such machines do not have a semiconductive coating. If the RTD or TC is installed adjacent to a phase end coil, the sensor is in a high electric field. Thus, most manufacturers install RTDs in slots containing only low-voltage coils.

Hydrogenerator PDA Monitor. The partial discharge analyzer (PDA) monitor system was developed in the late 1970s and was one of the first monitors that enabled nonspecialists to perform and interpret on-line PD tests [13.18]. The PDA monitor is intended specifically for salient pole machines (usually hydrogenerators) with a stator bore diameter exceeding about 3 m. At least two 80 pF capacitors are permanently installed per phase, at the ends of the circuit ring buses where the connections are made to the coils/bars (Figure 13.4). External noise is separated from stator PD on the basis of comparing the pulse arrival times from a pair of sensors at the PDA instrument, which in turn depends on the fact that pulses travel along the circuit ring bus at close to the speed of light: 3×10^8 m/s, or 0.3 m/ns.

A noise pulse from the power system will travel in both directions along the circuit ring bus. If the busses are the same length (which is rarely the case), the pulse will arrive at the two capacitors at the same time. With a bus length of 5 m, the pulse will arrive at the capacitors about 17 ns after it entered the stator. The coaxial cable on the low-voltage side of the capacitor will transmit the pulse to the PDA instrument. If the coaxial cables from both capacitors are the same length, the pulses will arrive at the pair of instrument inputs at the same time. Digital electronics will note the near simultaneous arrival of the pair of pulses, and digital logic will define the detected pulse as noise. (Since noise can sometimes indicate problems in the power system such as deteriorating electrical connections, modern PDAs separately display the noise.)

If a PD occurs at point A in Figure 13.4, the signal will be almost immediately detected by the capacitor near point A. Some signal also propagates around the circuit ring bus to the other sensor. The result is that the PDA detects the arrival of the pulse from the sensor at A many nanoseconds (34 ns in the example) before it detects the pulse at the other coupler. Since the pulse was detected by the PDA from sensor A well before the other sensor, the PDA digital logic determines that the pulse must be due to PD near coupler A. Similarly, a

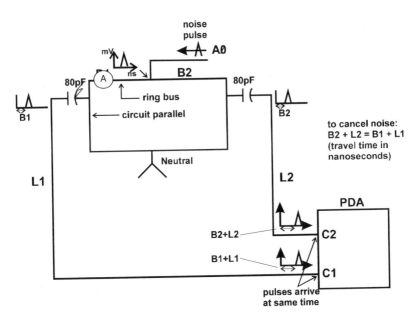

Figure 13.4. Location of capacitive couplers on the circuit ring bus of a hydrogenerator.

stator PD pulse occurring near the other sensor will be detected first by that sensor, and the PDA can determine that PD occurred in that portion of the winding. Thus, by measuring the relative time of arrival of pulses at the pair of sensors in a phase, the pulse can be identified as external noise or PD near each of the sensors. Since sequential PD pulses are usually many tens of microseconds apart and the two detected pulses arrive at most 40 or 50 ns apart, two different PD pulses are unlikely to confuse the instrument.

The time of arrival method separates stator PD from external noise, but it is still susceptible to "internal noise" from within the machine. The most likely source of this is slip ring sparking occurring on the rotor. Although such sparking initially has a fast pulse risetime at the slip rings, by the time the pulse is coupled across the air gap into the stator, its risetime has degraded. Thus, by measuring the risetime, "internal noise" can also be separated.

Motor and Turbine Generator Monitoring.

Most motors, most small turbine generators, and some hydrogenerators with "wave" windings have circuit ring buses that are too short to ensure reliable separation of noise and PD on the basis of pulse arrival times. In such cases, it has been found effective to permanently install a pair of 80 pF capacitors per phase on the bus connecting the machine to the power system (Figure 13.5). One capacitor is normally installed per phase as close as possible to the stator winding terminals, usually in the termination enclosure. The other capacitor in the pair is installed at the potential transformer cubicle (if present), or the circuit breaker.

A variation of the time of flight principle is used to distinguish PD from external electrical interference. A variation of the PDA, called the TGA-B, records the pulses from a pair of sensors [13.15]. A PD pulse from the stator will arrive at the M input of the instrument before it is detected by the capacitor further from the machine. Similarly, an external noise pulse will trigger the S input of the instrument before it is detected at the M input. Thus, digital logic can determine which pulse arrives at what input first if the pulse is external noise or from stator winding PD. In fact, it is also possible to determine if PD or noise is occurring on the bus between the capacitors.

Research with this system showed that it is often unnecessary to install the set of sensors at the switchgear if the connection between the machine and the switchgear is a power cable

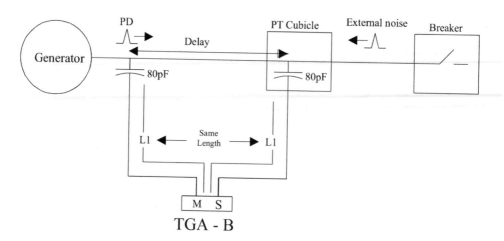

Figure 13.5. Installation of two capacitors per phase to separate PD from noise.

longer than about 30 m. It seems that fast risetime noise pulses will be both attenuated and distorted as they travel along the power cable to the machine sensor. The distortion involves the slowing of the pulse risetime: stator PD pulses will have a fast risetime, whereas noise that comes through the power cable will have a slower risetime. Thus, circuitry can separate the noise from the stator PD on the basis of risetime alone. Consequently, most motors, which often have power cable lengths in excess of 30 m, usually only require one sensor per phase. Generators usually have only short lengths of bus in the connection to the power system, and will not see sufficient pulse distortion along a short bus. Thus two sensors per phase are needed for most generators.

The pair of capacitive couplers mounted on the output bus has yielded false indications of stator winding problems in turbine generators rated at more than about 300 MW. The problem was high levels of relatively fast risetime pulses that were not associated with stator PD, but rather seemed to be associated with stator core lamination sparking. The solution was the development of the stator slot coupler (SSC) sensor [13.15]. Core sparking was found to yield slower risetime, highly oscillatory pulses from the SSC, as compared to the few nanosecond, unipolar PD pulses that are caused by PD in the slot or immediate endwinding. The SSC sensor only reliably detects PD in the slot it was installed in. A version of the TGA determines the pulse characteristics of each pulse to digitally discriminate between PD and all types of noise.

Data Acquisition. The PD sensors are permanently installed in the machine during a convenient machine shutdown. In most on-line PD monitoring systems, portable instrumentation is used to record the signals periodically. The measurements typically take about 1 hour. With the monitoring methods that use modern digital oscilloscopes and spectrum analyzers, permanent records can be made of the signals. Experts, at the time of the measurement or at a later time, use their knowledge to determine what the true PD activity is, as opposed to the noise. Usually, these types of measurements are performed by specialized staff from test service organizations or machine manufacturers. Similarly, portable PDA and TGA instruments can also be used to measure the PD activity from appropriate sensors on a regular basis. In this case, since noise is separated objectively from PD on a pulse-by-pulse basis, plant technicians can perform the testing.

For machines rated at 6 kV or more, experience indicates that the monitoring should be done at least every 6 months. With this measurement interval, PD signals can be detected 2 years or more before serious insulation deterioration with a consequent high risk of failure has occurred. For lower voltages, the warning time may be less—as short as a few weeks with the thermal deterioration process in 3 to 4 kV stators. In this situation, continuous PD monitoring is more cost-effective.

The PD activity in a machine strongly depends on several factors and is somewhat dependent on other factors. These include

- Stator voltage (increasing voltage strongly increases PD)
- Gas pressure (increasing hydrogen pressure strongly decreases PD)
- Humidity (increasing humidity will decrease any PD due to electrical tracking)
- Winding temperature (if the groundwall is delaminated, then as the temperature increases, the PD decreases as the internal voids shrink in size due to the components expanding)
- Machine load (if the coils/bars are loose in the slot, increasing load will increase magnetic forces on the coils/bars, increasing PD)

Thus, to obtain trendable results, all the above factors should be duplicated from test to test. Voltage and pressure are the most important factors to keep constant from test to test.

Continuous PD monitoring systems are now available that can ensure that PD is detected with the maximum amount of warning time. With suitable software, such continuous monitoring systems also ensure that the PD data is collected under the same machine operating conditions [13.19]. Such systems also enable the remote reading of PD results, with appropriate communications from the plant. This can significantly reduce monitoring costs in plants in remote areas.

On-line PD monitoring tends to produce high PD activity on newly installed windings, for two reasons. The epoxy or polyester in new windings is usually not fully cured when the stator is first operated, but gradually completely cures in its first few months of operation. Since partly uncured resin has a higher dielectric constant than cured resin, this will impose a higher stress on any small voids within the groundwall, resulting in greater PD activity until the resin is fully cured (Section 1.4.4). For nonglobal VPI stators with coils/bars having a semiconductive coating (Section 1.4.5), the coating may be isolated from the core in some locations along the slot. The isolation may be due to oxide on the core or insulating varnish on the surface of the semiconductive coating. The result is that the coating is not fully grounded along the slot and, under normal magnetic forces, may cause sparking as the coating makes and breaks contact with the core. Eventually, the coil/bar vibration will abrade any oxide or varnish coating, fully grounding the coating and preventing the sparking. Experience indicates that results from on-line PD monitoring taken within the first six months or so after installation of a new winding should not be immediately taken to imply that there are winding problems.

13.4.2 Interpretation

The PD monitoring instrumentation will record all the PD pulses that are detected. The earliest on-line monitors usually displayed the PD pulses on an oscilloscope with respect to the 50 or 60 Hz AC cycle (Figure 12.3). More recently, pulse magnitude analyzers enable the PD to be digitally recorded as a plot of the number of pulses versus the pulse magnitude (Figure 13.6a). Depending on the PD sensor and the recording apparatus, the magnitude can be measured in picoCoulombs, millivolts, microvolts, milliamps, or decibels. The two-dimensional graph in Figure 13.6a shows the number of PD pulses recorded per second at each magnitude interval. Modern PD monitors also display the pulse magnitude analysis as a function of the AC phase position (Figure 13.6b). Since both positive and negative PD pulses are recorded (Figure 12.3), the electronic pulse counters tend to record and display the two polarities separately (Figure 13.6a).

The fundamentals of on-line PD interpretation are given in IEEE Standard 1434 [13.20]. As mentioned in the section on off-line PD testing, the key result to interpret from any on-line PD monitor is the peak partial discharge magnitude. The highest PD magnitude recorded, sometimes referred to as Q_m, is important since the PD site associated with the peak PD will generally have the greatest volume of deterioration. That is, Q_m occurs at the most deteriorated site of the winding. It is most likely that if failure is to occur, it will occur at this site. There are many other quantities that can be derived from the PD data, but most of these are associated with the average condition of the winding, rather than the condition of the worst site. Since PDs are somewhat erratic, IEEE 1434 has defined Q_m to be the magnitude associated with a PD pulse repetition rate of 10 pulses per second (pps), as recorded by a pulse magnitude analyzer.

In principle, interpretation is comparative. That is, it is not generally possible to specify an acceptable level of Q_m or a level of Q_m at which there is a high risk of failure. The reasons

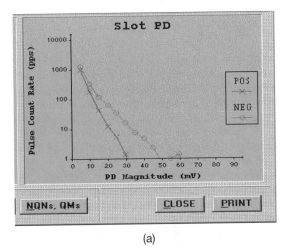

(a)

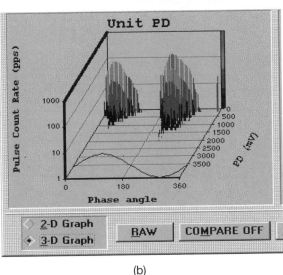

(b)

Figure 13.6. (a) Pulse magnitude and (b) pulse phase analysis plots from on-line PD detection. (Courtesy of Iris Power Engineering.)

are complex but relates to the inductive, capacitive, and transmission line nature of a stator winding, as well as the fact that PD is often only a symptom of the failure process, not a direct cause [13.21]. However, meaningful interpretation can occur by

- Trending the Q_m on the same machine over time, using the same monitoring method
- Comparing Q_m from several identical machines, using the same monitoring method
- With some restrictions, comparing Q_m from several "similar" machines, using the same monitoring method

The following assumes that the electrical interference has been separated from the stator winding PD.

Trend Over Time. This is the most powerful means of interpreting PD data, no matter which detection method is used. The idea is to get an initial fingerprint of the PD activity, such as the plots in Figure 13.6. The initial fingerprint is best when the winding is relatively new, i.e., after 6 months or so of operation. If the winding deteriorates, the volume of the worst defects will increase, which increases Q_m. Thus, if Q_m increases over time, then the deterioration is increasing, and there is a greater risk of failure. Doubling of Q_m every 6 months is an indication that rapid deterioration is occurring and, eventually, off-line tests or a visual inspection of the winding are warranted.

PD does tend to saturate after a strong increase over time; that is, the PD may increase rapidly over several years. But if the deterioration is significant, Q_m and other PD activity indicators tend to level off and stop increasing. This has implications for older machines that have been retrofitted with PD sensors. It may be that the winding is already in the "saturated" state, with no increasing trend in Q_m. Thus, without the initial PD data when the machine is in relatively good condition, one may not know that the winding may already be in poor condition. Consequently, although trend over time is the most powerful interpretation diagnostic, comparisons to other machines are still needed to ensure that saturation has not occurred.

Anecdotal observations have shown that if very high groundwall PD has occurred for many years, the trend is flat, and then Q_m plummets to 20% of previous measurements, winding failure may be imminent. Researchers believe the reason for the drop in PD before failure is that so much of the organic material has been degraded by the PD that the carbon formed tends to "short" out the voids, reducing PD.

For the trend to be meaningful, the trend plots should only show data collected under the same motor or generator operating conditions.

Comparisons to Similar Machines. The first measurement on a machine should be more than the initial fingerprint. It should also provide some indication of the relative condition of the insulation. Comparison to other machines is needed in the case of a monitoring system installed on an older winding where the trend is flat.

The best comparison occurs when all machines are identical, and the measurements are made with the same monitoring system under the same operating conditions. The machine with the highest Q_m will have the most severe deterioration and, thus, be most likely to fail. Because PD is erratic, differences of $\pm 25\%$ are not significant. However, if one machine has 10 times more activity than another identical machine, then the stator with the greatest activity should be subjected to further tests and/or inspections sooner than the other (Figure 13.7).

With the installation of 80 pF capacitive sensors on many thousands of machines, and the widespread use of compatible PDA and TGA instruments to test these sensors, statistical analysis has been possible on the tens of thousands of measurements obtained [13.22]. With the use of the 80 pF sensor, which tends to respond only to the high-frequency transmission line part of the PD signal, it seems that the size (capacitance) of the winding has only a minor impact on the detected PD magnitude for the same amount of deterioration.* The predomi-

---

*If PD measurements are made at lower frequencies, then the PD pulses "see" the entire capacitance of the winding, and the larger the winding capacitance, the lower the detected PD pulse. This effect can be partly compensated for by a "calibration," but the series inductance of the winding causes resonance that affects the PD magnitude in undefined ways [13.20].

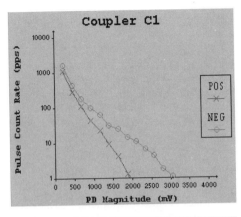

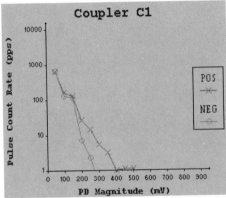

Figure 13.7. Comparison of PD activity in two stators. The stator with higher magnitude PD (on top) is most deteriorated. (Courtesy of Iris Power Engineering.)

nant characteristics that determine if machines are "similar" and thus if Q_m can be compared between them, are

- Voltage rating
- Surrounding gas pressure (that is, the hydrogen pressure in a hydrogen-cooled machine, or the altitude of an air-cooled machine)
- Monitoring system

These characteristics assume that the machine is operating at its normal full load and temperature.

For the 80 pF sensors with PDA/TGA instrumentation, Table 13.1 shows the cumulative probability of obtaining a Q_m of a certain magnitude as a function of voltage [13.22]. The table indicates that for 13.8 kV windings, 25% of the machines in the database had a Q_m of 9 mV or less, 50% had a Q_m of 79 mV or less, and 90% of measurements were less than 362 mV. Thus, if the first measurement on a 13.8 kV motor indicates a Q_m of 400 mV, the user knows that this machine has higher PD than 90% of all other similar machines. This high

Table 13.1. Distribution of Q_m for Air-Cooled Machines Using 80 pF Capacitive Couplers

Rated V	2–4 kV	6–8 kV	10–12 kV	13–15 kV	16–18 kV	> 19 kV
Average	89	88	121	168	457	401
Maximum	2461	1900	3410	3396	3548	3552
25%	2	6	27	9	145	120
50%	15	29	63	79	269	208
75%	57	68	124	180	498	411
90%	120	247	236	362	1024	912

reading would trigger further measurements and/or a visual inspection of the stator, since the stator is deteriorated in comparison to other stators.

The use of a database makes interpretation of an initial measurement both feasible and objective.

Identifying Failure Processes. If the trend or the individual reading is high, then the PD data can sometimes be further analyzed to determine the probable cause for the high activity. The first indication of possible deterioration processes comes from the "polarity" effect [13.20 and 13.23]. As shown in Figure 12.3, PD will normally produce both positive and negative pulses, resulting in positive and negative Q_m. If the ratio of $+Q_m/-Q_m$ is consistently

- > 1.5, then there is positive pulse predominance and the PD may be mainly occurring on the surface of the coils in the slot. This is an indication that the coils are loose in the slot, or the semiconductive or grading coatings are deteriorating.
- < 0.5, then there is negative pulse predominance. This indicates that the PD is occurring close to the copper conductors. Such negative predominance would occur if the coils were poorly impregnated or load cycling deterioration is occurring within the slot.
- About one, or changes from sensor to sensor, then either end-winding PD is occurring, or there is delamination of the groundwall conductors, usually caused by thermal deterioration.

Further information can sometimes be obtained by measuring the effect of various machine operating factors on the PD. As discussed above, in a machine with high activity, the PD can be affected by several operating factors such as load, winding temperature, and humidity. Although these influences make it more difficult to trend data over time, if controlled tests are done on the machine, the same influences can help to determine the likely deterioration mechanism. For example:

- If Q_m increases with increasing load (with constant temperature) then, perhaps, the coils/bars are loose in the slot
- If Q_m decreases with increasing temperature (and increasing load in many machines), then thermal deterioration or load cycling deterioration may have occurred
- If Q_m increases with increasing temperature (with load constant), then the silicon carbide coating may be deteriorating
- If Q_m decreases with increasing humidity, then PD is probably occurring in the end-winding, due to electrical tracking

If multiple failure processes are occurring, which is not uncommon on older windings, then isolating these various effects becomes difficult, even for an expert.

13.5 END-WINDING VIBRATION MONITOR

Stator end-winding vibration is an important cause of failure in two- and four-pole machines, and is a very important failure mechanism in large turbine generators (Section 8.12). End-winding vibration monitors have been developed to directly measure the presence of end-winding vibration.

13.5.1 Monitoring Principles

Vibration is detected by accelerometers, which produce an electrical output that is proportional to the magnitude of the acceleration that the sensor detects. With manipulation, this signal can also be calibrated in terms of vibration velocity or displacement. Most accelerometers are made from crystals that are enclosed in a metallic shell. Since these conventional accelerometers are metallic and operate near ground potential, they cannot be permanently installed in the end-winding of the stator. If they were, with normal pollution, it is likely that the grounded accelerometer would eventually initiate an electrical tracking failure.

End-winding vibration monitoring does not use conventional accelerometers, but rather uses fiber optic accelerometers. Such accelerometers contain no metal and are connected to the measurement instrument by means of fiber optic cable. Thus, the sensor does not introduce any change to the potential distribution in the end-winding.

Several different optical sensors have been developed. Westinghouse introduced the first in the early 1980s [13.24]. Light is transmitted along a fiber optic cable to a sensor head that contains a tiny cantilevered beam that vibrates with an intensity dependent on the displacement. The beam modulates the light, which is transmitted back to the instrument for analysis. This sensor is made resonant at 120 Hz, the main electromagnetic force frequency in a 60 Hz generator. Since this development was introduced, many types of fiber optic accelerometers have been introduced. Most of the recent versions will work over a broad range of frequencies, typically a few Hz to 1 kHz, and the output can be displayed on the same type of spectrum analyzer that is used with conventional accelerometers [13.25]. The output of many types of optical accelerometers depends on temperature, which can cause errors. This should be considered when selecting a brand of sensor.

The accelerometers measure vibration in one direction only. To obtain a proper idea of the end-winding vibration modes at any location, two sensors are mounted at each location, one to measure the vibration in the radial direction, the other to measure the vibration in the circumferential direction. A sensor for the axial position is not needed, since twice-power-frequency vibration in that direction is negligible.

Usually at least six pairs of accelerometers are installed on endwindings, three at each end. Choosing the optimum locations (i.e., the locations most likely to vibrate) for the pair of accelerometers needs careful thought. The best way to select locations is to perform the off-line impact tests described in Section 12.19. By impacting the endwinding in a variety of locations and measuring the response with temporarily installed conventional accelerometers, the locations most likely to vibrate are identified. Siting of the sensors is best done by an experienced technician.

If the monitoring system is installed in a hydrogen-cooled machine, the supplier should ensure that the fiber optic cable is compatible with hydrogen. Many optical fibers deteriorate in a hydrogen environment.

End-winding vibration monitoring systems are among the most expensive monitors available for rotating machines. To date, they are usually only installed if the user has other information indicating that failure due to endwinding vibration is reasonably likely.

13.5.2   Data Acquisition and Interpretation

The accelerometers can be read periodically or continuously monitored. The instrumentation outputs the peak acceleration, velocity, or displacement. The maximum acceptable end-winding vibration depends on the specific stator design and, thus, should be discussed with the OEM. If any of the sensors reads a displacement greater that about 0.25 mm peak-to-peak, then there is significant vibration. Displacements in excess of 0.4 mm peak-to-peak require attention in the very near future.

The trend in maximum displacement over the years is also meaningful. If the displacement is gradually increasing over the years, then this is an indication that the endwinding support system is loosening up.

13.6   SYNCHRONOUS ROTOR FLUX MONITOR

In synchronous machines, this monitor measures the magnetic flux near the rotor to determine if turn-to-turn shorts have occurred in the rotor winding. This monitor has been widely applied in large turbine generators and has found more limited application in hydrogenerators and critical synchronous motors. This is the most powerful means of monitoring the condition of round rotor windings on-line. In two- and four-pole machines, bearing vibration can also be an indicator of shorted rotor turns, although for different reasons (Section 13.10). Rotors equipped with average rotor temperature sensing (Section 13.1.2) can also detect shorted turns since a phantom lowering of the rotor temperature apparently (but not actually) occurs.

13.6.1   Monitoring Principles

A round rotor, in addition to producing the main radial magnetic flux that crosses from the rotor to the stator, produces a local field around each slot in the rotor. This fringing flux (also called a leakage flux) is caused by the current flowing through the turns in each slot (Figure 13.8). Any change in the total current within a slot due to shorted turns results in a change in the "slot leakage" magnetic flux associated with the affected slot. The first monitor to detect this ripple flux was developed by GE [13.26].

This monitoring system employs a "flux probe" and a digital oscilloscope to record the measured signals from the coil. The flux probe is usually permanently installed within the air gap to measure the slot leakage flux.* The sensor consists of a large number of turns of small-diameter magnet wire on a bobbin, with the axis of the coil oriented in the radial direction. To improve sensitivity to the leakage flux, the coil should be 2–3 cm from the rotor sur-

---

*The earliest versions of this monitoring system used a flux probe mounted on a rod that was pushed through a gland and manually inserted into the airgap during operation [13.26]. After the test, the coil was removed. Most users now opt for the permanent sensors since there is less risk of the flux probe hitting the rotor.

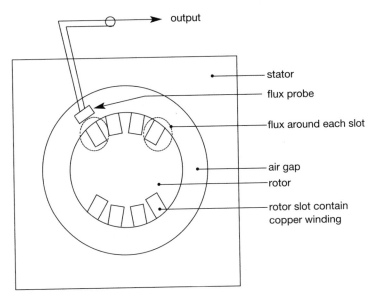

output

stator

flux probe

flux around each slot

air gap

rotor

rotor slot contain
copper winding

Figure 13.8. Drawing showing location of a flux probe in a turbogenerator airgap.

face. The flux probe is usually mounted at the turbine end of the stator core, at the 10 or 2 o'clock position (Figure 13.8) to minimize the possibility of damage to the coil when the rotor is removed.

During machine operation, the rotor flux from each slot will induce a current in the flux probe as the rotor moves past the flux probe. As each slot in the rotor passes, there will be a peak in the induced current caused by the slot leakage flux. The peaks in the current can then be recorded on a digital oscilloscope. Each peak of the waveform represents the leakage flux around one rotor slot. An interturn fault in a coil reduces the peaks associated with the two slots containing the faulted coil. Since there are fewer ampere-turns in these slots and, thus, less leakage flux, there is a reduced current induced in the flux probe. The recorded waveform data is analyzed to locate the coil(s) and the slot(s) containing the faulted turn and to indicate the number of turns faulted (Figure 13.9).

With the periodic monitoring system, signals from the flux probe are typically measured once per year or so, or when an increase in bearing vibration level has been detected (since this is also a sign of shorted turns). A laptop computer can be equipped with a digitizer to capture the signals and convert the data to spreadsheet format for analysis.

To maximize the sensitivity to shorted turns in all rotor slots, the signal from the flux probe needs to be measured under different load conditions, ranging from no load to full load.* Since the leakage flux is only one of three components of flux in the airgap (the others are the main flux from the rotor and the armature reaction flux due to the current flowing in the stator), greatest sensitivity to the leakage flux occurs when the other two fluxes are approximately zero. This occurs when there is a zero crossing of the total flux, which

*The first version of the test required the shutdown of the machine and an off-line test performed to measure the flux under stator winding short-circuit conditions. The modern version, which uses a number of different load conditions, has relatively minor impact on plant production.

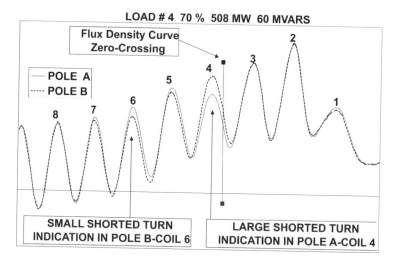

Figure 13.9. Flux probe output recorded by a digital oscilloscope. The smallest signal ripple is associated with the slot with a shorted turn. (Courtesy of GeneratorTech.)

in turn depends on the real and reactive load of the machine. The key idea is to adjust the load so that the flux zero crossing occurs for each slot pair in a pole pair. The number of load steps depends on the number of slot pairs in a pole pair. With the waveforms digitally recorded at each load step, specialized software then indicates if any specific load maneuvering is needed to line up the flux zero crossing with the slot. The software then determines the number of shorted turns in each slot and identifies the slots with the shorts [13.27]. This monitor has less sensitivity when applied to rotors with magnetic rotor slot wedges (older two- and four-pole turbine generators), since less leakage flux makes it into the air gap.

Most large turbine generator manufacturers provide the flux probes and monitoring services. There is at least one vendor of all components that provides sufficient training for users to perform and interpret their own monitoring. Continuous flux monitoring is also now possible [13.32].

13.6.2   Data Acquisition and Interpretation

Unlike in stator windings, in which turn shorts rapidly lead to failure, turn shorts in synchronous rotor windings do not in themselves lead to failure. One or more shorted turns can be sustained indefinitely, as long as the required excitation is adequate and the increased level of vibration that shorted turns produce is within acceptable limits (Section 1.5.2). However, if the shorted turns are the consequence of gradual aging, more and more shorted turns may occur over time. Sufficient shorted turns may occur so that high rotor vibration follows, or the machine will not be able to operate over the entire specified range of power factors. In two-pole rotors, this tends to occur if there are more than 5% shorted turns. In addition, if the number of shorted turns increases over time, there is a greater likelihood that a rotor ground fault may occur, which is much more serious.

A new rotor should have no shorted turns. If a shorted turn does develop over time, then increased monitoring frequency should be established. If more shorted turns appear in the same slot, then serious insulation degradation is occurring, sometimes increasing the risk of a ground rotor fault. Corrective action such as off-line testing and inspections should be considered. In addition, limitations on excitation current may be warranted until the root cause of the shorts is established. If increasing numbers of shorted turns are detected over time but are distributed through different slots, then further off-line tests and inspections are warranted, but usually with less haste. Some machines have been operating with shorted turns over their entire life with no adverse consequences.

13.7 CURRENT SIGNATURE ANALYSIS

Current signature analysis (CSA) monitoring has revolutionized the detection of broken rotor bars and cracked short-circuit rings in squirrel cage induction motor rotors (Section 9.4). The current on one of the power cables feeding the motor is analyzed for its frequency content. Specific spectral components in the current indicate the presence of defective rotor windings and/or airgap asymmetry during normal operation of the motor. It seems probable that future advances will also enable the detection of conditions such as problems in the driven load.

CSA monitoring works best on three-phase SCI motors rated at 30 HP and above that are operating at or near full load. The detection of broken rotor bars by CSA can sometimes also be accomplished or confirmed by bearing vibration analysis (Section 13.10.1).

13.7.1 Monitoring Principles

Current signature analysis was pioneered by Kliman [13.28] and Thomson [13.29] in the late 1970s. CSA finds the following problems, as a minimum:

- Cracked rotor bars
- Cast rotor bars with large internal voids
- Broken bar-to-short-circuit ring connections
- Cracked short-circuit rings

In simple terms, the current flowing in the stator winding not only depends on the power supply and the impedance of the stator winding, but it also includes current induced in the stator winding by the magnetic field from the rotor. That is, the stator winding is a probe or "transducer" for problems in the rotor. In this way, CSA is similar to the airgap flux monitor for synchronous machines (Section 13.6), in which instead of a separate flux probe to detect rotor problems, the entire stator winding acts as the flux probe. The key issue is separating currents that flow through the stator to drive the rotor from the currents that the rotor induces back into the stator if there is a problem. This separation is accomplished by measuring current components at frequencies other than power frequency, using high-resolution frequency spectrum analyzers.

In conventional squirrel cage motor theory, a three-phase symmetrical stator winding fed from a symmetrical supply will produce a resultant forward-rotating magnetic field at synchronous speed. If exact symmetry exists, there will be no resultant backward-rotating field. In practice, it is not possible to obtain exact symmetry and, consequently, there will be a small resultant backward-rotating field component. Any change in asymmetry of the power supply to the motor will cause an increase in the resultant backward-rotating field from the stator winding.

Reversing the normal situation, the rotor winding has a current that will induce a current in the stator. From conventional SCI motor theory, the frequency of the current in the rotor winding (f_2) is at slip frequency, and not at the supply (50 or 60 Hz) frequency; that is:

$$f_2 = sf_1$$

where s = per-unit slip and f_1 = supply frequency (50 or 60 Hz).

The rotor currents in a cage winding produce an effective three-phase magnetic field, which has the same number of poles as the stator field but is rotating at slip frequency with respect to the rotating rotor. If rotor current asymmetry occurs, then there will be a resultant backward-rotating (i.e., slower) field at slip frequency with respect to the forward-rotating rotor. Asymmetry results if one or more of the rotor bars is broken, preventing current from flowing through one or more slots. With respect to the stationary stator winding, this backward-rotating field at slip frequency (f_{sb}) with respect to the rotor induces a current in the stator winding at

$$f_{sb} = f_1(1 - 2s)$$

This is referred to as a twice-slip-frequency sideband due to broken rotor bars [13.28, 13.29]. The twice-slip-frequency sideband results from the low frequency (sf_1) in the rotor winding, combined with the slower than synchronous speed of rotation $(1 - sf_1)$.

The stator current is thus modulated at twice the slip frequency, which in turn causes a torque pulsation at twice slip frequency ($2sf_1$) and a corresponding speed oscillation that is also a function of the drive inertia. This speed oscillation can also be detected by bearing vibration analysis (Section 13.10). Since sf_1 is often close to about 1 Hz, this creates a hum at this low frequency that is easily recognized by knowledgable plant staff.

This speed oscillation can reduce the magnitude (amps) of the $f_1(1 - 2s)$ sideband but an upper sideband current component at $f_1(1 + 2s)$ is induced in the stator winding due to the rotor oscillation [13.30]. Broken rotor bars, therefore, result in current components being induced in the stator winding at frequencies given by

$$f_{sb} = f_1(1 \pm 2s).$$

CSA monitoring requires that the stator current be measured on one phase, usually via a current transformer. The current is analyzed with a spectrum analyzer or customized digital signal processing unit. A typical spectrum analyzer output is shown in Figure 13.10. If there are no broken rotor bars, then there will be no sidebands. If sidebands are present, then broken rotor bars are likely. Typically, the sidebands are only 1 Hz or so away from the very large power frequency component and the sideband currents are typically 100 to 1000 times smaller than the main power frequency currents. Consequently, exceptional dynamic range and frequency resolution is needed to accurately measure the sideband peaks due to broken rotor bars.

To detect broken rotor bars, the slip frequency must be accurately known. In early "broken rotor bar" detectors, s was measured with a stroboscope that directly detected the rotor speed (and thus allowed calculation of slip). Alternatively, slip can be detected from an axial flux probe near the rotor winding [13.28]. Present-day CSA monitors have proprietary means of estimating slip from the current itself. This greatly improves the ease of performing CSA. Some of these methods are effective, but many have been shown to produce errors for small

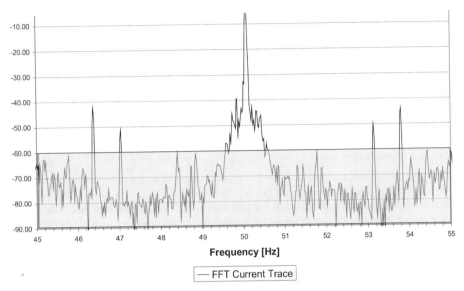

Figure 13.10. Spectrum analyzer output of the 50 Hz stator current, showing sidebands around 47 Hz associated with broken rotor bars. (Courtesy of Iris Power Engineering.)

motors or motors that have a large number of poles. Therefore, when purchasing a CSA monitor, users need to ensure that it will work (that is, measure the slip accurately) for the types of motors in their plant.

13.7.2 Data Acquisition

On small motors and those rated at 600 V or less, the stator current in one phase can be measured with a portable clamp-on current transformer placed on a phase lead either in the motor terminal box or at the motor control center (MCC) or breaker. For larger motors or those rated above 600 V, the current is usually measured by a clamp-on power frequency current transformer on the secondary of the CT that measures the supply current to the motor for protection purposes. These are almost always at the MCC. If the motor is in operation, extreme care is needed in installing the clamp-on CT since a shock hazard exists.

In principle, only a high-resolution, low-frequency spectrum analyzer is needed to measure the current spectrum. The spectrum analyzer suitable for vibration measurement (Section 13.10) can usually be employed for CSA. However, a means of calculating the slip frequency is needed.

The major suppliers of bearing vibration monitoring equipment, as well as some independent electrical test equipment manufacturers, either provide software to be used with a spectrum analyzer or have a specially built analyzer to measure the slip and determine if broken rotor bars are present.

The monitoring should be done about once per year, since it is unlikely that a healthy rotor would deteriorate to one that is ready to eject material in less than this time, unless a bizarre event has occurred. The measurements should be repeated whenever there is an increase in vibration at rotational speed times the number of rotor bars or a low-frequency hum is noted from the motor.

The testing is best performed at full load. At low load, there is insufficient current in the rotor to produce a significant reverse indication back into the stator. The minimum load for reliable interpretation is typically 50% of load.

To date, most CSA monitoring is done periodically. However, it is probable that it may eventually be incorporated in protective relaying devices, since overcurrent detection and CSA both require measurement of the stator current.

13.7.3   Interpretation

Basic interpretation requires comparison of the lower sideband with the power frequency stator current. Experience shows that if the sideband becomes larger than about 0.5% of the power frequency current, then broken bars are likely.* The greater the sideband current is (that is, the larger the fraction of the power frequency current), the more severe the rotor deterioration. As with most other monitors, it is best to trend the sideband magnitude over the years. If the sideband increases over time, then it is reasonable to expect that a greater number of bars have broken in more locations. At some point, there may be enough breaks in the rotor bars that the motor may fail to start, or some metal may fly off the rotor, destroying the stator. CSA may not detect bar breaks in large two- or four-pole motors if the breaks occur under end-winding retaining rings, since the retaining ring itself may allow the current to continue to flow.

Early CSA monitoring was prone to false indications (that is, indicating that a rotor had problems when it had none) and, less frequently, missing defective rotor windings. Thus, early users of this test had low confidence in the results. However, improvements in theory, software, and spectrum analyzer/digital signal processor resolution has made detection of rotor bar problems much more reliable.

There are many other circumstances that can cause frequencies other than the power frequency to be detected in the stator current:

- Power supply harmonics
- Eccentricity in the airgap due to bearing problems or unsymmetrical magnetic pull
- Oscillations in speed caused by the driven load (for example, reciprocating compressors, pulverizers, gear boxes, fluid couplings, and conveyors)

Although the latter two problems would be useful to identify, the CSA monitor needs to be able to distinguish them from broken rotor bars. Extensive research is underway to improve the reliability in detecting broken rotor bars by separately identifying other causes [13.30].

Broken rotor bars do not imply that the rotor must be rewound. Like turn shorts in synchronous machine rotors, motors have operated with broken bars for decades. However, if more broken bars are occurring over time, it is likely that aging is occurring. Consequently, there is greater risk that a motor may not start (due to insufficient starting torque), or chunks of the rotor winding may fly off the rotor. Also, broken rotor bars can lead to core burning due to the passage of rotor currents between bars through the core laminations (Figure 10.1).

---

* Since the initial CSA was performed with a spectrum analyzer, and such instruments normally record the output in db (rather than amps), conventional wisdom indicates that if the sideband is *less* than 45 db down from the 50 or 60 Hz current, then broken rotor bars are present.

13.8 AIRGAP MONITORING FOR SALIENT POLE MACHINES

Airgap monitoring systems (AGMS) measure the dynamic airgap between the rotor and the stator, most commonly in hydrogenerators and pump storage generators with salient pole rotors. This monitoring system normally measures various mechanical problems including:

- Movement of the stator frame, core, or parts of the core (if segmented) due to insufficient support, or movement of the cement foundations
- Distortion of the rotor rim due to high centrifugal forces combined with inadequate design
- Bearing deterioration or improper balancing

Most of these are beyond the scope of this book. However, the AGMS can help detect shorted turns in rotor winding.

13.8.1 Monitoring Principles

If a salient pole rotor has a turn-to-turn short, there will be unbalanced magnetic pull acting on the rotor. The pole with the shorted turn will have a weaker magnetic field than the other poles, and most particularly the pole that is 180° around the rotor from the faulted pole. This will create a "wobble" in the rotor that is manifested as an increase in bearing vibration, as well as an increase in airgap as the faulted pole revolves.

Two methods have been developed for measuring the airgap during normal operation of the stator: one is optical and the other uses capacitance. The capacitive approach has become dominant. With this technology, an electrically isolated partly conductive plate is glued to the bore of the stator (note that it cannot be fully conductive, otherwise it will block the magnetic field that is crossing the airgap) [13.31]. A high-frequency voltage (a few hundred kilohertz) is applied to the plate. A capacitor is formed between the plate and the grounded rotor poles. To a first approximation, the capacitance is given by Equation 1.5. Thus, the capacitance depends on the distance between the plate and the rotor. By measuring the current flowing into the plate, the capacitance can be calculated and, hence, the distance to the rotor. Since the distance can be measured every few milliseconds, the airgap in millimeters between the capacitor plate and the rotor can be measured thousands of times around the rotor periphery for each rotation of the rotor. During installation of the sensor, a measurement is performed to calibrate the measured current in millimeters. There is also a marker to generate a once-per-revolution signal so that poles with too small an airgap can be identified.

Usually, several capacitive sensors are located around the stator bore. Small rotors (< 8 m in diameter) usually have four sensors. Rotors >16 m in diameter may have up to 16 sensors. Algorithms have been developed to interpolate the airgap at any position around the stator bore, based on the discrete measurements at the sensors. The sensors can be measured occasionally with a portable instrument, or continuous monitoring systems are available. The latter are needed for some rapidly developing mechanical problems.

Interpretation to detect shorted turns on a rotor pole is straightforward. If a particular pole has enough shorted turns to weaken its magnetic field sufficiently so that the pole on the opposite side of the rotor has greater attraction to the stator, then the airgap will increase as this pole passes each sensor. The more shorted turns, the greater the dissymmetry in the airgap when compared to other poles. If the number of shorted turns increases over time, then the airgap increases.

13.9 VOLTAGE SURGE MONITOR

On-line voltage surge monitoring can determine if voltage surges from lightning, power system events, or inverter drives that can age the turn insulation in random-wound machines and machines with form-wound multiturn coils have occurred. Motors driven by modern IFDs benefit most from such monitoring.

13.9.1 Monitoring Principles

As discussed in Section 8.7, repetitive short-risetime voltage surges can gradually deteriorate the turn, ground, and phase insulation in motor stators. By monitoring the surge risetime and magnitude, as well as the number of surges, motors most at risk from this process can be identified. Unfortunately, since the magnitude and risetime of the surges is highly dependent on the power cable length to the IFD, the type of grounding, the cable insulation material, as well as the surge impedance difference between the cable and the motor, it is virtually impossible to calculate the surge environment. In addition, conventional voltage disturbance analyzers do not have the bandwidth to measure submicrosecond risetime voltage surges.

Specialized surge measurement systems have been developed especially for situations where fast risetime surges are possible [13.33]. The portable measurement system measures the magnitude of the surges, determines their risetime, and counts the total number of surges measured during the measurement interval. The monitor has the following specifications:

- It has wideband (50 Hz to 10 MHz) resistive or capacitive voltage dividers, capable of operating on motors rated up to 13.8 kV. The high bandwidth is needed to ensure that the peaks of the pulses with risetimes as fast as 50 ns are measured accurately.
- It is a portable electronic instrument that is temporarily placed near the motor for the duration of the measurement.
- It contains a laptop computer that downloads a summary of the measured surges recorded in the measurement interval for display or printout.

The system is normally connected to the motor during a short shutdown. The motor is then started and readings are collected at maximum speed and a number of lower speeds, to determine the surge characteristics under a variety of conditions. With pulse-width-modulated IFDs, the data at each speed can be collected in only a few minutes. However, if the system is to be used with conventional switchgear, then the system needs to be connected for enough switching events so that 25 or so surges are recorded to estimate the statistical variation of magnitudes and risetimes.

13.9.2 Interpretation

Modern surge monitors produce an output such as shown in Figure 13.11. In this three-dimensional plot, the horizontal axis is the surge risetime, the scale coming out of the page is the magnitude, and the vertical scale shows the number of surges in the recording interval at each risetime/magnitude combination. Surges with fast risetimes and higher magnitudes will cause faster aging of the insulation. Thus, a significant number of surges in the lower right of the plot are more dangerous.

At the time of printing, NEMA MG1, Part 31 has suggested that inverter duty motors *not* be exposed to voltage surges with a risetime of 100 ns or shorter, or to voltage surges greater than 3.7 pu (1 pu is peak line-to-ground rated voltage). Installing filters, changing power ca-

Motor 1 Phase V

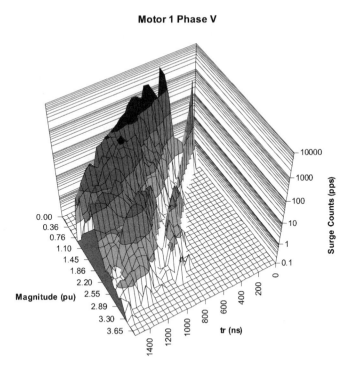

Figure 13.11. Surge repetition rate versus magnitude and risetime for a typical 600 V motor driven by a PWM-type IFD. (Courtesy Iris Power Engineering.)

ble length, or installing power cables with higher dielectric loss can reduce the severity of the surges.

13.10 BEARING VIBRATION MONITOR

Periodic bearing vibration monitoring is common on most important motors. Continuous bearing vibration monitoring is used for critical large motors and is standard on most large generators. Although the main purpose of such monitoring is to warn of problems with the bearings, such data can sometimes be used to indicate problems with the rotor windings and stator cores. Vibration monitoring can confirm the diagnosis from current signature analysis (Section 13.7) and rotor flux monitoring (Section 13.6).

13.10.1 Induction Motor Monitoring

If broken rotor bars are present in a SCI motor rotor, there will be an unbalanced magnetic pull on the rotor, since some of the rotor slots will not have the full current and, thus, will have less magnetic field in some portions of the rotor periphery. As indicated in Section 9.4, vibration monitoring can also identify fabricated rotor windings having some loose and some tight bars, or diecast windings with voids on the bar sections. If the stator core is loose in the frame, there will be a high radial vibration component that will be detected by bearing housing vibrations, if the bearings are mounted in brackets attached to the stator frame.

Sensors. Conventional accelerometers are often permanently or temporarily attached to the bearing housing. Only radial vibrations at two positions on the bearing housing, 90° apart, are usually measured in permanent installations. On the other hand, axial vibration levels are also measured in periodic monitoring, normally performed using a portable data logger, which includes a spectrum analyzer to which an accelerometer is connected. The signals from the accelerometers are broken down into magnitude versus frequency by a spectrum analyzer and are usually in velocity units of mm/sec and/or inches/second, peak.

Shaft displacement probes are mounted on the bearing housing to measure the relative movement of the shaft. This type of monitoring is used in larger motors with sleeve or tilting pad guide bearings because it gives an indication of how much of the bearing clearance is being used up when high rotor vibration levels occur. The units used for such vibration monitoring are usually mm or inches, peak-to-peak.

Interpretation of Results

- Breaks or air pockets in rotor cage windings are identified by a significant radial vibration frequency at rotational speed times the number of rotor bars. There will also be two side bands to this peak that are two times slip frequency above and below power frequency.

- Fabricated rotor windings with some loose and some tight bars in the slots, and diecast rotors with air pockets in the bar sections, will have significant radial vibrations at one time rotational speed frequency that change in magnitude with motor load. This problem is predominately found in two-pole rotors, which are less stiff radially than those in motors with four and more poles.

- Loose stator cores are identified by high vibration levels at two times power frequency.

13.10.2   Synchronous Machine Monitoring

Shorted turns in high-speed round rotors can induce rotor vibration that can be detected from accelerometers installed on the bearing housing or shaft displacement probes. A shorted turn in a rotor coil (which is installed in two slots), implies that the resistance of the coil will be smaller, since the current flows through a shorter path. Thus, the I^2R loss in the two slots where the coil is installed will be lower, resulting in a lower temperature in the affected slots. The lower temperature in the affected slots, combined with the higher temperature in the other slots, creates a "thermal bend." That is, the expansion of the coils and adjacent rotor body in the normal slots will be greater than the cooler slots, bending the rotor a small amount. This bend will result in a bearing vibration at the rotation speed. Thus, a peak in the frequency spectrum at rotational speed (rpm) divided by 60 (or 50) may indicate that shorts are present. The vibration tends to be greater in two-pole machines, since four-pole rotors tend to be stiffer. In two-pole machines, it is easier to detect shorted turns from the thermally induced vibration when the "small" coils contain the short (these are the coils away from the quadrature axis). Coils near the quadrature axis are in slots almost 180° apart and, thus, the thermal bend tends to cancel out. The bearing radial vibration will vary with excitation current, since the greater the rotor current, the greater will be the temperature differences between slots. Increasing vibration with increasing field current is a strong indicator of shorted turns.

Shorted turns can also give rise to unbalanced magnetic pull, especially in four-pole rotors. A short in one pole reduces the magnetic flux in this pole, whereas the flux is at full strength in the pole 180° around the rotor. This causes radial vibration at the rotational speed. Again, increasing the field current will increase the imbalance and, thus, the vibration.

REFERENCES

13.1   P. J. Tavner and J. Penman, *Condition Monitoring in Electrical Machines,* Research Studies Press, Wiley, New York, 1989.

13.2   A. J. Gonzalez, M. S. Baldwin, J. Stein, and N. E. Nilsson, *Monitoring and Diagnosis of Turbine-driven Generators,* Englewood Cliffs, NJ: Prentice Hall, 1995.

13.3   M. E. H. Benbouzid, "Bibliography on Induction Motor Faults Detection and Diagnosis," *IEEE Trans. EC,* 1065–1074, December 1999.

13.4   J. Dymond, R. Ong, and N. Stranges, "Instrumentation, Testing and Analysis of Electrical Machine Steady State Heating," in *Proceedings of IEEE Petroleum and Chemical Industry Conference,* Toronto, September 2001.

13.5   M. Bissonette, "Ten Case Studies of On-Line Monitoring of Hydroelectric Machinery," paper presented at Doble Client Conference, April 1999.

13.6   J. M. Braun, G. Brown, "Operational Performance of Generator Condition Monitors," *IEEE Trans. EC,* 344–349, June 1990.

13.7   S. Kilmartin, "Prefault Monitor for Air Cooled Generators," paper presented at EPRI Utility Motor and Generator Predictive Maintenance Conference, San Francisco, December 1993.

13.8   S. C. Barton et al., "Implementation of Pyrolysate Analysis of Materials Employing Tagging Compounds to Locate an Overheated Area in a Generator," *IEEE Trans. EC,* 1982.

13.9   D. M. Cartlidge, D. W. Casson, D. E. Franklin, J. A. Macdonald, and B. C. Pollock, "Machine Condition Monitoring Ozone Monitor for Air Cooled Generators," CEA Report 9134 G 864, February, 1994.

13.10 L. Rux, "Hydroelectric Generator Stator Winding Condition Assessment and Ozone Abatement Recommendations," paper presented at CIGRE/EPRI Colloquium on Maintenance and Refurbishment of Utility TurboGenerators, Hydrogenerators and Large Motors, Florence, Italy, April 1997.

13.11 G. C. Stone, "Importance of Bandwidth in PD Measurements in Operating Motors and Generators," *IEEE Trans. DEI,* 6–11, February 2000.

13.12 J. Johnson and M. Warren, "Detection of Slot Discharges in High Voltage Stator Windings During Operation," *Trans. AIEE,* Part II, 1993, 1951.

13.13 J. E. Timperly and E. K. Chambers, "Locating Defects in Large Rotating Machines and Associated Systems through EMI Diagnosis," CIGRE Paper 11-311, September 1992.

13.14 T. E. Goodeve et al., "Experience with Compact Epoxy Mica Capacitors for Rotating Machine PD Detection," in *Proceedings of the IEEE Electrical Insulation Conference,* Chicago, September 1995.

13.15 S. R. Campbell et al., "Practical On-Line Partial Discharge Test for Turbine Generators and Motors," *IEEE Trans. EC,* 281–287, June 1994.

13.16 K. Itoh et al., "New Noise Rejection Techniques on Pulse-by-Pulse Basis for On-Line Partial Discharge Measurement of Turbine Generators," *IEEE Trans. EC,* 585–594, September 1996.

13.17 S. R. Campbell and G. C. Stone, "Examples of Stator Winding Partial Discharges Due to Inverter Drives," in *Proceedings of IEEE International Symposium On Electrical Insulation,* April 2000, Anaheim, CA, pp. 231–234.

13.18 J. F. Lyles and T. E. Goodeve, "Using Diagnostic Technology for Identifying Generator Winding Maintenance Needs," *Hydro Review,* 58–67, June 1993.

13.19   B. A. Lloyd, S. R. Campbell, and G. C. Stone, "Continuous On-Line PD Monitoring of Generator Stator Windings," *IEEE Trans. EC,* 1131–1137, December 1999.

13.20   IEEE 1434-2000, "IEEE Trial Use Guide to the Measurement of Partial Discharges in Rotating Machinery."

13.21   G. C. Stone, "Calibration of PD Measurements for Motor and Generator Windings—Why It Can't Be Done," *IEEE Electrical Insulation Magazine,* 9–12, January 1998.

13.22   V. Warren et al., "Advancements In Partial Discharge Analysis to Diagnose Stator Winding Problems," in *Proceedings of the IEEE International Symposium on Electrical Insulation,* Anaheim CA, 497–500, April 2000.

13.23   M. Kurtz et al., "Application of PD Testing to Hydrogenerator Maintenance," *IEEE Trans. PAS,* 2148–2157. August 1984.

13.24   M. Twerdoclib et al., "Two Recent Developments in Monitors for Large Turbine generators," *IEEE Trans. EC,* 653–659, September 1988.

13.25   G. Pritchard and M. Couchard, "Fiber Optic Accelerometer for Use in Hostile Electrical Environments," *Hydro Power and Dams,* 1998.

13.26   D. R. Albright, "Interturn Short-Circuit Detector for Turbine-Generator Rotor Windings," *IEEE Trans. PAS,* PAS-90, 2, 1971:

13.27   D. R. Albright, D. J. Albright, and J. D. Albright, "Flux Probes Provide On-Line Detection of Generator Shorted Turns," *Power Engineering,* 28–32, September 1999.

13.28   G. B. Kliman et al., "Noninvasive Detection of Broken Rotor Bars in Operating Induction Motors," *IEEE Trans. EC,* 873–879, December 1988.

13.29   W. T. Thomson, "Diagnosing Faults in Induction Motors—Engineering Ideas," *Electrical Review,* 21–22, November 1984.

13.30   W. T. Thomson and M. Fenger, "Current Signature Analysis to Detect Induction Motor Faults," *IEEE Industry Applications Magazine,* 26–34, July 2001.

13.31   G. B. Pollock and J. F. Lyles "Vertical Hydraulic Generators, Experience with Dynamic Air Gap Monitoring," IEEE Winter Meeting Power Meeting, Paper 92 WM 042-2 EC, January 1992.

13.32   G. C. Stone and B. A. Lloyd, "On-Line Condition Monitoring to Improve Generator Availability," EPRI Steam Turbine Generator Workshop, August 2001, Baltimore.

13.33   M. Fenger, S. R. Campbell, and J. Pedersen, "Motor Winding Problems Caused by Inverter Drives," *IEEE Industry Applications Magazine,* 22–31, July 2003.

CHAPTER 14

CORE TESTING

This chapter describes the main tests that are commercially available for assessing rotor and stator core tightness and stator core insulation condition. All of the conventional forms of these tests require the motor or generator to be taken out of service and to have its rotor removed.

The tests covered are:

- Knife test—for stator or rotor core tightness
- Rated and reduced flux (EL-CID) tests—for core insulation local and general condition
- Core loss test—for general core insulation condition

For each test, the purpose is described, as well as the types of machines it is useful for. The theory of the test and its advantages and disadvantages are also covered. Finally, practical guidelines on performing the test and interpreting the results are given.

There are some technologies available to check the condition of large synchronous generator stator cores without removing the rotor. The EL-CID test can sometimes be performed without rotor removal. In addition, on-line monitoring of the stator core is possible with temperature sensors placed on the core (Section 13.1) and condition monitors to detect stator core overheating in hydrogen-cooled generators (Section 13.2).

14.1 KNIFE TEST

This is a simple, commonly used test that requires no specialized equipment.

14.1.1 Purpose and Theory

Well-designed and constructed cores should remain tight for the life of the machine in which they are installed. The laminations in such cores should be impenetrable with a sharp

Electrical Insulation for Rotating Machines. By Stone, Boulter, Culbert, and Dhirani
ISBN 0-471-44506-1 © 2004 Institute of Electrical and Electronics Engineers

object. On the other hand, if the core is poorly designed or constructed, it will relax with service and the laminations in the whole core or in sections will become loose (Chapter 10). In a loose core, a sharp object, such as a winder's knife, can be inserted between adjacent laminations. The "knife test" is designed to evaluate the tightness of a laminated stator or rotor core.

14.1.2 Test Method

Before starting this test, remove from the surface of the core any varnish or other coatings that may be loose. If the core is a large segmented type, loose areas may be indicated by dust from relative movement between core laminations.

This test involves trying to insert a standard winder's knife blade, with a maximum thickness of 0.25 mm, between the laminations at several locations around the core bore (stator), or core outside diameter (rotor). When performing this test, care must be taken to avoid breaking off the tip of the knife while it is in the core. That is, do not rock the knife blade as it is pushed between the core laminations (Figure 14.1). Instead, insert the knife (if possible) by holding the knife handle in one hand and carefully applying pressure to the back of the knife blade with the palm of other hand.

14.1.3 Interpretation

If the blade penetrates the section of core being tested by more than 5 mm, then it is assessed to be loose. If there are many places where the knife can be inserted, repairs as described in Section 10.5 may be advisable.

Figure 14.1. Photo of a knife being inserted between the laminations of a loose stator core.

14.2   RATED FLUX TEST

The rated core flux test (also called the "ring flux", "full flux", or "loop" test) is the traditional method of determining the insulation integrity of any type of laminated stator core in an AC motor or generator. The test can assess the severity of damage, detect hidden damage, and, for any type of damage, provide information to indicate whether repair is required. The major disadvantage of this test is that it requires machine disassembly and removal of the rotor before it can be performed. As the test is normally carried out at or near the rated back-of-core flux, it may aggravate an existing problem if core temperatures are not carefully monitored.

14.2.1   Purpose and Theory

The equipment required for this test depends on the size of the core to be checked. For small and medium-size machines that can easily be transported to a motor service shop, a commercial core tester is normally used. On the other hand, large generators have to be tested on-site using heavy cables and local large 50 Hz or 60 Hz power supplies.

This test is performed by installing an excitation winding around the stator core, as illustrated in Figure 14.2. For a commercial core tester, this winding normally consists of one or two turns of heavy cable, whereas, for large cores, the number of turns and cable size will be much greater. The excitation winding must have an appropriate number of turns and be insulated for the voltage to be applied across its ends. Voltage should be sufficient to produce a back-of-core flux that will give approximately the rated operating flux density in the core area behind the winding slots to induce normal axial voltages between laminations. When this axial voltage is applied, any defective areas of core or tooth insulation will show up as "hot spots" in that they will become significantly hotter than those with "healthy" insulation. These hot spots are created by the currents that are induced between shorts in the core insula-

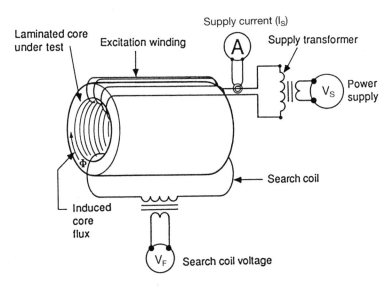

Figure 14.2. Test arrangement for rated core flux test.

tion. Surface defects are indicated by hot areas of core that become evident soon after the application of the excitation current. On the other hand, in large cores, deep defects may take more than 30 minutes to show as high temperatures on the observed surfaces, because the surrounding "healthy" sections of core act as a heat sink.

This is the most appropriate test for determining the need to perform core insulation repairs and for determining the effectiveness of repairs. For motors, it is important to perform this test before burning out stator windings that are to be replaced and after winding burnout to confirm that this winding removal process has not caused significant core insulation deterioration. This test is often used to confirm the seriousness of core defects detected by the EL-CID test described in Section 14.4.

14.2.2 Test Method

If a commercial core tester is used, then it usually has an on-board computer that calculates the required back-of-core flux and number of excitation winding turns from stator winding information and stator core dimensions.

For large machines that cannot be tested with a commercial device, guidelines on the design of the excitation winding required to induce flux in the core are given in IEEE Standard 432 [14.1]. A 50 or 60 Hz power supply of sufficient capacity is needed to induce the required level of excitation in the core. If possible, the power supply should have the capability of raising the excitation level gradually to avoid transients that may damage the core lamination insulation during energization. Commercial core tester excitation winding power supplies have this capability. For large machines that have to be tested on-site, there is usually a 3.3 or 4.16 kV supply available from the plant distribution system. Experience has shown this to be generally adequate for testing most large machines such as turbogenerators and hydrogenerators. However, a variable autotransformer of suitable rating is difficult to obtain for testing of large turbine generators. Therefore, sudden application of the supply voltage to the core excitation winding is often unavoidable for large stator core testing. In order to establish the required capacity of the supply, it is first necessary to determine the excitation level needed to produce rated or near rated flux in the stator core. As indicated in Reference 14.2, this is calculated as follows.

Normal operating flux per pole is

$$\phi = \frac{V_{ph} \cdot N_p}{4.44 \cdot f \cdot K_d \cdot K_p \cdot N_{TP}}$$

where:
V_{ph} = rated phase-to-phase voltage (rms)
N_p = number of stator winding parallel paths per phase
f = machine supply (motor), or output (generator) frequency
K_d = winding distribution factor (usually = 0.955)
K_p = winding cording factor = sin ½ (180° · coil slot pitch/coil full pitch)
N_{TP} = number of turns per phase in the stator winding

For the flux test, the excitation winding and its power supply are designed such that only half the flux per pole is induced in a circumferential direction in the stator core yoke. Therefore, the flux volts required to produce rated flux in the stator yoke are:

$$V_F = 4.44 \cdot f \cdot (\phi/2)$$

Once this is known, and the supply voltage is known, the number of turns for the excitation winding is determined by:

$$N_T = \frac{V_S}{V_F}$$

For example, if the calculated flux volts were 1050 and V_s is 4160 V, then N_T would be 4160/1050 = 3.96. For the test the number of turns would be four, since the turns have to be a whole number and four turns would not overflux the core area behind the teeth. The excitation level in this example would be about 99% of rated flux.

To correctly size the cable and determine if the supply is adequate, it is necessary to know how many amperes this excitation winding will draw. This is done using a *B–H* curve for the core laminations, which gives the relationship between flux density (*B*) in Tesla and magnetic field intensity (*H*) in ampere-turns/m for the material used in the stator core. If such a curve is not available, a generalized curve for silicon steel laminates for rotating machines should be used. The procedure for supply current estimation is as follows.

Determine the stator core back of teeth flux density:

$$B = \frac{(\phi/2)}{\text{area}}$$

where, per Figure 14.3,

$$\text{area} = \text{core length (cl)} \cdot \text{back of core depth (W}_y) \cdot \text{stacking factor}$$

$$\text{Length of flux path (lfp)} = [(\text{stator bore} + 2) \cdot (\text{slot depth} + W_y)] \cdot \pi$$

The stacking factor is typically 0.95, and allows for the lamination insulation in the core.

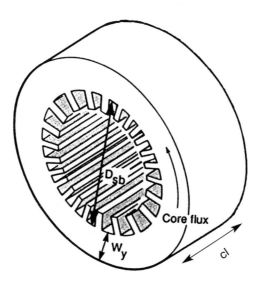

Figure 14.3. Dimensions for calculating core area.

On a *B–H* curve for the type of core steel in the stator, find the calculated value of *B* and the corresponding value for *H* (ampere-turns per meter). The total ampere-turns required to induce *B* in the stator back of core area is

$$MMF = H \cdot \text{lfp ampere-turns}$$

Therefore, the supply current would be

$$I_S = \frac{MMF}{N_T} \text{ (amperes)}$$

For small and medium-size cores, it is usually possible to induce a back-of-core flux of 100% to 105% of the rated value [14.1]. In large core testing, if I_S is too high for the supply available or the resultant cable size too large, then the excitation level may be reduced by adding another turn. This process of recalculation can be repeated until the supply is capable of providing adequate excitation for the test. For good results, the excitation level should induce at least 75% of the rated back-of-core flux.

Core saturation effects and leakage fluxes result in the actual flux not being directly proportional to the applied ampere-turns. Also, estimations from approximate *B–H* curves introduce a certain percentage of error. For these reasons, it is advisable to start with extra turns on the excitation winding to ensure personnel safety and equipment security. A one-turn search coil can be used to more accurately measure the flux volts, V_F, directly [14.4].

Practical considerations when setting up this test are as follows:

- The excitation winding should be a flexible cable suitably insulated and sized for the supply voltage and expected current capacity. Where possible, the excitation winding should be at the axis of the stator bore (Figure 14.4).

- The excitation winding should not obscure any areas of core having obvious damage and care should be taken to protect the core from damage when assembling the excitation winding.

- Installed stator core thermocouples or RTD's should also be monitored during testing in addition to infrared scanning.

- The power supply for large machine core tests is usually obtained using two phases of an adequately sized three-phase 3.3 or 4.16 kV breaker. Breaker protection (overcurrent and ground fault) should be used and properly calibrated for the expected load. A remote breaker trip switch should be installed at the test site to allow immediate shutdown of the excitation winding power supply in the event of rapid core heat up due to shorted insulation

Commercial core testers have built-in instrumentation to monitor the voltage applied, the current drawn, and the power absorbed by the excitation winding and core system. If an excitation system has to be fabricated, then current and voltage transformers will be required to allow the connection of an ammeter and a voltmeter to monitor excitation levels. A wattmeter may also be connected into the circuit to measure the power drawn.

Note: When testing large turbine generators, care should be exercised since high voltages can be induced between laminations and high magnetic fields exist in and around the core.

Figure 14.4. Photo of the excitation winding through the axis of a turbogenerator stator. (Courtesy Ontario Hydro.)

Shorted laminations will create high temperatures when excited near full flux. A thermal imaging camera (infrared scanner) is used to quickly survey the whole core, detect the location of faults, and measure actual core temperatures. A special infrared mirror with nonreflecting glass, which is movable axially and rotatable, is inserted into the bore of long turbogenerator stators to better monitor developing hot spots with an infrared camera. For motor stator core testing, it is also beneficial to measure the power absorbed by the excitation winding, as described in Section 14.3. This is particularly relevant if previous power readings have been taken with the same induced core flux. A significant increase in the absorbed power compared to previous readings indicates higher core losses due to deterioration in core lamination insulation.

It is recommended that, unless severe damage is detected, the duration of this test should be at least 30 minutes for small and medium-size machines and up to 2 hours for large machine cores, with temperature measurements taken every 15 minutes. This will ensure that deep-seated core faults are detected. The ambient air temperature should also be monitored to allow a comparison between it and core temperatures. When large generator cores with embedded thermocouples are being tested, it is advisable to also monitor the temperatures indicated by these sensors since they may help confirm the presence of good or poor core insulation. With the full flux test, it is necessary to allow time for the core to cool down before the test can be repeated.

Caution is needed not to apply a full flux test without comprehensive thermal monitoring. Since the core is not being rapidly cooled by forced air or hydrogen during the test, thermal runaway is possible.

14.2.3 Interpretation

Core temperatures should be monitored from the instant the test flux is applied. This is necessary because the rate of temperature rise in areas of the core with damaged insulation can give a good indication of their location. If the fault is near the core surface, hot spots will appear rapidly, whereas deep-seated damage will be indicated by a more gradual increase in temperature.

Most cores with healthy insulation will still have areas that are a few degrees above the average core temperatures obtained from this test. These are due to flux concentrations. Consequently, insulation damage is not likely unless hot spot temperatures are at least 10°C above the coolest areas of the core for motors and 5°C for large generators. Temperature hot spots of up to 15°C above ambient core temperature may be acceptable if attempts to remove local core insulation shorting are unsuccessful. Cores with this characteristic should be more frequently tested since the condition could deteriorate with time, requiring corrective action.

14.3 CORE LOSS TEST

This is a simpler version of the rated flux test described in Section 14.2. During this test, a wattmeter is connected to the excitation winding power supply to measure the power absorbed. The magnitude of this power and changes in its level between tests give an indication of general core insulation condition. Most repair shops have commercial core loss testers adequate for motor stator cores.

14.3.1 Purpose and Theory

The core loss test gives an indication of the general condition of the core insulation, and, for a given core, the higher the losses per mass of core, the poorer the condition. As core insulation condition deteriorates, the currents that flow between the laminations increase, thereby increasing the amount of power needed to reach a certain flux level. This test is most useful for detecting a deterioration in core insulation condition over time from increased power losses for the same excitation winding configuration and applied voltage. It is also useful to perform this test before and after winding burnouts (Section 10.1) to detect any significant core insulation deterioration due to the burnout.

14.3.2 Test Method

The test setups for different sizes of machines are the same as those described in Section 14.2.2. As indicated, a wattmeter measurement is already incorporated in commercial core testers. If the core excitation winding system described in Section 14.2 is used, then a wattmeter is required to measure the loss. The wattmeter must be connected into the current and voltage transformers used to obtain current and voltage readings. The weight of the core should be known.

14.3.3 Interpretation

For small and medium-size motor and generator stators:

- The core loss should not exceed about 10 W/kg. Cores with lamination insulation in good condition typically have a core loss less than 6 W/kg.
- The increase in core loss from a previous test should not exceed 5%.

For large motor and generator stators:

- The cause of core loss increases of more than 5% above those from previous tests should be investigated and, if possible, eliminated.
- The core loss should not exceed 6 W/kg.

14.4   LOW CORE FLUX TEST (EL-CID)

The traditional rated flux test described in Section 14.2 requires a significant amount of time and resources to set up if it is to be used for testing large generator stator cores. The alternative low-flux test described in this section was devised by the Central Electricity Research Laboratories in the United Kingdom in the 1980s and is now being routinely performed on large motors and generators around the world by both machine manufacturers and utilities. The main advantage of this test is that it requires a much smaller capacity power supply for the excitation winding, since only 3 to 4% of rated flux is induced in the core. In fact, for most tests, the power supply can be obtained from a 120 V or 220 V AC wall outlet. Also, this test normally takes much less time to perform than the rated flux test. Detailed information on this test is contained in Reference 14.5. In recent years, developments in robotic techniques have allowed this test to be performed in large turbogenerators without removing the rotor [14.6].

14.4.1   Purpose and Theory

The electromagnetic core imperfection detector (EL-CID) identifies the presence of faulty core insulation by an electromagnetic technique. It operates on the basis that currents will flow through failed or significantly aged core insulation even when a flux equivalent to a few percent of rated voltage is induced in it. The EL-CID system uses a special pickup known as a "Chattock coil" (or Maxwell's worm) to obtain a voltage signal that is proportional to the eddy current flowing between laminations. This solenoid coil is wound with a double layer of fine wire on a U-shaped form. When this coil is placed in a position bridging two core teeth, the voltage induced in it by an axial fault current I_f between the laminations is approximately proportional to the line integral of the alternating magnetic field along its length (l).

Applying Ampere's law for any closed loop of integration, the line integral of the magnetic field is equal to the enclosed current, in this case, the fault current. Hence, if the effects of the field in the core are ignored, the Chattock coil will give a voltage output proportional to the current flowing in the area encompassed by it, the two teeth it spans, and the core behind these teeth.

The output from the Chattock coil cannot be used directly to give an indication of core insulation quality. This is because there is another component of voltage induced in it by the circumferential magnetic field generated by the excitation winding that produces the test flux in the core. Consequently, the voltage induced in the Chattock coil is the mean potential difference between teeth due to both currents (I_f) flowing between laminations and those induced circumferentially in the laminations by the excitation winding field (I_W). Fortunately, there is a phase shift between the excitation winding flux and that produced by axial fault currents since these currents are at 90° to one another. Thus, if the voltage output from the Chattock coil is fed to a signal processor, the portion generated by the excitation winding can be eliminated to produce a voltage that is proportional to the axial component of current.

This is done by measuring the amplitude of voltage from the Chattock coil at the instant when the voltage component induced by the excitation winding is zero. This zero voltage time is determined from a voltage input obtained from a reference coil attached to a different section of the core.

The Chattock coil head used for large generators is designed to give better sensitivity by having a compensating coil fitted to it. This compensating coil is located about 50 mm above the ends of the U-shaped Chattock coil. The voltage induced in this coil is approximately the same as that induced in the Chattock coil by the excitation winding flux. By connecting the compensating coil in series with and in phase opposition to the Chattock coil, the voltage fed to the signal processor contains a very low excitation winding voltage component. This improves the resolution of the voltage component generated in the Chattock coil by eddy currents flowing axially between laminations.

The signal-processing unit gives an output in milliamperes that is a function of the component of Chattock coil voltage generated by axial eddy currents. If there is faulty insulation in the core or there are interlamination shorts at the core surfaces, they will be indicated by relatively high milliampere readings. The manufacturer of this equipment states that for 5 V/m induced voltage along the axis of the core, output readings above 100 mA indicate significant core plate insulation shorting. If the V/m were lower than this, the 100 mA limit should be reduced in proportion; e.g., if the V/m were 4, then the mA limit would be 4/5 · 100 = 80 mA.

14.4.2 Test Method

The EL-CID test kit comes with a preformed excitation winding and 1 turn trace winding for testing small to medium-size stator cores. This consists of a multistrand cable with plugs at either end that, when connected together, form the winding. A multiturn excitation winding has to be constructed for large stator cores. It is important for the excitation winding to pass through the center of the stator bore.

The setup of equipment for a stator core test is illustrated in Figure 14.5. Specific details of how to calibrate and set up the instrumentation and select the number of turns, applied voltage, etc. for the excitation winding are given in References 14.3 and 14.5.

The voltage applied to the excitation winding and the current flowing through it should be recorded at the start and end of the test. This will verify whether the excitation was relatively constant throughout the test and provides good reference information for future tests.

At the start of the test, the V/m along the bore of the core should be measured with a voltmeter and recorded. This value is used to determine the EL-CID signal processor current value limit for good insulation, as described in Section 14.4.1.

Once all the test equipment has been set up, the test is performed by axially traversing each core slot from one end to the other at a speed of less than one meter every 2 seconds using an appropriately sized Chattock coil from the EL-CID equipment kit. During the test, the mA readings from the signal processor should be continuously monitored and at least maximum values should be recorded. The locations and polarity of any reading close to, or above the mA calculated limit for a good core should be marked with a nonconductive substance and examined for signs of defects.

As indicated above and in [14.6], the cores in large synchronous machines can be tested without removing the rotor. The Chattock coil is attached to a trolley whose movements can be controlled to make it traverse each slot axially. For this setup, the rotor itself is used as part of a one-turn coil that induces flux in the stator core.

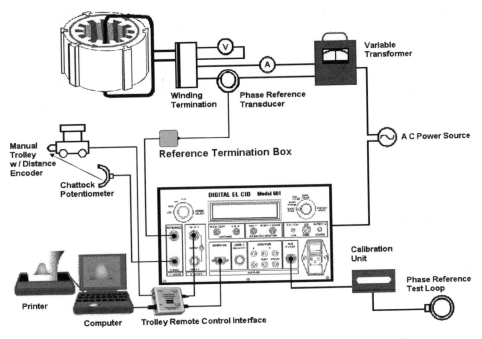

Figure 14.5. EL-CID test setup for a stator core. (Courtesy Adwel International.)

14.4.3 Interpretation

The polarity of the signals can give an indication of a fault location. This is important when there is a winding in a core, since the fault may not be visible [14.3]. Some guidelines on interpretation of the test results are as follows:

- If general high mA readings are obtained, the core insulation is either in poor condition, or the signal processor has not been properly calibrated; e.g., it was calibrated on a magnetic steel surface.
- If readings above the established threshold value of about 100 mA are obtained, the core in the area of such readings should be carefully examined for defects. If none are found, it is advisable to perform a rated flux test, as described in Section 14.2, to assess the severity of the core insulation damage.
- Using the guidelines given in Reference 14.5, the polarity of mA readings around a fault location can give an indication of position and depth of the damaged core [14.3].

One advantage of this test is that the equipment can be left connected and is easily reenergized to check repair work done on a section of core containing shorted laminations. With the full flux test, it is necessary to allow time for the core to cool down before it can be repeated. The EL-CID test does require some skill and experience on the part of the operator to determine if a fault exists and where it is located.

There is considerable controversy on the usefulness of this test in hydrogenerators in which the stator core contains one or more sections. The most likely area for core problems is at the interface between sections. In this area, the magnetic field can be very unusual, causing the EL-CID to read incorrectly, unless knowledgeable technicians perform the test.

REFERENCES

14.1   IEEE Standard 432, Guide for Insulation Maintenance for Rotating Electric Machinery (5 HP to less than 10,000 HP).

14.2   M. G. Say, *Performance and Design of Alternating Current Machines,* Pitman Publishing, 4th ed., 1976.

14.3   G. Klempner, "Experience and Benefit of Using EL-CID for Turbine Generators," paper presented at EPRI Motor and Generator Predictive Maintenance and Refurbishment Conference, Orlando, Florida, November 28–30, 1995.

14.4   J. E. Timperly and A. F. Bristar, "Machine Stator Iron Evaluation through the Use of Resonant Circuits," *IEEE Trans. EC, 1,* 3, 94–100, September 1986.

14.5   S. J. Sutton, "Theory of Electromagnetic Testing of Laminated Stator Cores," *Insight, 36,* April 1994.

14.6   J. W. Shelton and D. Paley, "Robotic Delivery Systems for Generator Air Gap Tests," paper presented at EPRI Motor and Generator Predictive Maintenance and Refurbishment Conference, Orlando, Florida, November 28–30, 1995.

CHAPTER 15

ACCEPTANCE AND SITE TESTING OF NEW WINDINGS

This chapter describes the main tests that are available for prequalifying manufacturer's insulation systems and assessing the quality of insulation in new windings, both in the manufacturer's plant and on-site. All of these tests are designed to help give assurance that stator and rotor windings in new machines and new replacement windings will give reliable service. Many of these tests are described in Chapter 12 and these are referenced where appropriate.

Guidelines on the acceptance and pass/fail criteria for each test are also given. It is important that such information be given in user technical specifications and be accepted by the winding supplier before placing a purchase order.

15.1 STATOR WINDING INSULATION SYSTEM PREQUALIFICATION TESTS

For stator winding insulation systems rated at 3 kV and above, it is important to gain some assurance that a high-quality insulation system that will give long life can be provided by the manufacturer. This is best done by subjecting sample coils to one or more high-voltage tests and dissecting the test coils for inspection afterward. Such coil design and manufacturing qualification tests and inspections should be identified in a technical specification and be performed before a new machine or replacement winding is purchased. This technical specification should indicate pass/fail criteria, and only those companies whose coils have passed all of the tests should be asked to bid for such work.

15.1.1 Power Factor Tip-Up Test

This test, described in Section 12.9, makes use of the fact that the power factor or dissipation factor of a coil or bar ground insulation will change with applied AC voltage. The magnitude of this change gives an indication of how void-free the groundwall insulation is. The test is applicable for coils/bars rated at 3 kV and above.

Electrical Insulation for Rotating Machines. By Stone, Boulter, Culbert, and Dhirani
ISBN 0-471-44506-1 © 2004 Institute of Electrical and Electronics Engineers

This test can also be performed on new coils, prior to accelerated aging tests (Sections 15.1.4, 15.1.5), to give an indication of the original condition of the ground insulation. If the test is then repeated during and after the accelerated aging process, changes in power factor or dissipation factor will give an indication of degradation of the groundwall insulation from aging effects such as loss of insulation bonding from thermal degradation.

Test Method. To prepare coils for testing, a ground plane of aluminum foil or tinned copper mesh must be applied to the coil or bar straights. If the coils represent a global VPI insulation system, then they should be fitted with steel or aluminum plates on two sides to simulate a stator slot; these plates can be used as the ground plane. If there is a silicon carbide stress-control material applied at the ends of a semiconducting layer on the coil straights, this must be guarded out and grounded to eliminate the effects of this material on the test readings. Guarding is commonly done by wrapping aluminum foil on top of this stress-control material and directly grounding the foil.

A variable-voltage AC power source and a power-factor or dissipation-factor measuring instrument are required. Measurements of power factor or dissipation factor are then taken at two voltages, normally 25% and 100% of rated line-to-ground voltage, and the tip-up calculated. The values obtained serve as the initial levels of the power factor for the insulation system under test. For the detection of aging effects, it is only necessary to take measurements at rated line-to-ground voltage since degradation of insulation condition will be detected by the changes in values between the initial and the final readings.

Acceptance Criteria. For new unaged coils or bars, with well-bonded and well-impregnated ground insulation, many suggest that the power factor tip-up should be < 0.5% and certainly no greater than 1.0%

As ground insulation ages in any aging test (Sections 15.1.4 and 15.1.5), these values will increase, and some judgment and experience are needed to relate the test values to coil insulation condition. However, at the end of the accelerated aging process, coil dissection findings can be related to test values.

15.1.2 Partial Discharge Test

Off-line partial discharge testing of complete windings is discussed in Section 12.10. This test is similar to the tip-up test in that it detects any voids in winding groundwall insulation resulting from poor coil impregnation as well as problems with the semiconductive or stress relief coatings. The test can also detect any delamination due to thermal and other types of aging tests. The test can be applied to any coil or bar with a semiconductive coating.

Test Method. As with the power factor tip-up test, a ground plane has to be provided on the coil straights. The test requires an AC power source capable of supplying PD-free 50 or 60 Hz variable voltage to at least the rated line-to-ground voltage of the insulation system under test. A power-frequency separation filter must be installed between the PD measuring device and the test coil to block power frequency voltages and their harmonics. PD measurements are taken for each test coil in voltage steps up to the rated line-to-ground voltage of the insulation system to measure the discharge inception voltage (DIV) and the discharge extinction voltage (DEV). The first set of readings is taken before the commencement of any accelerated aging tests that may be required. The positive and negative PD magnitudes are measured and recorded. This test is repeated at various times throughout the accelerated aging test and the results are trended to look for increases in PD activity. Reductions in DIV and DEV values provide another indication of deterioration in the condition of the coil ground insulation.

Acceptance Criteria. New unaged coils or bars with well-bonded and impregnated ground insulation will often have the following characteristics:

- The peak PD magnitude at rated line-to-ground voltage should be < 50 mV, if measured with an 80 pF capacitor feeding a 50 ohm load, or less than 50 pC with other types of PD detectors.
- DIV should be 0.5 times rated line-to-ground voltage.
- The location of the deterioration can be determined by comparing the positive and negative PD pulses as follows:
  a) If the positive and negative PD pulses remain approximately equal, the voids are present in the bulk groundwall insulation.
  b) If the negative PD pulses are higher than the positive ones, then it is likely that the coil ground insulation is separated from the conductor stack or the coil is poorly impregnated near the copper conductors.
  c) If the positive PD pulses are higher than the negative ones, it is likely that the coil or bar semiconductive or grading material is defective.

15.1.3 Impulse (Surge) Test

As indicated in Section 12.12, this test is designed to check the integrity of turn insulation in form- and random-wound multiturn coils. It is not applicable to Roebel bar windings. Weak turn insulation will fail when a fast risetime, high-voltage surge is applied to the coil.

Test Method. The test procedure is described in IEEE 522. The surge test unit should have a risetime of 0.2 μs or shorter. The test is performed by increasing the peak voltage in steps, until the recommended peak voltage is reached. IEEE 522-1992 recommends a peak voltage for new coils of 3.5 pu, where 1 pu is the rated peak, line-to-ground rated voltage. Suggested steps are 25%, 50%, 75%, and 100% of the maximum peak voltage.

Acceptance Criteria. There should be no change in the surge voltage waveform between low voltage and high voltage. If a change occurs, the turn insulation has failed.

15.1.4 Voltage Endurance Test

High-voltage stresses are an important aging mechanism for stator windings rated at 3 kV and above since partial discharge is created (Section 1.4.4). It is well known that increasing the voltage across the coil ground insulation will reduce the time to failure if partial discharges are present. Voltage endurance testing uses this phenomenon to identify any weakness in the ground insulation materials used and their method of application. The general principles of voltage endurance testing are discussed in Section 2.4. Any coils/bars subjected to this test must be discarded.

The manufacture of electrical coils is a highly specialized and complex process. A large number of requirements must be satisfied before a coil can be produced that will provide long-term reliable service under various operating conditions. Some of the important requirements include material properties, material handling, shaping and forming, processing temperatures and pressures, and quality control at every stage.

Inevitable changes over time in materials, their suppliers, manufacturing methods and processes, and the staff in the design office and the shop add further challenges to obtaining a consistent high quality of the end product. The voltage endurance test provides an important means of detecting weaknesses in any of the numerous activities involved in the design and manufacture of electrical coils.

Consistent application of the test helps vendors recognize potential problems and to avoid significant costs of redesign and retooling involved in correcting mistakes. Customers benefit from a higher level of assurance of long-term reliable service from the machines as well as having a number of qualified vendors available for competitively supplying their motor and generator needs.

This test is designed to verify that the coil ground insulation and, where appropriate, semiconducting and grading materials are adequate both in material properties and in the method of manufacturing. It can reveal manufacturing defects such as:

- Reduced insulation thickness at coil corners (due to too high a tape tension during application or skipped layer of tape)
- Wrinkles in insulation
- Poor application of semiconducting and grading materials
- Voids (due to inadequate or improper bonding resin application)

In a prequalification test, two coils or four bars are subjected to a much higher than normal voltage stress between the conductors and ground. The above-mentioned defects will cause excessive insulation electrical aging and coil failures before the end of the test. The temperature of the insulation is raised to the expected operating temperature of the coils, usually 90 to 135°C, depending on the type of machine and the thermal rating of the insulation system. For the most widely applied version of the voltage endurance test, the voltage stress normally used is 3.75 to 4.4 times the normal operating line-to-ground voltage.

Test Method. IEEE Standard 1043-2000, "Recommended Practice for Voltage Endurance Testing of Form-Wound Stator Bars and Coils for Large Generators," describes the test method. At least four test bars or two coils are fitted with steel or aluminum platens clamped in place on either side of the straight sections. Thermostatically controlled electric heaters are then fitted to the outside of the platens to raise the coil temperature to the desired level for the test. The test voltage is then applied between the coil conductors and the steel platens. At the start of the test, this voltage should be gradually increased to the required value so that no significant overshoots or spikes result from the voltage application process. The voltage is maintained for a predetermined time period, e.g., 250 to 400 hours or until a coil failure occurs [15.1, 15.2].

Acceptance Criteria. If all coils pass the test and coil or bar visual examinations and dissections reveal no manufacturing or material defects, then the insulation system can be considered to be of good quality and should give long life if maintained properly. For multiturn coils where two coils will provide four legs, three out of four legs surviving is normally considered as a pass criterion. Similarly, three out of four bars surviving the voltage for 250 or 400 hours constitutes a pass.

Dissection of coils that fail the test can often result in identifying the material and manufacturing improvements required to achieve satisfactory test results. In particular, coil construction at the points of failure should be carefully examined to determine the cause of failure.

15.1.5 Thermal Cycling Test

As indicated in Section 8.2, stator winding insulation in machines with long stator cores that see rapid starts and stops, or rapidly changing loads, is susceptible to deterioration from thermal cycling. This can affect both groundwall and turn insulation [15.3] in multiturn coils. It is with this knowledge that IEEE Std. 1310 was developed to simulate accelerated thermal cycling duty on stator winding insulation to determine the effects on particular large generator bars and multiturn coils. This test is normally only performed on coils or bars rated at 3 kV or more, and is intended for use in machines that see full load changes in less than a few minutes. Bars and coils subjected to this test cannot be placed in service.

The purpose of this test is to qualify coil design and manufacturing processes for their relative ability to resist deterioration due to rapid heating and cooling due to load cycling. The test is primarily intended for machines in which the windings are indirectly cooled by air. This test does not include a simulated core and therefore does not identify any deterioration of the groundwall surface or structure from movement relative to the core.

This test makes use of I^2R heating, from passing AC or DC current through the test coils, and forced-air cooling to produce fairly rapid thermal cycling. Temperature variations depend on the class of insulation being tested; e.g., for a Class F insulated winding, a typical temperature test profile would begin at 40°C, increase to 155°C, then decrease to 40°C. The test coils are typically subjected to 500 thermal cycles. As soon as the maximum coil temperature is reached, cooling is immediately initiated.

The rate of insulation deterioration in the test coils can be determined by performing nondestructive diagnostic tests prior to the commencement of the thermal cycling, at predetermined intervals during the test, and after the test is completed. The diagnostic tests used include dissipation factor and power factor tip-up, partial discharge, physical measurements, and a tap test to detect groundwall insulation delamination. The surge test (Section 15.1.3) can be used for multiturn coils to determine if significant aging of the turn insulation is occurring from thermal cycling.

Test Method. Both IEEE 1310 and IEC 60034- Part 18, Section 34, describe test procedures for complete coils and bars. However, the IEEE procedure is more closely defined and, thus, results from different manufacturers will be more comparable if IEEE 1310 is specified. Detailed test procedures are contained in IEEE 1310 and Reference 15.3 and are summarized as follows:

1. Obtain the test specimens from the coil manufacturer. The minimum number of specimens should be four bars or two coils having four legs. This does not include one or more specimens to be used as temperature control bars, which have a temperature measuring device embedded in them to monitor coil surface and conductor temperatures.

2. Visually examine the coil specimens to ensure they are in good condition prior to starting the test. Apart from detecting any obvious manufacturing defects, the intent is to ensure the absence of any damage during shipment.

3. Perform baseline tests such as capacitance, dissipation factor, and partial discharge on each test coil. An IEEE Std. 522 impulse voltage test can also be included if turn insulation degradation due to thermal cycling is to be evaluated.

4. Instrument the temperature control bars with temperature measuring devices such as thermocouples, thermistors, or fiber optic sensors.

5. Connect the coils into a test circuit that allows AC or DC current to be passed through them to induce heating.

6. Install a chamber, with cooling air fans, around the test coil assembly.

7. Install a temperature-controlled circuit that energizes the coils during the heating cycle and the fan(s) during the cooling cycle.

8. Start the test with a predetermined number of thermal cycles, typically 50.

9. After 50 cycles, stop the test, remove the cooling chamber, disconnect the coils, and perform nondestructive tests, as defined in step 3, on the coils.

10. Reconnect the coils, reinstall the cooling chamber, and recommence cyclic testing. Repeat these nondestructive tests at predetermined intervals, typically 100, 250, and 500 cycles.

11. At the end of the thermal cycling test, some additional proof tests, such as AC or DC hipot to breakdown, turn insulation impulse to breakdown, and voltage endurance, may be performed to determine the voltage-withstanding capability of the aged insulation.

12. Coil dissection should be performed to obtain engineering information that could improve coil materials and manufacturing techniques. Specific attention should be paid to the condition of insulation at hipot test failure points.

This is an involved and expensive procedure. It should be required only for large, critical machines subject to fast, frequent load changes.

Acceptance Criteria.   The coils tested represent the insulation systems to be qualified for use in motors or generators. The particular insulation system is considered to be suitable for cycling duty provided the test coils display the following characteristics:

- No significant increases are observed in dissipation factor, capacitance, and partial discharge values between the start and finish of the test.
- Coils can withstand hipot proof testing at the end of the test.
- Multiturn coil turn insulation can withstand the IEEE Std. 522 or IEC 60034-15 impulse test voltage at the end of the test.
- Dissected bars or coils do not show significant delamination of groundwall insulation.
- Dissected multiturn coils do not show significant deterioration of bonding between conductor strands or turn insulation degradation.
- There are no visible signs of significant degradation of semiconducting layer or grading material.

15.1.6 Thermal Classification Tests

As discussed in Sections 2.1.1 and 2.3, all insulation systems are given a thermal classification. This classification defines the normal maximum operating temperature for the winding, although the expected life is not defined. For example, a Class F or Class 155 system defines a normal maximum operating temperature, at the hottest spot, of 155°C.

The classification is done in accordance with test methods specified by IEEE and IEC. For example, IEEE 275 and IEC 60034, Part 18 detail the test procedure for form-wound groundwall insulation systems. Such procedures require determining the thermal life of a new insulation system in model coils under accelerated aging tests, and compare the results to an insulation system with proven performance in service (Section 2.3). Thus, the expected service life at the temperature class is not stipulated. To have meaning, purchasers of windings need to not only specify the thermal class of a winding, but also specify the expected service life (for example, 10, 30, or 40 years). Virtually all insulation systems in present-day use were compared to older insulation systems that were operated at many tens of degrees below the thermal class. Thus, the satisfactory service life experienced with the older insulation systems usually does not imply that the same life can be expected if operated at an operating temperature defined by the thermal class. For example, it is very unlikely that a Class F winding would operate for more than 5 or 10 years or so at a copper hot spot temperature of 155°C.

The thermal classification of a stator (or rotor) winding is normally only evaluated when a new insulation system is introduced. Thus, it would be very unusual for a winding purchaser to specify a thermal classification test be done to qualify a vendor's insulation system, since the test is very expensive. However, it is not unreasonable to ask the manufacturer for the test data of their present system, and what service-proven system it was compared against.

15.2 STATOR WINDING INSULATION SYSTEM FACTORY AND ON-SITE TESTS

There are a number of tests that can be performed to give a good indication of the quality of a new coil/bar or a new winding that is intended for service. Again, it is important to have a technical specification for new windings that states what tests are to be performed and the acceptance criteria for the results. This is especially true for tests that give quantitative results rather than those that have pass/fail acceptance criteria.

It is important to note that no single test can give complete assurance of insulation system quality, so a number of different tests that evaluate the different components are required. The factory tests described below also apply to large hydrogenerator and turbogenerator windings that need to be performed on-site.

15.2.1 Insulation Resistance and Polarization Index Tests

These tests, as described in Section 12.1, are designed to determine whether coils/bars or windings are clean and dry and do not have major ground insulation defects. The tests should always be done before any high-voltage AC or DC hipot tests (Section 15.2.2) or surge tests (Section 15.2.3).

Factory Tests. The IR/PI tests should be performed at the following stages of new machine manufacture:

- After manufacture of the bars or coils.
- During coil or bar winding. Usually, sections of the winding are tested after installation in the slots and wedging.
- After all the bars/coils are installed, wedged, braced, and connected. For global VPI windings, these tests should be performed prior to VPI to ensure the winding is clean and dry.

- After the winding is fully processed, but before assembly.
- After the machine is completely assembled. It should be noted that such tests should be performed on directly water-cooled stator windings before the cooling system is filled with coolant.

Once the winding is fully processed, the DC test voltage and minimum IR and PI values given in IEEE 43-2000 should be met. Note that older versions of IEEE 43 had considerably less stringent acceptance levels. For coils and bars that have not been impregnated yet, each manufacturer will determine the acceptance levels.

Site Tests. Baseline tests should be performed before new or rewound stators are put into service and before any on-site high-voltage tests such as an AC hipot test.

For motors and air-cooled generators, such tests are most easily performed from the local MCC or breaker to also verify that the cabling to the machine is clean, dry, and free from significant ground insulation defects. If low readings are obtained, the power supply cables or bus must be disconnected from the stator windings so that each can be tested separately to determine which component is causing the low IR and PI values. If the machine has space heaters, it is important that these be energized for at least two days before such tests are performed to ensure that there is no moisture on the winding surfaces that could affect the results.

To obtain satisfactory readings from water-cooled stator windings, the winding must be completely drained of water and a vacuum pump used to obtain very low moisture levels in the water-cooling circuits.

15.2.2 AC and DC Hipot Tests

These tests are performed to determine the integrity of the winding ground insulation during the installation of new or replacement windings and before a winding is placed in service. As indicated in Sections 12.2 and 12.4, the magnitude of the test voltage and the use of an AC or DC test is a function of where and when the test is performed. In general, AC hipot tests are more revealing than DC tests and, thus, AC tests are preferred [15.4]. If a hipot test failure occurs, a repair or rewind is required. It is important that windings pass an IR and PI test before any high-voltage testing is done.

Factory Tests. The following tests should be performed at various stages of winding installation:

1. After coil or bar manufacture. The manufacturer will determine what the levels should be, depending on whether the coils are impregnated or "green" [15.4].
2. During coil or bar winding and after IR and PI tests, sections of winding are hipot tested following installation in the slots. This procedure is designed to detect faulty coils or bars at a stage at which they can be easily removed and replaced. The test level, as with all hipot tests before complete assembly, are generally set by the manufacturer. There are wide variations among manufacturers [15.4]. The magnitude of the hipot voltage is a function of whether the coils are "green" for a global VPI or fully processed. For green coils, a test value of 50 to 60% of the final complete winding, one-minute hipot test value is a minimum test voltage. On the other hand, values of 1.33 (or more) times the final hipot test value are used for fully processed turbine generator and hydrogenerator bars to increase the assurance of success of subsequent hipot tests.

3. After all the bars/coils are installed, wedged, braced, and connected. For global VPI windings, these tests should be performed prior to VPI to ensure the winding is clean and dry. The minimum hipot voltages used for this test would be the same as those described in step 1.

4. After the winding is fully processed, but before assembly, a full-voltage one-minute AC or DC hipot test should be performed, i.e., $2E + 1$ for an AC test and $(2E + 1) \times 1.7$ for a DC test. These are set by standards such as NEMA MG1 and IEC 60034.

5. If the winding is a sealed type, a water immersion or spray test per NEMA MG1-2000, Section 20.18 or IEEE 429 should be performed. This includes a one-minute hipot test at 1.15 times of the winding rated line-to-line voltage while the stator is immersed in or after it has been soaked for 15 minutes with water containing a wetting agent. IR and PI tests are required both before and after this hipot test.

6. After the machine is completely assembled, the test described in step 4 should be repeated. For very large generators this test may be performed on-site. It should be noted that on direct-water-cooled stator windings, such tests should be performed before the cooling system is filled with coolant, especially if a DC hipot test is to be performed.

On-Site Tests. On-site tests should be performed after the installation of a new machine or rewound stator. For motors and air-cooled generators, this test is normally performed from the local breaker or MCC. Larger hydrogen-cooled generators are normally tested by removing the isolating links connecting them to the insulated phase bus (IPB) and testing from the generator-side bus. Removal of the neutral connection is required to test each phase separately.

Most on-site testing of new stators in North America is done with AC [15.4]. On-site tests in Japan and Europe are usually AC hipot tests. As discussed in Sections 12.2 and 12.4, hipot test levels for windings having seen service are typically around 75% of the value for a new-winding factory test. If a new winding has been installed on-site, it should be subjected to a full $2E + 1$ AC hipot test before placing it in service. If the winding is direct-water-cooled, this test should be performed before its cooling system is filled with water.

15.2.3   Impulse (Surge) Tests

As described in Section 12.12, the integrity of multiturn coil turn insulation can be checked by impulse or surge testing. This is most important as a quality check of new coils prior to and after their installation in the stator slots. The test described below is not relevant for Roebel bar windings since there is no turn insulation. Conventional lightning impulse (1.2 μs/50 μs) tests are rarely performed on the groundwall insulation since the AC and DC hipot tests are better indicators of the groundwall condition.

Factory Tests. Surge testing per IEEE Std. 522 or IEC 60034-15 should be performed at the following stages of winding installation:

- Prior to installing coils in slots
- After coil installation, wedging, and bracing but before coil interconnections are made
- After winding connection to verify that this has been correctly done; incorrect connection will show up as a phase impedance imbalance when a surge comparison test is done

When the test is done on complete windings, high-voltage interturn stress only occurs in the first few turns or coils. Thus, only the turn insulation in this part of the winding is tested. As noted in Section 12.12, determining if turn insulation puncture has occurred can be especially difficult in complete form-wound windings. Surge tests done according to IEC 60034, Part 15 are usually less stringent since this standard allows testing with a 0.5 μs risetime surge. This longer risetime means there is less voltage developed across the turn insulation.

On-Site Tests. On-site tests are assumed to be on complete windings. As discussed above, a surge test on a complete winding will only test the integrity of the turn insulation in the coils connected to the phase terminals. Also, especially in form-wound stators, it can be difficult to determine when the turn insulation has been punctured. Thus, on-site testing is relatively rare, at least for form-wound machines.

15.2.4   Strand-to-Strand Test

This test is performed on new stator bars during manufacture to check the integrity of the strand insulation, which is normally a double layer of Dacron glass or a single layer of Dacron glass over a varnished or enameled surface. This is particularly applicable to Roebel bars that have transposed strands in which the mechanical forces required to bend the strands at the transition may also cause insulation damage. For liquid-cooled, stators this test must be performed before the bar nozzles are installed and brazed to each end of the bar.

The test involves performing a one-minute AC hipot test between all adjacent strands in each bar. This is a good quality check on the strand insulation and typically 110–220 V AC is used. If the strand insulation fails this test, the bar should be scrapped or repaired if feasible.

15.2.5   Power Factor Tip-Up Test

This is a well-established test to determine if form-wound coils and complete windings have been properly impregnated. It is mainly a QC test for individually impregnated coils. The test is described in Section 12.9.

Standards such as IEEE 286 and IEC 60894 do not provide acceptance levels. With modern insulation systems, many users require the tip-up test on all individual coils or bars (with the silicon carbide coating guarded out) to be below 1% (test at 25% and 100% of rated line-to-ground voltage). However, it is typical to require about 75% of the bars or coils to have a tip-up of 0.5% or less. Acceptance levels on complete windings have not been widely accepted since the silicon carbide coatings can not be guarded out, resulting in a tip-up that depends both on the quality of the insulation and the characteristics of the silicon carbide coating. Thus, acceptance levels for complete windings can only be set in comparison to other similar windings.

Factory Tests. The test is performed on individual coils after impregnation, but before the hipot tests. Some manufacturers measure the tip-up on all coils and bars, whereas others do tests on a statistical sampling basis. The test is not relevant for green (unimpregnated) coils and bars. Although some measure the tip-up on complete windings, the test can normally only serve as a baseline measurement for future trending. The off-line PD test is more relevant.

On-Site Tests. Since on-site tests at site are assumed to be on complete stators, any tip-up test will only serve as a baseline for trending, rather than an acceptance test.

15.2.6 Partial Discharge Test

The test can be used as a quality check for the impregnation and bonding of ground insulation in new windings, similar to the power factor tip-up test (Section 15.2.5). However, the PD test can also help to determine the root cause of any poor manufacturing. The off-line PD test is also better than the tip-up test at finding problems in complete windings, such as global VPI stators. The test is only relevant for machines rated at 3.3 kV and above, although lower-voltage windings to be operated from an inverter drive may benefit from the test [15.5]. Off-line PD tests (Section 12.10) require specialized equipment that includes a PD-free power source or a means of separating machine winding PD from power source noise.

Factory Tests. Specification of acceptance criteria for factory tests will ensure that a well-impregnated and bonded ground insulation system has been provided. Peak PD magnitude acceptance limits are somewhat dependent on:

- The instrument used to measure PD
- The type of sensors used
- The machine voltage
- The cooling gas, i.e., air or hydrogen

The acceptance criteria should, therefore, be based on comparison to the results of tests on well-consolidated windings of similar voltage rating and the same type of cooling system, and measured with the same particular instrument and sensors [15.6]. There are no broadly agreed upon limits to PD, although each manufacturer may have its own in-house limits. Acceptance levels could be set using statistical process control principles, by which coils, bars, or windings with PD greater than, say, the mean plus 3 standard deviations are not accepted. The mean refers to the manufacturer's total database of results for similar bars, coils, or windings using the PD measurement method used in the plants [15.6].

Where possible, PD tests should be performed on individual coils or bars as well as on fully impregnated individual phases. As described in Section 12.10.3, the PD patterns can sometimes indicate if the problem is at the copper, within the groundwall, or on the coil or bar surface.

On-Site Tests. PD tests are assumed to be on complete windings. Acceptance levels can only be set in comparison to similar machines, or by specifying that there will be little increase in PD during the warranty period of the machine [15.6].

15.2.7 Capacitance Test

The capacitance test (Section 12.5) can aid manufacturers in determining if the groundwall has been fully impregnated and cured. Since it is only a relative test, it is rarely used as an acceptance test. There are no industry standards for the test method.

Factory Tests. Capacitance can be measured during global VPI winding processing, as an indication that resin impregnation is complete. A low-voltage or high-voltage capacitance bridge is connected to the winding during the resin impregnation process. The capacitance will increase with time. If capacitance versus time is monitored, complete impregnation is indicated when the capacitance stops increasing. This method is not foolproof since flattening out of the capacitance versus time curve may be an indication that no more resin can penetrate the winding, but that there may still be voids in it. As the coil or winding is cured, the capacitance will decrease and level out.

Capacitance tip-up can also be used as an acceptance criterion for new windings (Section 12.6). However, there appears to be no widely accepted pass/fail levels and the silicon carbide coating distorts measurements on complete windings.

On-Site Tests. A baseline complete phase or winding test can be performed on-site to allow periodic testing to detect winding groundwall degradation. This test should be performed from the stator winding terminals.

15.2.8   Semiconductive Coating Test

This test, as described in Section 12.14, is used to check the condition of the semiconductive coating on coils and bars, usually only found on windings with voltage ratings of 6 kV and above. This test can be performed on new windings as a quality check on the semiconductive coating applied. It is also a useful on-site check to obtain a baseline to determine if the semiconductive coating has deteriorated in service.

The test is performed by measuring the resistance between each bar or coil leg semiconductive coating and the core at each end, as described in Section 12.14. The semiconductive coating on new windings can be considered to be well made if resistance values are all below 2000 ohms per square.

15.2.9   Wedge Tap

This test is used to determine the tightness of stator slot wedges that are installed to prevent radial displacement of the bar or coil in the slot. In general, this test is not applicable to stators with global VPI windings since the wedges should be tightly "glued" or bonded in place to prevent coil or bar movement.

Factory Test. It is important that the wedges in a new or rewound stator are initially tight since they loosen under the influence of electromagnetic forces acting on the coils or bars during machine operation. It is therefore advisable, especially for large high-voltage machines, to specify that this test be performed before assembly, in accordance with one of the procedures described in Section 12.16.2, and that the results meet the following acceptance criteria:

- No more than two adjacent wedges in the same slot are to be loose.
- No end wedges are to be loose.
- No more than 25% of the wedges in the complete winding are to be loose.
- None of the installed wedges should be cracked.

On-Site Test. For large generators, this test is sometimes performed on-site before insertion of the rotor. If a stator is rewound on-site, this test should be performed with acceptance criteria the same as indicated above. It is important to obtain baseline test values for comparison with subsequent test results after the generator has been placed in service. For new machines, a wedge test after a year of service is critical for determining the scope of warranty work and for arresting any damage due to loose bars after the initial period of settling in the slot.

15.3 FACTORY AND ON-SITE TESTS FOR ROTOR WINDINGS

This section deals with appropriate factory and on-site tests for all insulated rotor windings. It also covers tests for specific types of windings. As a general rule, all of these tests can be performed during winding manufacture and installation. On the other hand, it is not practical to perform some of these tests on-site on a complete winding. For smaller motors and generators that have been in service, it is often more convenient to transport them to a local repair company to perform such tests.

15.3.1 Tests Applicable to All Insulated Windings

Insulation resistance (IR), polarization index (PI), and hipot tests can be performed on all three types of insulated windings covered in this section. These tests can be performed both during manufacture or rewinding and on-site, since the equipment required to perform them is portable.

IR and PI Tests. As indicated in Section 12.1, these tests are performed to confirm whether the winding is clean, dry, and free from major flaws such as cracks. These tests should be performed in accordance with IEEE Std. 43-2000, which gives test voltages and minimum acceptable IR and PI values for each type of winding. These tests should always be a prerequisite to a hipot test, because low IR values increase the risk of failure. It should be noted that for random-wound rotor windings, PI is usually not very meaningful.

Tests during Winding Manufacture. During winding manufacture these tests should be performed at the following stages:

- After coil installation, but before interconnecting coils to form a winding
- Before resin or varnish treatment or curing to ensure that the winding is dry so that no moisture is trapped in it
- After final winding processing, but before rotor installation
- After the machine is completely assembled.
- Prior to any hipot tests

On-Site Testing. On-site, these tests should be used to confirm that the winding is clean, dry, and free from serious ground insulation defects before it is placed in service.

Hipot Tests. As for stators, rotor winding hipot testing is performed to check the integrity of the winding ground insulation. The DC hipot test (Section 12.2) is used for synchronous machine rotor windings since they operate with an applied DC voltage. For wound-rotor windings, either an AC hipot test (Section 12.4) or a DC hipot test (Section 12.2) can be performed.

Factory Tests. Hipot tests should be performed at the following stages of manufacture, but prior to final winding processing. The test voltage should be less than the final test value since complete winding integrity is not present until that time.

- After coil installation, but before interconnecting coils to form a winding
- Before resin or varnish treatment or curing to ensure that the winding is dry and that no moisture is trapped in it

- After final winding processing, but before rotor installation
- After the machine is completely assembled

On-Site Tests. Periodic hipot testing of rotor windings on-site is not commonly done. The main reason is that, compared to the stator, rotor operating voltages are much lower and, if their IR and PI are acceptable, the groundwall insulation is not likely to fail. Hipot levels for windings that have seen service are usually about 75% of the acceptance level.

15.3.2 Round-Rotor Synchronous Machine Windings

One of the most critical components in this type of winding is the turn insulation. Consequently, it is important to check the integrity of turn insulation during manufacture, during repairs, and if the machine operating performance indicates the presence (or not) of turn shorts. The following tests can be used to detect shorted turns, although they are not effective for finding weak (but unfailed) turn insulation:

- RSO test (Section 12.21)
- Surge test (Section 12.21)
- Voltage drop test (Section 12.20)
- Airgap flux probe test (Section 13.6)

The RSO and surge tests can be performed on the rotor by itself or while it is installed in the stator. The airgap flux probe test is performed either during a spin-pit test or when the machine is put into service. On the other hand, the rotor has to be removed from the stator bore to perform the voltage drop test.

At operating speed, radial conductor loading, due to centrifugal forces, can induce turn shorts that otherwise have a high resistance without these forces. To simulate this at standstill, checks for turn shorts should be performed at four rotor positions 90° apart. Using this technique, the coils in the 4 to 8 o'clock positions experience radial loading due to gravity, which may induce turn shorts similar to those due to radial conductor loading at speed.

Factory Tests. Tests for shorted rotor turns are performed as a quality check to detect defective turn insulation after the rotor coils are installed and wedged. It is advisable to perform the first test before the retaining rings are installed since removal of coil turns to repair defective turn insulation is much easier at this stage of assembly. A second test should be performed after the retaining rings are fitted. The radial forces exerted on the end-windings by the assembly process for these rings can convert high-resistance contacts in weak insulation to turn shorts. If it is possible to run the rotor in a spin pit, or with the machine assembled and the stator winding shorted, the airgap flux test can be performed to check for shorted turns before shipment [15.7]. This is the best test since full radial conductor loading, due to centrifugal forces, is present.

On-Site Tests. As already indicated, the RSO and surge tests are easiest to perform on-site and they can be conducted with the rotor installed. If the machine has a brushless exciter, it has to be disconnected to allow direct access to the two ends of the winding. Such tests are performed to give baseline data for comparison with those from future tests to detect the condition of the rotor turn insulation.

If an airgap flux probe is fitted, baseline test data should, if possible, be obtained from a short-circuit test [15.7].

15.3.3 Salient Pole Synchronous Machine Windings

It is important that these windings be checked for shorted turns during manufacture, if performance indicates that such a fault may exist and as a routine predictive maintenance tool. Strip-on-edge type windings are most susceptible to shorted turns, due to the fact that most of their conductors are not insulated on the edges (see Chapter 1).

Factory Tests. These windings should be checked for shorted turns by performing a surge test or voltage drop test at the following stages of manufacture:

- After coil winding, but before consolidation with varnish or resin
- After coil consolidation, but before connection to form a winding
- After winding connection
- After the machine has been run

Surge testing (Section 12.21) or voltage drop testing (Section 12.20) are the best methods for checking this type of winding for shorted turns in the factory.

On-Site Tests. The voltage drop test is the most widely applied test to detect shorted turns in salient pole rotors on-site.

15.3.4 Wound Induction Rotor Windings

As with stator windings, the three-phase rotor windings in these machines are susceptible to shorted turns. In addition, bar-type windings are prone to having high-resistance brazed or soldered clip connections.

Surge testing (Section 12.21) is the most effective method for detecting shorted turns and incorrect winding connections. It can also detect the presence of high-resistance winding connections. This should be performed at the following stages of manufacture:

- After winding and wedging of individual coils or coil groups, but before connection
- After winding connection, but before resin or varnish impregnation
- After winding impregnation
- After the motor has been run

Another effective method of detecting high-resistance connections is to pass an AC or DC current through the winding and scan the connections with a thermovision camera. High-resistance connections will be indicated by much higher temperature, as seen by the thermovision camera, than those that are properly connected.

15.3.5 Squirrel Cage Rotor Windings

The main concerns with this type of winding both during manufacture and in service are:

- Cracking of the bars and short-circuit rings in fabricated rotors
- Air pockets in diecast aluminum windings
- Poorly brazed or welded bar-to-short circuit connections

Factory Tests. It is important that quality checks to detect flaws in rotor cage windings be performed during manufacture. These should be done after the winding is installed. The most common tests used are:

- Growler test (Section 12.22)
- Fluorescent die penetrant test (Section 12.23)
- Rated flux test (Section 12.24)
- Single phase rotation test (Section 12.25)
- Bar-to-short circuit ring ultrasonic tests

In addition, for diecast rotors, a plug reversal and vibration test or X-ray examination is often performed to detect the presence of air pockets. The plug reversal and vibration test involves running the motor at no load, measuring vibration levels, then disconnecting and reversing two leads to make the rotor stop and run in the opposite direction. This has the effect of heating up the rotor winding. If air pockets are present, the motor vibration levels will increase significantly after a few cycles, due to nonuniform bar expansion, which bends the rotor. The test is particularly good for two-pole motors, which are most sensitive to rotor imbalance.

On-Site Tests. By far the most effective test to check for rotor cage winding breaks or air pockets is stator current signature analysis as described in Section 13.7. This involves using an instrument that measures stator current and performs a spectrum analysis for side bands that are two times slip frequency removed from the fundamental 50 or 60 Hz current frequency. A baseline test, for new rotor windings, should be performed with the motor loaded to at least 50% of rated load if this type of monitoring is to be used to obtain early detection of rotor cage winding breaks. The number of rotor bars should also be noted.

15.4   CORE INSULATION FACTORY AND ON-SITE TESTS

These tests can be performed on new cores either in the factory or on-site. However, some are more difficult to perform on-site. This section provides details on when the tests should be performed and how easy it is to perform them on-site.

15.4.1   Core Tightness Test

As indicated in Section 14.1, this test involves trying to insert a standard winder's knife blade, with a maximum thickness of 0.25 mm, between the laminations at several locations around the core bore (stator) or core outside diameter (rotor). If the blade penetrates the section of core being tested by more that 5 mm, then it is assessed to be loose.

This test should be used to check the tightness of new cores before the windings are installed in them. This would normally be done in the manufacturer's plant. It can also be used on large hydrogenerators cores that are built on-site.

15.4.2   Rated Flux Test

The rated core flux test (also called the "ring flux" or "loop" test) described in Section 14.2 is the traditional method of determining the insulation integrity of any type of laminated stator core in an AC motor or generator.

Factory Tests. For the following reasons, this is the best test for checking the integrity of the interlaminar insulation in new stator cores:

- It gives an indication of the quality of the core insulation by means of the power required to induce rated flux in the back of the core. If a commercial core flux tester is used, these losses are usually expressed in watts per kilogram or watts per pound of core.
- It gives an indication of the severity and location of core insulation damage (see Section 14.2.3) by the magnitude of hot spot temperatures and the time they take to develop.

This test should be performed before the stator windings are installed, since repairs of core insulation damage is much easier at this stage of manufacture. Since there is a danger that core insulation surface shorting could be induced during the installation of the slot wedges, it is advisable to repeat the test after the wedges are installed.

On-Site Tests. The main reasons for performing this test on a new stator on-site would be:

- If the machine is a hydrogenerator for which the core was built on-site.
- If a baseline EL-CID test indicated a significant core insulation defect.

15.4.3 EL-CID Test

The EL-CID test described in Section 14.4 is most suited for on-site testing since the test equipment is portable and easy to install.

Factory Tests. The main reason for performing a factory test on a new stator core is to obtain baseline data for comparison with future site test results. This would allow identification of developing general and local core insulation deterioration.

On-Site Tests. On-site EL-CID tests on new cores are normally only performed if a baseline test was not performed in the manufacturer's plant or if the core was built on-site.

REFERENCES

15.1 G. C. Stone, B. K. Gupta, J. F. Lyles, and H. G. Sedding. "Experience with Accelerated Aging Tests on Stator Bars and Coils," paper presented at IEEE International Symposium on Electrical Insulation, Toronto, 356–360, June 1990.

15.2 IEEE 1553-2003, "Draft Standard for Voltage Endurance Testing of Form Wound Coils and Bars for Hydrogenerators."

15.3 G. C. Stone, J. F. Lyles, J. M. Braun, Ontario Hydro, Canada, and C. L. Kaul, Virginia Power, Richmond, VA, "A Thermal Cycling Type Test for Generator Stator Winding Insulation," *IEEE Transactions on Energy Conversion, 6,* 4, 707–713, December 1991.

15.4 B. K. Gupta, G. C. Stone, and J. Stein, "Use of Machine Hipot Testing in Electric Utilities," in *Proceedings of IEEE Electrical Insulation Conference,* Cincinnati, October 2001, pp. 323–326.

15.5 G. C. Stone and S. R. Campbell, "Examples of Stator Winding Discharge Due to Inverter Drives," in *Proceedings of IEEE International Symposium on Electrical Insulation,* Anaheim CA, April 2000, pp. 231–234.

15.6 G. C. Stone, "Using the Partial Discharge Test as a Winding Acceptance Test," paper presented at Iris Rotating Machine Conference, Scottsdale, AZ, June 1999.

15.7 J. W. Wood and R. T Hindmarsh, "Rotor Winding Short Detection," *IEEE Proceedings, 133,* pt. B, 3, May 1986.

CHAPTER 16

MAINTENANCE STRATEGIES

Previous chapters have provided information on the types of materials used in motors and generators, as well as the design of the insulation systems in windings and cores. The aging processes of these insulation systems were also described, along with various on-line and off-line diagnostic tests that can be used to determine if specific deterioration mechanisms are occurring. This chapter will discuss how the above information, together with the results of visual inspections, can be applied to maintenance strategies to assess the insulation condition in any particular rotating machine.

Even for experts, there are considerable difficulties in trying to estimate the condition of a winding. Tests or inspections that have a likelihood of destroying all or part of the insulation should normally be avoided. In addition, available diagnostic tests are not always good enough to predict insulation condition for all possible aging mechanisms. Also, nondestructive visual inspections cannot find all possible problems, but can often identify insulation system degradation in its early stages before on- or off-line diagnostics can do so.

This chapter will give guidelines on procedures to assess the insulation and indicate the relative condition of any particular machine. In many cases, enough information is given to indicate if insulation failure is imminent, or whether the winding can be expected to survive until the next major outage. The emphasis is on assessing the general condition. Since suitable tests and expertise are not available for all likely deterioration mechanisms, it is not considered feasible to accurately estimate insulation condition without performing some machine disassembly to allow a visual inspection. Information on the insulation condition is necessary to decide whether the winding may readily be repaired, thus arresting the degradation, or whether a major rewind or reconstruction must be considered.

16.1 MAINTENANCE AND INSPECTION OPTIONS

The specific application and type of business will influence the inspection and maintenance strategy adopted by a user of electrical machines. Any continuous process operation, such as a utility generating station or oil refinery, usually justifies some form of planned maintenance program since the consequential costs of unscheduled downtime can significantly exceed the required investment.

The extent of an inspection and maintenance program should be geared to a machine's application, consequences of failure (including lost production and revenue), redundancy, complexity, and value.

The inspection and maintenance options are as follows and, in practice, a cost-effective strategy would contain a mixture of these options:

- Breakdown or corrective maintenance
- Time-based or preventative maintenance
- Condition-based (or predictive) maintenance

16.1.1 Breakdown or Corrective Maintenance

This is essentially a "no maintenance" option and may be cost-effective for low-value equipment for which it can be expected that outages from winding failures would have no safety or serious economic consequences to plant system operation. In carrying out this option, it must be recognized that a failure during service may extensively damage the machine to the extent that it cannot economically be repaired and so it must be replaced.

Decisions on whether to repair or replace a winding following breakdown will usually be determined by cost. Assessment of insulation failures using the guidelines listed in Chapters 8, 9, and 10 may indicate that changes are required either in the installed equipment or in maintenance strategy, to avoid future in-service failures. A simple example is the replacement of an open-type motor enclosure by a totally enclosed equivalent when repeated failure is attributed to winding contamination and the cost of routine cleaning is not practical or cost-effective. In another case, stator winding failures due to end-winding movement could lead to a rewind with a bracing system specifically designed for repeat starts. Similarly, sustained overheating (overloading) may indicate the need for a machine with a higher rating or rewinding with a higher temperature class of insulation.

The advantage of this option is that there is no investment in any planned program, although it requires a larger spares inventory, with significant associated costs, to cover the inevitable failures of critical machines. A disadvantage is the constant exposure to a breakdown that can have significant repercussions; for example, significant production system downtimes and repair costs.

16.1.2 Time-Based or Preventative Maintenance

This option involves performing inspections and repairs during scheduled outages or turnarounds. The timing of the outages may not, however, be optimal for electrical machines since outages are often dictated by other considerations. With this maintenance option, inspection intervals vary from 1 to 10 years, depending on the type of machine and if outages/turnarounds are dictated by other equipment in the plant.

A key ingredient in developing time-based maintenance programs is experience with the machines. It is necessary to build up confidence with a specific design to assess when maintenance is required. Where schedules permit, more frequent inspections should be conducted on a new machine until such time as the results give the assurance to increase the intervals. In addition, old machines usually also require more frequent inspections and tests.

Although the aging mechanisms described in Chapters 8, 9, and 10 justify routine inspection and maintenance activities throughout a machine's life, the following additional concerns merit conducting an initial inspection shortly after the machine is placed into service:

- Failures in machines of similar design
- New design features
- Incorrect installation
- Settling in/premature degradation of windings
- Actual versus specified environment
- Actual versus specified operating regime

Any inspection and maintenance program should be flexible and sensitive to sudden changes in insulation condition in response to new operating requirements. For example, changing a generating unit's mode of operation from base load to maneuvering or peaking duty can cause winding loosening and abrasion due to thermal cycling and acceleration of other generic insulation degradation mechanisms in both the turbine generator and large motors in auxiliary systems.

A manufacturer normally introduces design changes following an appropriate development program. This work may have resulted from technological change (e.g., replacement of asphalt insulation by an epoxy system), cost reduction (e.g., changing to a global stator winding VPI system from a nonglobal VPI winding), or the need to eliminate a known problem (e.g., an inadequate end-winding support arrangement). Verification of these new features comes from initial operation of production units incorporating the changes. An early inspection either detects incipient design weaknesses before significant damage occurs or gives some assurance that generic design problems do not exist. Such reassurance may be applicable to a population of similar machines following a careful examination of the first machine of that type. Nonetheless, some forms of degradation require extended time to become apparent. Therefore, a continued program of inspection of the prototype machines over several years is prudent.

Although cores and windings are assembled to be tight when shipped from the factory, they may settle during initial operation under the influence of vibration and thermal cycling. As indicated in Chapters 8, 9, and 10, many electrical failures are actually mechanical in origin, i.e., the original insulation capability is degraded due to movement or mechanical failure to a point where dielectric breakdown occurs. For this reason, it is essential to arrest any progressive deterioration in the winding support systems at an early stage. Experience has shown that windings retightened after a year or two in service will still be tight and in good condition when examined again following 5 to 10 years of operation.

Equipment is purchased assuming a certain operating environment, including atmospheric temperature, humidity, type of operation, voltage surges, etc. If the actual environment differs from the specified environment, the insulation may age more rapidly than anticipated.

Required maintenance activities during an outage will be determined by the results of diagnostic tests, inspections, and assessment techniques described in Chapters 12, 13, and 14. An example of this option is to inspect a turbine generator initially after one year of service and every five to six years thereafter.

The advantage of the time-based option is that outages can be planned in advance and resources allocated accordingly; where the flexibility exists to adjust the outage timing based on previous inspection results, it can be very effective. However, there are several disadvantages to this option:

- Machines that are in good condition will be removed from service when not necessary, incurring unnecessary labor and outage costs. In addition, the machine is exposed to the risk of inadvertent damage or incorrect reassembly.
- The outage or turnaround scheduling may be dictated by other requirements.
- A preventative inspection may, on occasion, be deemed expendable in the event of manpower limitations or other corporate considerations.
- Failure mechanisms that develop in a relatively short time can still occur between inspections.

Since most machine windings are very reliable, most planned outages are not needed.

16.1.3   Condition-Based or Predictive Maintenance

An ideal maintenance program would initiate corrective activities on an electrical machine only when a problem is known to exist. With the advent of more advanced on-line diagnostic techniques, condition-based programs, rather than preventative or corrective maintenance programs, have become very popular in recent years.

Assessment of a machine's insulation condition requires the provision of appropriate on-line monitoring devices plus the application of selected off-line diagnostic tests, as described in Chapter 13. The normal (healthy) signatures from each monitoring device must be known so that any deviation provides useful intelligence. Therefore, baseline test data, as indicated in Chapter 15, should be obtained.

Installed monitoring devices might include rotor and stator winding temperature detectors, accelerometers, condition (or core) monitors, airgap flux probes, and partial discharge detectors. Routine monitoring of these sensors, together with auxiliary system status and actual operating load, will give an indication of machine deterioration. Outages can be scheduled to confirm any problems and to effect repairs or maintenance when significant deviations from baseline are detected.

The advantage of this maintenance option is that it will maximize the equipment's availability while minimizing resources invested in routine maintenance activities. To be successful, this type of program may require a significant investment in monitoring equipment, in personnel to perform the monitoring and assess the results, and in the integration of sensors into the machine, either during original manufacture or disassembly of the machine during an outage. It also requires a good knowledge of the design and construction of the machine and expertise in interpreting the significance of deviations from normal. However, confidence must be built up concerning the relevance, accuracy, consistency, and reliability of the monitoring. Unfortunately, it is not known how to monitor all the failure processes that can occur in a machine. As a consequence, sufficient on-line monitoring that incorporates all these functions is, as yet, not available to effectively implement this option, unless at least some short-term inspection outages are taken.

Even with an effective monitoring program in place, the opportunity should always be taken to conduct a visual inspection any time a machine is opened for other maintenance purposes (Section 11.3). However, it should be recognized that a considerable amount of effort is required to dismantle the parts to carry out an effective and meaningful inspection, with a risk of damage to the machine in the process. For example, the rotor is often removed to perform a good visual examination of the stator winding. Sometimes, the amount of disassembly can be reduced by making use of technologies such as video probes, cameras and robotic devices that can gain access for visual inspections through small spaces, e.g., the EL-CID test described in Section 14.4. Furthermore, excessive application of this option to all elements of the power plant could lead to unrealistic amounts of downtime through lack of coordination.

16.1.4 Inspections

The success of any maintenance strategy is dependent on the effectiveness of the associated inspections. Additionally, the planning of these inspections is governed by the particular strategy adopted.

An accurate assessment of insulation condition requires an outage, and planning should be directed at minimizing the time for which the machine is not available. A key ingredient of any planned inspection is to identify the necessary resources ahead of time and have them available when required; these resources can encompass manpower, equipment, and spare parts.

After an In-Service Failure. An inspection at this time is not a planned inspection since the timing is not within the user's control. However, the adoption of a breakdown maintenance strategy inherently accepts the risk of sudden unplanned outages and general provisions should be made to handle them. Electrical machine breakdown due to insulation failure leads to a repair/replace decision rather than an assessment of condition. However, an inspection after a failure can be instrumental in determining the cause, in assessing the extent of corrective action, and in implementing a plan to inspect similar equipment to prevent similar failures in the future.

For motors, complete spares can usually be justified for low ratings and applications in which a failure causes significant financial loss to the user if there is no spare motor immediately available. The selection of spares to be carried should also be based on the population and failure rate of similar machines in the plant. Following replacement of a failed motor, economics will dictate whether it be repaired or scrapped. The availability of competent repair contractors should be established in advance; qualified organizations should be identified and listed together with their specific capabilities concerning insulation systems, size limits, and turnaround time.

For larger, more complex electrical machines, an assessment of required spare parts should be conducted after some consultation with the manufacturer. Decisions on which components to purchase and stock will be influenced by their criticality, probability of failure, cost, shelf life, number of machines of the same design, and procurement time. Approved repair techniques and replacement winding insulation systems can be documented together with sources for required consumables such as insulating tapes, resins, etc. Another advantage of carrying spare components, such as complete stator and rotor assemblies for large machines, is that these assemblies can be cycled through the machines to allow an ongoing refurbishment program to be implemented.

As mentioned earlier, failure causes should be monitored for signs of a trend that would lead to a review of the application and possibly equipment or component changes.

Years of Service or Operating Hours. Where a time-based maintenance strategy is employed, outage planning can initially be based on recommendations from the equipment supplier, modified by experience with similar machines. A common practice with large generators is to alternate minor and major inspections. A major inspection would include withdrawing the rotor or inspecting it using in-situ tests and robotic techniques. The initial inspection often occurs early in a machine's life and may be identified as a contractual requirement tied to the expiration of the warranty period.

Prior to an inspection, any repair parts that may be required should be identified and acquired. Past experience and in-service tests are invaluable for determining the appropriate parts to be acquired. Examples are rewedging materials, insulating tapes and resins, core tightening wedges, and stress grading paint. If a major rewedging job is anticipated, the availability of specialized craftsmen, on a standby basis, should be determined. The results of previous diagnostic tests should be reviewed and compared with the new values to identify any trends. Hence, although a specific test value may in itself still be within an acceptable range, deterioration may be evident that should influence future inspection intervals. Once experience has been gained with a particular machine, the timing of inspections can be adjusted based on previous results; this could lead to either longer or shorter intervals.

Outage planning will often be based on the requirements of many pieces of equipment, so the timing for a particular machine may have to be a compromise. Naturally, more flexibility exists for systems incorporating redundant components. As an example, where three 50% capacity pump-motor drives are installed in an auxiliary service water system for a large generating unit, one motor can be scheduled for inspection/maintenance without derating or shutting down the generating unit.

Even though future inspections have been planned based on operating time, the results of routine on-line diagnostic tests may indicate that earlier attention is required. For example, a partial discharge test may raise concern for loose generator stator bars, in which case an inspection and any required remedial action should be scheduled as soon as possible.

Trend Monitoring. Outage or turnaround inspections can be initiated based on excursions in an on-line monitor. Effective plant condition monitoring during normal operation may lead to two significant benefits:

1. Reduced maintenance costs and increased availability via elimination of unnecessary inspection and preventative maintenance for healthy equipment
2. Identification of potentially serious states prior to the occurrence of damage

Vibration monitoring has become accepted as a valuable method of tracking the mechanical health of rotating machines. A normal signature, or fingerprint, is established and vibration limits identified at which specific actions should be initiated; these may range from investigation to equipment shutdown. Likewise, insulation condition can be monitored via assessment of appropriate sensor trends and diagnostic test results. Assessments that can be made from well-proven trend monitoring systems are discussed in Chapter 13. Unfortunately, not all insulation system deterioration mechanisms can be detected with available monitoring systems. For example, deterioration of rotor winding groundwall insulation is difficult to detect, since on-line monitors have not been developed that can detect the processes that lead to weakened electrical or mechanical capability of the insulation.

Where a monitored characteristic shows a progressive trend away from the normal value, an inspection can be planned for the first available outage. It is important that operating staff appreciate the significance of "abnormal" trends such that they can evaluate the risks of deferring inspections due to other priorities.

16.2 MAINTENANCE STRATEGIES FOR VARIOUS MACHINE TYPES AND APPLICATIONS

Maintenance strategies need to be custom-designed on the basis on the following factors:

- Machine criticality and outage costs
- Machine replacement/repair costs
- Redundancy
- Required manpower resources and cost benefit
- Safety requirements (for nuclear power plants and certain chemical plants)
- Machine accessibility (for nuclear power plant inside containment applications or hazardous locations in chemical plants)
- Public health considerations, such as drinking water and waste water processing

In this section, maintenance strategies will be discussed under the following headings

- Turbogenerators
- Hydrogenerators
- Motors

16.2.1 Turbogenerators

There is no doubt that the criticality, outage costs, replacement/repair costs, and cost benefits warrant a well-designed comprehensive winding maintenance program for this type of machine. This should include the following elements.

On-Line Monitoring

- On-line continuous stator winding temperature monitoring is particularly important for water-cooled stator windings since it will provide early detection of blockages in conductor strands if individual stator bar inlet and outlet water temperatures can be monitored. Such monitoring can detect blockages from debris in the stator coolant, copper oxide build-up in the conductor strands due to poor water chemistry, and loss of hydrogen-cooler water supply.
- On-line monitoring for hydrogen in stator coolant will give an early warning of conductor strand breakage in water-cooled stators since the hydrogen is always at a higher pressure than the water.
- Periodic or continuous airgap flux probe monitoring is recommended for machines that may be susceptible to or have a history of turn insulation failures, as it is capable of detecting the development of a single turn short.
- Continuous vibration monitoring can often give an early indication of developing rotor winding turn shorts.
- Continuous or periodic stator winding partial discharge monitoring is essential to allow early detection of winding ground insulation deterioration problems as described in Section 13.3.

Off-Line Tests. These are required if no on-line winding monitoring is being performed, to detect insulation degradation that cannot be identified by on-line techniques, or to verify conditions indicated by the on-line techniques identified above. The number of off-line tests can be kept to a minimum if the on-line monitoring described above is performed. As discussed in Chapters 12 and 14, some of the techniques identified below do require major disassembly of the generator.

- IR and PI—These tests, described in Section 12.1, should be conducted before any high-voltage AC or DC tests are performed and if winding surface contamination is present or suspected.

- AC or DC hipot tests—There are some companies that believe in performing one of these tests on stator windings (Section 12.2 or 12.4) on a periodic basis to give some assurance that the winding will survive until the next major outage. Such tests should also be performed after any winding repairs, bar replacements and bushing repairs, or replacements to verify the integrity of such work.

- Partial discharge test—Off-line PD tests (Section 12.10) are performed to verify the results of on-line tests or if on-line monitoring devices are not fitted. Also, the location of high PD can often be verified by the PD probe test described in Section 12.11.

- Winding capacitance and dissipation factor tests—These tests, described in Sections 12.5, 12.6 and 12.8, are alternatives to the partial discharge test and, in some cases, are used in conjunction with the PD test to determine stator winding ground insulation degradation by trending the results of periodic tests.

- Capacitance mapping—As indicated in Section 12.5, this type of testing should be performed on certain water-cooled generator stator designs that have clip joints that are susceptible to crevasse corrosion cracking. This type of degradation eventually leads to stator-cooling water seeping into the bar ground insulation, causing it to delaminate.

- Wedge tap—As indicated in Section 15.2.9, it is important that a baseline stator wedge tap be performed and a wedge map be obtained before a turbine generator goes into service. This test should be repeated before a new winding warranty runs out and periodically throughout the life of the winding. To reduce outage times, more use is now being made of robotic wedge tap devices that allow this test to be performed without withdrawing the rotor. Time savings from this approach are only achieved if the results indicate that no rewedging is required.

- RSO or surge test—These rotor tests, as described in Section 12.21, are normally only performed if running vibration data indicate the presence of turn shorts and to verify airgap flux probe data indicating such defects.

- Pressure and vacuum decay tests—Section 12.27 covers these tests, which are performed to verify whether there are any leaks in the stator cooling systems in water-cooled stators. These tests should be performed to help confirm leaks from crevasse corrosion cracking and after any maintenance that required the stator coolant to be drained.

- EL-CID test—The results of this test are trendable and, like all such tests, it is important to obtain baseline results before the generator is placed in service to determine the initial core insulation condition. This test, which is relatively easy to perform, should be repeated during major outages when the rotor is removed, or when a wedge tap is performed with the rotor in-situ. In the latter case, there are a number of companies that have a robotic device that can crawl along the airgap to perform a wedge tap and an EL-CID test.

Visual Inspections. As discussed in Section 16.1.4, it is important to perform visual inspections of both stator and rotor insulation system components at every opportunity. When performing such inspections, particular attention should be paid to the following:

1. Stator
   - Slot wedges for cracks
   - End-winding bracing for looseness
   - End-windings for evidence of partial discharges, especially between line-end bars in different phases; puffiness in ground insulation; insulation cracking; and flow of asphalt in windings that use this as a ground insulation-bonding agent
   - Stator core for looseness, surface damage, and localized overheating. Such inspections should include both the stator bore and as much of the back of core area as possible
   - Phase connections for evidence of cracked insulation, overheating, and loose blocking
   - Terminal bushings for evidence of cracks in the porcelain, surface contamination associated with tracking
   - Removal of some hoses or waterbox lids to allow inspection of bar ends for evidence of strand bore oxidation, blockage from debris, and wall thinning from erosion

2. Rotor and sliprings
   - If there is a need to remove the end-winding retaining rings for inspection, an opportunity should be taken to inspect the end-windings for copper dusting, loose blocking, displaced turn insulation, and conductor distortion.
   - Visible areas of the slot portions of the windings should be inspected through radial gas ducts for evidence of turn insulation and slot packing migration and copper dusting.
   - If there is a history of up-shaft lead or radial pin insulation cracking, this insulation should be periodically inspected. It should be possible to conduct such inspections without removing the rotor.
   - If the rotor has sliprings, their surfaces should be inspected for pitting, grooving, and fingerprinting.
   - Check radial pin insulation for cracks or other damage.
   - If the rotor has slip rings, inspect for pitting, grooving, or fingerprinting.

16.2.2 Salient Pole Generators and Motors

As with turbine generators, the criticality, outage costs, replacement/repair costs, and cost benefits warrant a well-designed comprehensive winding maintenance program for this type of machine. This is particularly true for hydraulic generators that produce electrical power at low cost. This should include the following elements.

On-Line Monitoring

- On-line continuous stator winding temperature monitoring is particularly important to detect blockage of cooling air passages and extensive contamination. If the machine has an air-to-water heat exchanger, such monitoring will also give an early warning of a cooling water supply loss and cooler tube blockages by debris or silt in the water supply.

- Continuous or periodic stator winding partial discharge monitoring is most beneficial for machines with voltage ratings of 6.0 kV and above to allow early detection of winding ground insulation deterioration problems by PD trend monitoring, as described in Section 13.4.
- Continuous vibration monitoring can often give an early indication of developing rotor winding turn shorts.
- Airgap monitoring for mechanical problems and to detect rotor winding shorted turns and various mechanical problems (Section 13.8).

Off-Line Tests. These tests are required if no on-line winding monitoring is being performed, to detect insulation degradation that cannot be identified by on-line techniques, or to verify conditions indicated by the on-line techniques identified above. The number of off-line tests can be kept to a minimum if the on-line monitoring described above is performed. As discussed in Chapters 12 and 14, some of the techniques identified below do require major disassembly of the generator.

- IR and PI—These tests, described in Section 12.1, should be performed before any high-voltage AC or DC tests are performed and if winding surface contamination is present or suspected.
- AC or DC hipot tests—There are some companies that believe in performing one of these tests on stator windings (Section 12.2 or 12.4) on a periodic basis to give some assurance that the winding will survive until the next major outage. Such tests should also be performed after any winding repairs or bar replacements to verify the integrity of such work.
- Partial discharge tests—Off-line PD tests (Section 12.10) are performed to verify the results of on-line tests. Also, the location of high PD can often be verified by the PD probe test described in Section 12.11.
- Winding capacitance and dissipation factor tests—These tests, described in Sections 12.5, 12.6, and 12.8, are alternatives to the partial discharge test and in some cases are used in conjunction with the PD test to determine stator winding ground insulation degradation by trending the results of periodic tests.
- Wedge tap—As indicated in Section 15.2.9, it is important that a baseline stator wedge tap is performed and a wedge map obtained before a nonglobal VPI winding goes into service. This test should be repeated before a new winding warranty runs out and periodically throughout the life of the winding.
- Pole drop test (Section 12.20) to detect rotor shorted turns.
- EL-CID test—The results of this test are trendable and, like all such tests, it is important to obtain baseline results before the generator is placed in service to determine the initial core insulation condition. This test, which is relatively easy to perform, should be repeated during major outages when the rotor is removed.

Visual Inspections. As discussed in Section 16.1.4, it is important to perform visual inspections of both stator and rotor insulation system components at every opportunity. When performing such inspections particular attention should be paid to the following:

1. Stator
   - Slot wedges for cracks
   - End-winding bracing for looseness
   - End-windings for evidence of partial discharges, especially between line-end bars in different phases; puffiness in ground insulation; insulation cracking; and flow of bitumen in windings that use this as a ground insulation-bonding agent
   - Stator core for looseness, surface damage, and localized overheating. Such inspections should include both the stator bore and as much of the back of core area as possible, as well as at core splits (if present).
   - Phase connections for evidence of cracked insulation, overheating, and loose blocking
   - If fitted, circuit ring bus and output bus supports for surface contamination associated with tracking

2. Rotor and Sliprings
   - Rotor windings should be periodically inspected for insulation cracking and pole winding bulging from inadequate support.
   - Winding top and bottom washers should be inspected for cracks and migration.
   - Interpole connections should be inspected for conductor cracking and insulation damage.
   - Interpole "V" braces should be inspected for looseness and cracking of the insulation between them and the winding poles.
   - Amortisseur winding bars and interconnections should be checked for cracks, distortion, and migration out of the laminations.
   - If the rotor has sliprings, their surfaces should be inspected for pitting, grooving, and fingerprinting.
   - The condition of the visible portions of the insulation on the leads, between the sliprings or brushless exciter and field winding, should be checked.
   - If the rotor is a high speed, strip-on-edge type, with bolted-on pole tips, the bolts should be inspected for looseness and cracking.

16.2.3 Squirrel Cage and Wound-Rotor Induction Motors

Since these types of motors are used in a wide range of applications, from very critical to noncritical, developing a cost-effective maintenance program involves assessing each application to determine the impact of in-service failures. This is best illustrated by the following example for a large plant with many motors.

Maintenance Strategy Development. The following strategy has been found to be cost-effective for plants containing a large number of motors.

1. Make a list of all applications using motor-driven equipment.
2. Analyze the list and assign one of the following categories to each motor application:
   - Critical—loss of production costs high, major safety concerns or environmental hazards occur if a motor fails.
   - Important—high repair costs, loss of redundancy or environmental impact if a motor fails.

- Noncritical—no impact on plant production, safety, or environment if motor fails; repair or replacement costs are low.

3. Develop a maintenance program for each motor category based on allocating the greatest resources to the critical motors and the least to the noncritical ones. This should include selection of appropriate on-line monitoring, off-line tests, and visual inspections for each category of motor and how frequently they should be performed.

Typical Maintenance Activities for each Motor Category. For completeness, the following maintenance recommendations include both winding and motor mechanical component monitoring techniques.

Critical Motors

1. On-line monitoring
   - Continuous or periodic shaft and/or bearing housing vibration monitoring and trending. Periodic readings should be taken monthly.
   - Continuous or periodic temperature monitoring and trending of stator windings (if fitted with RTDs or thermocouples). Periodic measurements should be taken monthly.
   - Continuous or periodic on-line stator winding partial discharge monitoring for all motors rated at 3.3 kV and above.
   - Continuous or periodic temperature monitoring and trending of bearings. Continuous monitoring requires that the bearings be fitted with RTDs or thermocouples. Monthly bearing housing temperature monitoring can be performed with a thermovision camera if bearing temperature detectors are not fitted.
   - Stator current signature analysis for motors with fabricated rotor windings that are started frequently or are driving high-inertia equipment such as induced- or forced-draft fans. Periodic readings should be taken annually and more frequently after indications of broken bars are found.
   - Periodic visual inspections for evidence of unusual noises, bearing lubricant leaks, oil levels, blocked air inlets, etc. The maximum interval for these should be one week.
   - Periodic oil analysis for oil lubricated bearings (typically every 6 months).
   - Periodic thermography for indications of overheated connections, restricted cooling, etc.
2. Off-line tests
   - Periodic IR, PI, and winding conductivity (Section 12.3) measurements (typically every 2 to 4 years).
   - Periodic DC step voltage or ramp tests (typically every 2 to 4 years).
   - Periodic partial discharge (PD), capacitance, or dissipation factor tests if on-line PD tests indicate an upward trend.
   - If bearings are insulated, an IR value should, if possible, be taken every 2 to 4 years. The practicality of this depends on how the motor bearings are insulated.
   - IR of wound-rotor windings.
3. Visual inspections. At least one motor from each application should be partially or completely disassembled every 6 to 8 years to allow a detailed inspection of internal

components for generic degradation. The motor selection should be based on total running hours, number of starts, and the results of on-line tests and off-line inspections. Such inspections, which should also be performed when a motor is disassembled for repair, should cover the following components:

- Stator core and windings.
- Rotor core, windings, end-winding retaining rings (if fitted), sliprings (wound-rotor motors), shaft cooling air fans, bearing journals/mounting surfaces, and shaft.
- Bearings, housings, and oil-to-water coolers (if fitted) for signs of tube bore erosion/corrosion.
- Air-to-water coolers for evidence of tube erosion/corrosion resulting in wall thinning.
- Physical condition and functional checks of all accessories including wiring and terminal blocks.

If such inspections reveal what could be generic defects, a program to inspect all motors of the same or similar design should be developed.

Important Motors
1. On-line monitoring
   - Continuous or periodic shaft and/or bearing housing vibration monitoring and trending. Periodic readings should be taken monthly.
   - Continuous or periodic temperature monitoring and trending of stator windings (if fitted with RTDs or thermocouples). Periodic measurements should be taken monthly.
   - Continuous or periodic temperature monitoring and trending of bearing windings. Continuous monitoring requires that the bearings be fitted with RTDs or thermocouples. Periodic bearing housing temperature monitoring can be performed with a thermovision camera if bearing temperature detectors are not fitted. This should be done monthly.
   - Continuous or periodic on-line PD monitoring on motors rated 6 kV or more.
   - Stator current signature analysis for motors with fabricated rotor windings that are started frequently or are driving high-inertia equipment such as fans. Periodic readings should be taken annually and more frequently after indications of broken rotor bars are found.
   - Periodic oil analysis for oil lubricated bearings (typically every 6 months).
   - Periodic thermography for indications of overheated connections, restricted cooling, etc.
   - Periodic visual inspections for evidence of unusual noises, bearing lubricant leaks, blocked air inlets etc. The maximum interval for these should be one week.
2. Off-line tests
   - Periodic IR, PI, and winding conductivity measurements (typically every 2 to 4 years).
   - Periodic DC step voltage or ramp tests (typically every 2 to 4 years).
   - If bearings are insulated, an IR value should, if possible, be taken every 2 to 4 years. The practicality of this depends on how the motor bearings are insulated.

- IR of wound rotor windings (typically every 2 to 4 years).
- Off-line PD (every 2 to 4 years).

3. Visual inspections. At least one motor from each application should be partially or completely disassembled every 6 to 8 years to allow a detailed inspection of internal components for generic degradation. The motor selection should be based on running hours and results of on-line tests and off-line inspections. Such inspections, which should also be performed when a motor is disassembled for repair, should cover the following components:

- Stator core and windings.
- Rotor core, windings, end-winding retaining rings (if fitted), sliprings (wound-rotor motors), shaft cooling-air fans, bearing journals/mounting surfaces, and shaft extension.
- Bearings, housings and oil-to-water coolers (if fitted) for signs of tube bore erosion/corrosion.
- Air-to-water coolers for evidence of tube erosion/corrosion resulting in wall thinning.
- Physical condition and functional checks of all accessories including wiring and terminal blocks.

If such inspections reveal what could be generic defects, a program to inspect all motors of the same or similar design should be developed.

Noncritical Motors. These would receive very little maintenance since they could be run to failure. Typically, they would receive:

- Periodic bearing lubrication (grease-lubricated bearings) or replacement (oil-lubricated bearings).
- Periodic vibration monitoring (say, every 3 months).

INDEX

Electrical Insulation for Rotating Machines. By Stone, Boulter, Culbert, and Dhirani   **365**
ISBN 0-471-44506-1 © 2004 Institute of Electrical and Electronics Engineers

ABOUT THE AUTHORS

Greg C. Stone is an electrical engineer who has developed new tests to evaluate stator insulation systems, many of which have become IEEE and IEC standard test methods. He received his doctorate from the University of Waterloo, Canada, and has worked for the Research Division of Ontario Hydro, the largest electrical power utility in North America. Dr. Stone is the principal author of *Handbook to Assess the Insulation Condition of Large Rotating Machines* and numerous technical papers and articles. Dr. Stone currently works at Iris Power Engineering and is a Fellow of the IEEE.

Edward A. Boulter, Lt. Commander (Ret.), USN Reserves, is a consulting engineer serving a wide variety of clients including General Electric Company, Von-Roll-ISOLA, utilities in several states, and manufacturers. Previously, he spent nearly 40 years working as a project and senior engineer and technical team leader at General Electric.

Ian Culbert is a graduate of Dundee Technical College, Dundee, Scotland. He has worked as a specialist at Ontario Hydro, providing technical support to power station project engineering, as well as operations and maintenance staff on all types of motors and standby generators and designer of motors for Reliance Electric and Parsons Peebles. Currently, Ian is a rotating machine specialist at Iris Power Engineering.

Hussein Dhirani is a senior design engineer in the Engineering Services Division of Ontario Power Generation Inc. in Toronto, Canada. He received a BE (electrical) degree from the University of Poona, India